Introduction to the Quantum Yang–Baxter Equation and Quantum Groups: An Algebraic Approach

Mathematics and Its Applications

Volume 423

Introduction to the Quantum Yang–Baxter Equation and Quantum Groups: An Algebraic Approach

by

Larry A. Lambe

CAIP, Rutgers University,
Piscataway, NJ, U.S.A. and
University of Wales, Bangor
Bangor, Gwynedd, U.K.

and

David E. Radford

University of Illinois at Chicago,
Chicago, Illinois, U.S.A.

SPRINGER-SCIENCE+BUSINESS MEDIA, B.V.

A C.I.P. Catalogue record for this book is available from the Library of Congress.

ISBN 978-1-4613-6842-7 ISBN 978-1-4615-4109-7 (eBook)
DOI 10.1007/ 978-1-4615-4109-7

Printed on acid-free paper

To Mary Kay: for all the years of support and understanding (LAL).

To Robert G. Heyneman: my thesis advisor who introduced me to the subject of Hopf algebras (DER).

Contents

Foreword

Chapter 1

The algebraic prerequisites for the book are covered here and in the appendix. This chapter should be used as reference material and should be consulted as needed. A systematic treatment of algebras, coalgebras, bialgebras, Hopf algebras, and representations of these objects to the extent needed for the book is given. The material here not specifically cited can be found for the most part in [Sweedler, 1969] in one form or another, with a few exceptions.

A great deal of emphasis is placed on the coalgebra which is the dual of $n \times n$ matrices over a field. This is the most basic example of a coalgebra for our purposes and is at the heart of most algebraic constructions described in this book.

We have found pointed bialgebras useful in connection with solving the quantum Yang–Baxter equation. For this reason we develop their theory in some detail. The class of examples described in Chapter 6 in connection with the quantum double consists of pointed Hopf algebras. We note the quantized enveloping algebras described elsewhere are pointed Hopf algebras. Thus for many reasons pointed bialgebras are objects of fundamental interest in the study of the quantum Yang–Baxter equation and quantum groups.

Chapter 2

The discussion of the quantum Yang–Baxter equation begins here. Various forms of the equation are defined and discussed. We treat the constant form and the one-parameter and two-parameter forms of the equation. The fundamental notations for describing the equation are developed and the basic algebraic structures associated with the equation are motivated. In particular the FRT [Faddeev et al., 1990], [Faddeev et al., 1988] and reduced FRT constructions [Radford, 1994b] are first presented in this chapter.

Computer algebra can play an important role in the solution of the equation and development of the subject. We describe two computer experiments for solving the equation in special cases in this chapter. The second [Lambe, 1996], described in Section 2.11.2, suggests that more general algebraic constructions can be associated with the quantum Yang–Baxter equation. We develop these ideas in Chapters 3 and 8.

Chapter 3

We explore various categorical settings for the study of constant form of the quantum Yang–Baxter equation. The most fundamental setting is the category of left QYB-modules over a bialgebra which is slight variation on the basic setting of [Yetter, 1990], the prebraided monoidal category described in the Preface. See also [Majid, 1990b]. The results here have very little overlap with the results of [Yetter, 1990] as the focus here is quite different. The material related to the category of left QYB-modules is taken from [Lambe and Radford, 1993] and generally follows this reference very closely.

Chapter 4

We give a universal mapping property for the FRT construction which is found in [Radford, 1993c]. The reduced FRT construction and a universal mapping property for it, and the notion of M-reduced, were given in [Radford, 1994b].

Important constant solutions to the quantum Yang–Baxter equation have matrices that are upper triangular in a basis. In this case the reduced FRT construction is a pointed bialgebra. Conditions for the reduced FRT construction to be a pointed bialgebra, or pointed Hopf algebra, are addressed in Section 4.4 which includes as its first theorem a slight variation of [Radford, 1994b, Theorem 3].

Chapter 5

We motivate quantum groups and quantized enveloping algebras through the study of the most basic example, quantum $\mathrm{SL}(2, k)$. Most of the material from this chapter is a slight revision of material from [Lambe and Radford, 1993, Section 9].

Chapter 6

We discuss finite-dimensional algebras which give rise to solutions to the quantum Yang–Baxter equation through their representations in a natural way. These are the quasitriangular Hopf algebras and the quasitriangular algebras. Probably the most important example of a finite-dimensional quasitriangular Hopf algebra is the quantum, or Drinfel'd, double [Drinfel'd, 1987].

The treatment of quasitriangular Hopf algebras follows the second author's work on minimal quasitriangular Hopf algebras found in [Radford, 1993b]. The families of quasitriangular Hopf algebras discussed in Section 6.6 is found in [Radford, 1994a] and our exposition draws heavily from Section 5 of that paper.

It is very interesting to note that every finite-dimensional Hopf algebra is the reduced FRT construction of some solution to the quantum Yang–Baxter equation. This is a consequence of quantum double construction. See Exercise 6.4.5.

Chapter 7

We examine the notion of coquasitriangularity, the notion dual to the notion of quasitriangularity discussed in Chapter 6. See [Larson and Towber, 1991], [Majid, 1990b], and [Schauenburg, 1992a]. At center stage are the FRT construction and the reduced FRT construction again. We define the notion of coquasitriangular coalgebra and construct the free quasitriangular bialgebra on a coquasitriangular coalgebra. Left comodules over a coquasitriangular bialgebra A form a category of left QYB A-modules [Radford and Towber, 1993] but they are *locally finite* as left A-modules. The FRT construction and the reduced FRT construction are both coquasitriangular bialgebras. Some of these results are generalized to the case of one-parameter QYBE solutions in [Cotta-Ramusino et al., 1993] and we extend those results in the last section of this chapter.

A new method for producing one-parameter QYBE solutions from given ones is presented in Section 7.5.

Chapter 8

We use the techniques developed thus far to find solutions to quantum Yang–Baxter equation in this chapter. The role of the reduced FRT construction is illustrated in the solution of the QYBE in the upper triangular two-dimensional case. Computer algebra methods which determine all solutions in the two-dimensional case [Hietarinta, 1993a] are mentioned. The analysis of the upper triangular two-dimensional case follows the discussion in [Radford, 1994b]. We believe that these methods can be generalized to systematically yield higher dimensional solutions.

The notion of "patching solutions" of Section 8.4 can be found in [Lambe and Radford, 1993, Section 10].

The examples of Chapter 2 which were discovered by computer methods are analyzed in a theoretical framework in Section 8.5.

We give an example of the ρ-perturbation method from Section 7.5.2 applied to the one-parameter solution associated to the XXX-magnet model.

Chapter 9

This is intended to be an introduction to the work found in [Joyal and Street, 1991a], [Joyal and Street, 1991b], [Joyal and Street, 1991c], [Joyal and Street, 1993], [Schauenburg, 1992a], [Schauenburg, 1992b], [Pareigis, 1996] and the references found therein.

The reader will realize the FRT construction as a coend [MacLane, 1988] in this chapter. More generally, quantum groups can arise as the kinds of coends found here. See [Majid, 1991b] [Yetter, 1990] for early material in this direction. An older specific example may be found in [Pareigis, 1981].

Appendix A

In the appendix we discuss various notations and conventions used in the text as well as develop in rather careful detail the linear algebra used in Chapter 1. The linear algebra should be referred to as needed, and the discussion on notation and conventions should be consulted before the reader gets too far into the text. The various forms of notation we use in the book reflect what the reader will find in the literature related to the quantum Yang–Baxter equation and quantum groups.

Preface

We would like to highlight some of the results contained in this book which we have not been able to find in the literature.

- A good part of our treatment of the constant quantum Yang–Baxter equation takes place in the category of left QYB-modules which is a minor variation of the pre-braided monoidal category of left crossed bimodules over a bialgebra. The category of weak QYB-modules of Section 3.1.4 is one of several generalizations of the category of QYB-modules which was motivated by experiments in computer algebra. This material was motivated by the class of examples in Section 2.11.2.

- A bialgebra formally gives rise to three others by twisting the multiplication, comultiplication, or twisting both. In Section 4.3 we consider these variations for the FRT construction and show they are themselves FRT constructions.

- In Section 4.4 we find necessary and sufficient conditions for the reduced FRT construction to be a Hopf algebra in the pointed case (with one trivial exception the FRT construction is never a Hopf algebra).

- The quantum double in the finite-dimensional case is known to be a minimal quasitriangular Hopf algebra. We motivate the construction of the quantum double in what may be a novel way in the context of minimal quasitriangular Hopf algebras in Section 6.4.

- Definitions are given for quasitriangular algebras and coquasitriangular coalgebras in Section 7.2. The free coquasitriangular bialgebra on a coquasitriangular coalgebra is defined and studied in Section 7.4. We follow this with some new observations based on previous work which extends the algebra to the one-parameter case of the QYBE. With this, we present a method for "perturbing" a given one-parameter solution of the QYBE by a given family of parameterized comodules.

- The category of weak QYB H-modules is studied when $H = k[G]$ is the group algebra of a cyclic group G over a field k in Section 8.5. The motivation for ensuing theory was the computer algebra experiment described in Section 2.11.2 for the special case when $G = \mathbb{Z}_2$. Of great interest is whether or not a natural tensor product exists for the category. This is systematically treated. We give some computer calculations that yield a classical one-parameter solution and use the perturbation method of Theorem 7.5.1 to transform it.

Acknowledgments

Much of the first author's work on this book was carried out during visits to the Department of Mathematics of Stockholm University during which he was generously supported. He wishes to thank the Department for providing a very stimulating environment, and he also wishes to thank the Wenner–Gren Foundation of Stockholm for its kind support. During this period, the first author also enjoyed the hospitality of the Department of Mathematics at the University of Wales and received on occasion support from the London Maths Society and EPSRC for which he is very grateful.

The second author's work on this book was done at the University of Illinois at Chicago and during a short visit to Stockholm University in May of 1995. He wishes to thank in particular the Mathematics Department of Stockholm University for its kind hospitality during his visit there.

The symbolic computational system and language AXIOM [Jenks and Sutor, 1992] was used for all computer calculations and experiments described in the book.

INTRODUCTION

The quantum Yang–Baxter equation has roots in statistical mechanics [Yang, 1967], [Akutsu et al., 1989], [Zamolodchikov and Zamolodchikov, 1975], [Baxter, 1982], [D'Ariano et al., 1985].

A purely algebraic derivation of quantum Yang–Baxter equation based on some physical assumptions can be obtained as follows. Let $\{e_i(u)\}$ represent a collection of particles. Here u represents a parameter that depends on the particle's mass and (relativistic) momentum (called the rapidity). Assume that the scattering of two such particles is described by a probability amplitude (scattering amplitude)

$$R_{i,j}^{k,l}(u-v)$$

(here the first particle depends on u and the second on v). One often sees diagrams of the form

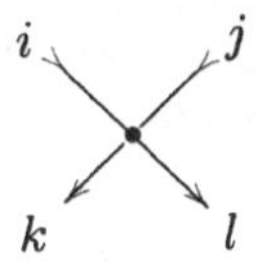

to represent this situation. Now some standard theories assume that in multiple particle scattering, only two particles scatter at a time. This has the natural implication that if we formally write

$$e_i(u)e_j(v) = R_{i,j}^{k,l}(u-v)e_l(v)e_k(u), \tag{I.1}$$

we can use this commutation relation in order to calculate the scattering amplitudes

$$R_{i_1,\ldots,i_n}^{j_1,\ldots,j_n}(u_1-u_2,\ldots,u_{n-1}-u_n) \tag{I.2}$$

for n-particle interactions. In other words, one uses (I.1) to calculate

$$e_{i_1}(u_1)\cdots e_{i_n}(u_n)$$

and consequently the coefficients (I.2). But, indeed, consider the free algebra generated by symbols $e_i(u)$ where $u \in X$ and X is some ordered group. Suppose one imposes the relations (I.1). If this operation is to be associative, the matrices $R_{i,l}^{j,k}(u)$ will have to satisfy a certain equation because there are two different ways to transform $e_i(u)e_j(v)e_k(w)$ into linear combinations of $e_c(w)e_b(v)e_a(u)$ (which we encourage the reader to work out). This is the "one-parameter Quantum Yang–Baxter equation" which we will present in Chapter 2. The algebra just defined is called the *Zamolodchikov Algebra.*

A quantum version of the inverse scattering method [Faddeev and Takhtadzhan, 1987] can be found in [Faddeev, 1990], [Kulish and Sklyanin, 1982], [Sklyanin, 1982], [Sklyanin, 1991], [Takhtajan, 1984], [Takhtajan, 1990b] and the references therein. At the heart of the quantum inverse scattering method (QISM) is an algebra $A(R)$ with generators $t_i^j(x)$, $1 \le i, j \le n$, $x \in X$ and relations

$$R_{s_1,\, s_2}^{a,\, b}(x)t_i^{s_1}(xz)t_j^{s_2}(z) - t_{s_2}^{b}(z)t_{s_1}^{a}(xz)R_{i,\, j}^{s_1,\, s_2}(x)$$

where $R_{i,j}^{k,l}$ satisfies the same equation as above, i.e. the one-parameter Quantum Yang–Baxter equation (see Chapter 2 for details). It turns out that $A(R)$ is naturally a bialgebra. It will play a crucial role in this text, particularly in Chapters 4, 7, and 8.

The actual term "quantum group" was introduced in [Drinfel'd, 1987] and was motivated by work on the QISM. Since the appearance of this paper, there has been am explosion of work in the area of quantum groups and the Quantum Yang–Baxter equation.

In addition to connections with the QISM and statistical mechanics, the QYBE has important connections with knot theory and invariants of 3-manifolds. The braid equation is in the construction of knot and link invariants. The braid equation is equivalent to the QYBE. We will not deal with this aspect of the theory in this book in any detail, but refer the reader to [Akutsu et al., 1989], [Kauffman, 1991] [Yang and Ge, 1989], [Yang and Ge, 1994], and the references found there for details. The interplay with 3-manifold theory as can be found in [Kauffman and Lins, 1994], [Kauffman and Radford, 1995], [Reshetikhin, 1991], and [Reshetikhin and Turaev, 1991] to cite very few examples.

We suggest the fundamental paper [Drinfel'd, 1987] for an overview of the QYBE and the theory of quantum groups and the survey papers [Majid, 1990a], [Majid, 1990b] as well as [Takhtajan, 1990a] and [Takhtajan, 1993]. Besides the references mentioned above, the reader may find the following works and the references therein very useful compliments to this book: [Chari and Pressley, 1995], [Kauffman, 1991], [Shnider and Sternberg, 1993]. This book is intended to fill a gap in the existing literature and is not intended to be comprehensive.

1 ALGEBRAIC PRELIMINARIES

In this chapter we develop the theory of coalgebras, bialgebras, Hopf algebras, and their associated representations which is needed for our treatment of algebraic structures related to the quantum Yang–Baxter equation. Many exercises of various degrees of difficulty are provided to help the reader understand the concepts introduced. We provide a complete and self contained treatment of the topics mentioned above for our purposes. There are general references on coalgebras, bialgebras, and Hopf algebras which the reader may find useful to consult: [Abe, 1980], [Heyneman and Sweedler, 1969], and [Heyneman and Sweedler, 1970], [Montgomery, 1993], and [Sweedler, 1969].

1.1 Coalgebras

Coalgebras are structures which are formally dual to algebras. Coalgebras are important in their own right and have a local finiteness which algebras generally do not possess. The algebras of interest to us in this book have a coalgebra structure which plays an important part in their structure theory.

In this section we define coalgebra over a field k and related notions, describe some very basic examples, and prove a few elementary results on coalgebras. The theoretical role of the coalgebra $\mathrm{C}_n(k)$, which is the counterpart of the algebra $\mathrm{M}_n(k)$ of $n \times n$ matrices over k, will begin to emerge.

To understand the dual nature of coalgebras we formulate the axioms for an algebra in terms of linear maps. An algebra with unit over the field k is a triple (A, m, η), where A is a vector space over k and $m : A \otimes A \longrightarrow A$ and $\eta : k \longrightarrow A$ are linear maps with images denoted by $m(a \otimes b) = ab$ and $\eta(1_k) = 1$, such that $a1 = a = 1a$ for all $a \in A$. The map m is referred to as the multiplication map and η is referred to as the unit map. Using the canonical k-linear identifications $A \otimes k \simeq A \simeq k \otimes A$, this last pair of equations is equivalent to commutativity of the diagrams

$$\begin{array}{ccc} A \otimes A & & \\ {\scriptstyle 1_A \otimes \eta} \uparrow & \searrow {\scriptstyle m} & \\ A \otimes k & \xrightarrow[\simeq]{} & A \end{array}$$

and

$$\begin{array}{ccc} A \otimes A & & \\ {\scriptstyle \eta \otimes 1_A} \uparrow & \searrow {\scriptstyle m} & \\ k \otimes A & \xrightarrow[\simeq]{} & A \end{array}$$

Observe that if (A, m, η) is an algebra over k then so is (A, m^{op}, η), where $m^{op}(a \otimes b) = ba$ for all $a, b \in A$. In terms of the twist map we have $m^{op} = m\tau_{A,A}$.

Definition 1.1.1 *An algebra (A, m, η) is* commutative *if $m = m^{op}$.*

We will frequently denote an algebra (A, m, η) by A.

Definition 1.1.2 *The algebra (A, m^{op}, η), denoted by A^{op}, is the* opposite algebra.

Definition 1.1.3 *An algebra A is* associative *if $a(bc) = (ab)c$ for all $a, b, c \in A$.*

Taking the canonical identification $A \otimes (A \otimes A) \simeq (A \otimes A) \otimes A$ for granted, associativity translates to commutativity of the diagram

$$\begin{array}{ccc} A \otimes A \otimes A & \xrightarrow{m \otimes 1_A} & A \otimes A \\ {\scriptstyle 1_A \otimes m} \downarrow & & \downarrow {\scriptstyle m} \\ A \otimes A & \xrightarrow[m]{} & A. \end{array}$$

Note that if (A, m, η) is an associative algebra then A^{op} is associative also. From this point on we will assume that A is associative, unless otherwise specified.

Definition 1.1.4 *A coalgebra with counit over the field k is a triple (C, Δ, ϵ), where C is a vector space over k together with linear maps $\Delta : C \longrightarrow C \otimes C$ and $\epsilon : C \longrightarrow k$, such that the diagrams*

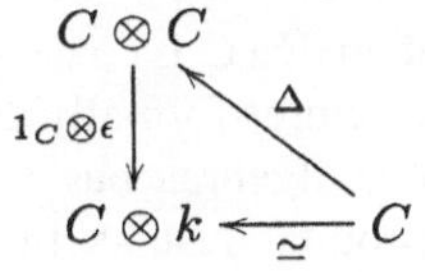

$$
\begin{array}{ccc}
 & C \otimes C & \\
{\scriptstyle 1_C \otimes \epsilon}\downarrow & & \nwarrow{\scriptstyle \Delta} \\
C \otimes k & \underset{\cong}{\longleftarrow} & C
\end{array}
$$

and

$$
\begin{array}{ccc}
 & C \otimes C & \\
{\scriptstyle \epsilon \otimes 1_C}\downarrow & & \nwarrow{\scriptstyle \Delta} \\
k \otimes C & \underset{\cong}{\longleftarrow} & C
\end{array}
$$

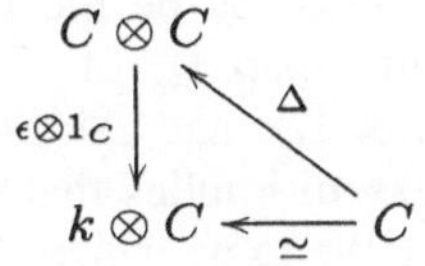

commute.

The map Δ is referred to as the comultiplication map, or the coproduct, and ϵ is referred to as the counit map. The modified Heyneman-Sweedler (H-S) notation for expressing $\Delta(c)$ is $\Delta(c) = c_{(1)} \otimes c_{(2)}$. See Section A.3.2 for a discussion of this terminology. The commutativity of the two diagrams above translates to

$$
c_{(1)}\epsilon(c_{(2)}) = c = \epsilon(c_{(1)})c_{(2)}
$$

for all $c \in C$ in this notation. We will frequently denote the coalgebra (C, Δ, ϵ) by C.

Definition 1.1.5 *A coalgebra C is* coassociative *if the diagram*

$$
\begin{array}{ccc}
C \otimes C \otimes C & \overset{\Delta \otimes 1_C}{\longleftarrow} & C \otimes C \\
{\scriptstyle 1_C \otimes \Delta}\uparrow & & \uparrow{\scriptstyle \Delta} \\
C \otimes C & \underset{\Delta}{\longleftarrow} & C
\end{array}
$$

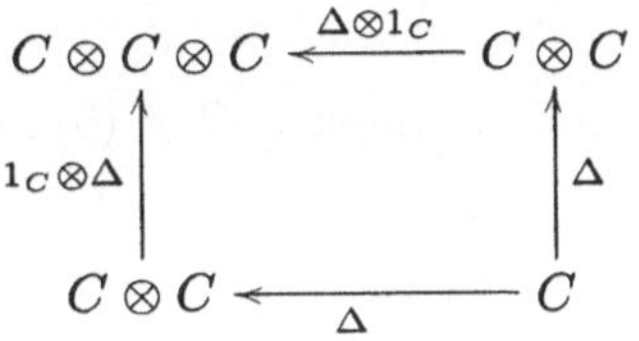

commutes.

In terms of the H-S notation for $\Delta(c)$ the commutativity of the diagram is expressed as

$$
c_{(1)(1)} \otimes c_{(1)(2)} \otimes c_{(2)} = c_{(1)} \otimes c_{(2)(1)} \otimes c_{(2)(2)}
$$

for all $c \in C$.

Suppose that (C, Δ, ϵ) is a coalgebra over k. Then $(C, \Delta^{cop}, \epsilon)$ is a coalgebra where $\Delta^{cop}(c) = c_{(2)} \otimes c_{(1)}$ for all $c \in C$. In terms of the twist map $\Delta^{cop} = \tau_{C,C}\Delta$.

Definition 1.1.6 *The coalgebra $(C, \Delta^{cop}, \epsilon)$, denoted by C^{cop}, is the* opposite coalgebra.

Observe that C^{cop} is a coassociative coalgebra whenever C is.

Definition 1.1.7 *A coalgebra C is* cocommutative *if* $\Delta = \Delta^{cop}$.

We will assume from this point on that C is coassociative unless otherwise specified. Coalgebras can be regarded as structures formally dual to algebras – or vice versa.

The coassociative axiom for coalgebras has implications for the evaluation of iterated applications of the coproduct just as the associative axiom for algebras has implications for the evaluation of repeated multiplications. For $1 \leq j \leq i$ let $\Delta^{(i,j)} = 1_C \otimes \cdots \otimes \Delta \otimes \cdots \otimes 1_C$ be the tensor product of i linear operators, all of which are the identity map 1_C of C except for the jth which is Δ. Thus $\Delta^{(1,1)} = \Delta, \Delta^{(2,1)} = \Delta \otimes 1_C$, and $\Delta^{(2,2)} = 1_C \otimes \Delta$. It is a nice exercise to show the coassociative axiom implies that for fixed n all of the composites $\Delta^{(n,j_n)}\Delta^{(n-1,j_{n-1})} \ldots \Delta^{(1,1)}$ are the same, where $1 \leq j_i \leq i \leq n$. Let $\Delta^{(n)}$ denote the common value of the composites. We drop the summation sign in the Heyneman-Sweedler notation for expressing $\Delta^{(n-1)}$ and write

$$\Delta^{(n-1)}(c) = c_{(1)} \otimes c_{(2)} \otimes \cdots \otimes c_{(n)}$$

for all $c \in C$.

Definition 1.1.8 *A subspace D of a coalgebra C is a* subcoalgebra *of C if* $\Delta(D) \subseteq D \otimes D$.

Thus a subcoalgebra D of C is a coalgebra in its own right with the restriction $\Delta|_D$ as comultiplication for D and with $\epsilon|_D$ as counit. Notice that C and (0) are subcoalgebras of C.

Definition 1.1.9 *A coalgebra C is* simple *if $C \neq (0)$ and (0) and C are the are the only subcoalgebras of C.*

Observe that the intersection of subcoalgebras of C is a subcoalgebra of C by part e) of Exercise A.4.4. Among all the subcoalgebras containing a given subspace V of C there is a unique minimal one.

Definition 1.1.10 *This unique subcoalgebra just described is the* subcoalgebra of C generated by V.

Notice that k has unique coalgebra structure. At this point we consider other basic examples, some of which we will revisit as we consider more complex structures.

Example 1: Let S be a set and $C = k[S]$ be the vector space over k with basis S. Then C has a coalgebra structure over k determined by

$$\Delta(s) = s \otimes s \quad \text{and} \quad \epsilon(s) = 1 \tag{1.1}$$

for all $s \in S$. If S is empty we set $C = (0)$.

Definition 1.1.11 *Let C be a coalgebra over k. An element $s \in C$ which satisfies (1.1) is a* grouplike element *of C.*

The reader should note that $s \in C$ is a grouplike element if and only if $s \neq 0$ and $\Delta(s) = s \otimes s$.

We denote the set of grouplike elements of C by $G(C)$. It follows that $k[G(C)]$ is always a subcoalgebra of C by virtue of the following:

Lemma 1.1.1 *Let C be a coalgebra over the field k. Then $G(C)$ is linearly independent.*

Proof: If $G(C)$ is not linearly independent then there is a dependency relation among some $s_1, \ldots, s_r \in G(C)$. Assume that r is as small as possible and $\sum_{i=1}^{r} \alpha_i s_i = 0$, where $\alpha_i \in k$, is a dependency relation. Then the s_i's are distinct, $r > 1$, and each $\alpha_i \neq 0$. Thus we have a relation

$$s_r = \sum_{i=1}^{r-1} \beta_i s_i$$

where $\beta_i \in k$ is not zero for all $1 \leq i < r$. By our choice of r the set $\{s_1, \ldots, s_{r-1}\}$ is linearly independent. Applying Δ to both sides of the last equation we obtain

$$(\sum_{i=1}^{r-1} \beta_i s_i) \otimes (\sum_{j=1}^{r-1} \beta_j s_j) = \sum_{i=1}^{r-1} \beta_i s_i \otimes s_i.$$

Expanding the left hand side of this equation and comparing terms with the right hand side we see that $\beta_i \beta_j = 0$ whenever $i \neq j$. Therefore $r - 1 = 1$. Since $r = 2$, the dependency relation is $s_2 = \beta_1 s_1$. Applying ϵ to both sides of this equation yields $\beta_1 = 1$. Consequently $s_2 = s_1$, a contradiction. Hence there is no dependency relation in $G(C)$. ∎

If $C = k[S]$ is the coalgebra of Example 1 then $S = G(k[S])$ by virtue of Lemma 1.1.1. Through this particular coalgebra structure on $k[S]$ we can recover the set S.

When S has an algebraic structure there may be other ways of defining a coalgebra structure on $k[S]$.

Example 2: Suppose that S is a (multiplicative) semigroup with identity e such that each $s \in S$ has at most finitely many factorizations $s = ab$, where $a, b \in S$. Then the vector space $C = k[S]$ over k with basis S is a coalgebra over k where

$$\Delta(s) = \sum_{ab=s} a \otimes b \quad \text{and} \quad \epsilon(s) = \delta_{e,s}$$

for all $s \in S$.

We note that the requirement for S in Example 2 is satisfied when S is a finite group or S is the semigroup of non-negative integers N under addition. The latter arises quite often in practice is usually described in different terms. A linear basis for $C = k[N]$ is frequently denoted $\{c_0, c_1, c_2, \ldots\}$ and the coproduct and counit are described by

$$\Delta(c_n) = \sum_{i=0}^{n} c_{n-i} \otimes c_i \quad \text{and} \quad \epsilon(c_n) = \delta_{0,n}$$

for all $n \geq 0$.

Example 3: Let S be a finite non-empty set and $C = k[S \times S]$ be the vector space over k with basis the Cartesian product $S \times S$. Then C is a coalgebra over k where

$$\Delta((i,j)) = \sum_{\ell \in S} (i,\ell) \otimes (\ell, j) \quad \text{and} \quad \epsilon((i,j)) = \delta_{i,j}$$

for all $(i,j) \in S \times S$.

These examples essentially depend only on the cardinality of S. For $n = |S|$ we denote the coalgebra $C = k[S \times S]$ by $\mathrm{C}_n(k)$ and describe it as the vector space over k with basis $\{e^i_j\}_{1 \leq i,j \leq n}$ whose coalgebra structure is determined by

$$\Delta(e^i_j) = e^i_\ell \otimes e^\ell_j \tag{1.2}$$

and

$$\epsilon(e^i_j) = \delta^i_j \tag{1.3}$$

for all $1 \leq i, j \leq n$. Set $\mathrm{C}_0(k) = (0)$.

Definition 1.1.12 *The coalgebra* $\mathrm{C}_n(k)$ *is a* comatrix coalgebra. *Equations 1.2 and 1.3 are the* comatrix identities.

Definition 1.1.13 *Let C be a coalgebra over k. A subset of elements of C described by* $\{e^i_j\}_{1 \leq i,j \leq n}$ satisfies the comatrix identities *if they satisfy (1.2) and (1.3).*

Observe that the span of a set of elements of C satisfying the comatrix identities is a subcoalgebra of C.

Definition 1.1.14 *Any linear basis* $\{e^i_j\}_{1 \leq i,j \leq n}$ *for* $\mathrm{C}_n(k)$ *which satisfies the comatrix identities is a* standard basis for $\mathrm{C}_n(k)$.

Suppose that (A, m_A, η_A) and (B, m_B, η_B) are algebras over k and $f : A \longrightarrow B$ is a linear map.

Definition 1.1.15 *The linear map $f : A \longrightarrow B$ is an* algebra map *if $f(ab) = f(a)f(b)$ for all $a, b \in A$ and $f(1) = 1$.*

The conditions of the definition are expressed by the commutativity of the diagrams

$$\begin{array}{ccc} A\otimes A & \xrightarrow{f\otimes f} & B\otimes B \\ {\scriptstyle m_A}\big\downarrow & & \big\downarrow{\scriptstyle m_B} \\ A & \xrightarrow[f]{} & B \end{array}$$

and

$$\begin{array}{ccc} A & \xrightarrow{\quad f \quad} & B \\ & {\scriptstyle \eta_A}\nwarrow \qquad \nearrow{\scriptstyle \eta_B} & \\ & k & \end{array}$$

Suppose $(C, \Delta_C, \epsilon_C)$ and $(D, \Delta_D, \epsilon_D)$ are coalgebras over k and $f : C \longrightarrow D$ is a linear map. Then dually:

Definition 1.1.16 *The linear map $f : C \longrightarrow D$ is a* coalgebra map *if the diagrams*

$$\begin{array}{ccc} C\otimes C & \xrightarrow{f\otimes f} & D\otimes D \\ {\scriptstyle \Delta_C}\big\uparrow & & \big\uparrow{\scriptstyle \Delta_D} \\ C & \xrightarrow[f]{} & D \end{array}$$

and

$$\begin{array}{ccc} C & \xrightarrow{\quad f \quad} & D \\ & {\scriptstyle \epsilon_C}\searrow \qquad \swarrow{\scriptstyle \epsilon_D} & \\ & k & \end{array}$$

commute.

Thus to require $f : C \longrightarrow D$ to be a coalgebra map is to require

$$\Delta(f(c)) = f(c_{(1)}) \otimes f(c_{(2)}) \quad \text{and} \quad \epsilon(f(c)) = \epsilon(c)$$

for all $c \in C$. A coalgebra map is an isomorphism if it is one-one and onto.

Let $f : C \longrightarrow D$ be a coalgebra map. Then $\epsilon f = \epsilon$ means that $\epsilon(\mathrm{Ker} f) = (0)$. Since $\Delta f = (f \otimes f)\Delta$ we conclude that $\Delta(\mathrm{Ker} f) \subseteq \mathrm{Ker}(f \otimes f)$. As $\mathrm{Ker}(f \otimes f) = (\mathrm{Ker} f) \otimes C + C \otimes (\mathrm{Ker} f)$ it follows that $\Delta(\mathrm{Ker} f) \subseteq (\mathrm{Ker} f) \otimes C + C \otimes (\mathrm{Ker} f)$.

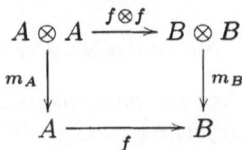

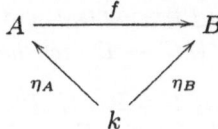

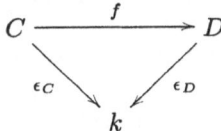

Definition 1.1.17 *A subspace I of a coalgebra C which satisfies $\epsilon(I) = (0)$ and $\Delta(I) \subseteq I \otimes C + C \otimes I$ is a* coideal *of C.*

We leave it as an exercise for the reader to establish the fundamental homomorphism theorem for coalgebras.

Theorem 1.1.1 *Suppose that C is a coalgebra over the field k.*

a) *Let I be a coideal of C. Then the quotient space C/I has a unique coalgebra structure so that the linear projection $\pi : C \longrightarrow C/I$ is a coalgebra map.*

Suppose D is also a coalgebra over k and that $f : C \longrightarrow D$ is a coalgebra map. Then:

b) $\mathrm{Ker} f$ *is a coideal of C, and if I is a coideal of C such that $I \subseteq \mathrm{Ker} f$ then there is a unique coalgebra map $F : C/I \longrightarrow D$ such that $F\pi = f$.*

c) *Suppose that f is onto and let $I = \mathrm{Ker} f$. Then there is a unique isomorphism $F : C/I \longrightarrow D$ of coalgebras such that $F\pi = f$.*

∎

We end this section by showing that a finite-dimensional coalgebra over k is the homomorphic image of $C_n(k)$ for some $n \geq 0$.

Proposition 1.1.1 *Suppose that C is an n-dimensional coalgebra over the field k. Then C is the homomorphic image of $\mathrm{C}_n(k)$.*

Proof: We may as well assume that $n > 0$. Choose any basis $\{m_1, \ldots, m_n\}$ for C. For each $1 \leq j \leq n$ we write

$$\Delta(m_j) = m_i \otimes c^i_j$$

where $c^i_j \in C$. It is an easy but important exercise to show that the c^i_j's satisfy the comatrix identities. The desired map is $\pi : \mathrm{C}_n(k) \longrightarrow C$ defined by $\pi(e^i_j) = c^i_j$. ∎

There are close connections between algebras and coalgebras as one might suspect given their dual nature. We will explore some of them in the next two sections.

Throughout the following exercises C and D are coalgebras over the field k.

Exercise 1.1.1 Let $C = k[S]$ be the coalgebra of Example 1.

a) Show that the difference $s - s'$, where $s, s' \in S$, spans a coideal of C.

b) Show that any coideal of C is spanned by certain differences $s - s'$, where $s, s' \in S$.

Exercise 1.1.2 Suppose that $f : C \longrightarrow D$ is a coalgebra map.

a) Show that $f(G(C)) \subseteq G(D)$.

b) Show that if $\mathcal{C}$ is a subcoalgebra of C then $f(\mathcal{C})$ is a subcoalgebra of D.

c) Show that if I is a coideal of C then $f(I)$ is a coideal of D.

d) Show that if J is a coideal of D then $f^{-1}(J)$ is a coideal of C.

e) Give an example to show that if $\mathcal{D}$ is a subcoalgebra of D then $f^{-1}(\mathcal{D})$ is not necessarily a subcoalgebra of C.

Exercise 1.1.3 Suppose $g, h \in G(C)$. Set $P_{g,h}(C) = \{c \in C \mid \Delta(c) = g \otimes c + c \otimes h\}$.

a) Show that any subspace of $P_{g,h}(C)$ is a coideal of C.

b) Show that $\mathcal{C} = U + kg + kh$ is a subcoalgebra of C, where U is any subspace of $P_{g,h}(C)$.

Definition 1.1.18 *The elements of $P_{g,h}(C)$ are* skew primitive *(*g:h-skew primitive *when more precision is required).*

We let $P_g(C)$ stand for $P_{g,g}(C)$.

Exercise 1.1.4 Show that $C \otimes D$ is a coalgebra, where

$$\Delta(c \otimes d) = (c_{(1)} \otimes d_{(1)}) \otimes (c_{(2)} \otimes d_{(2)})$$

and

$$\epsilon(c \otimes d) = \epsilon_C(c)\epsilon_D(d)$$

for all $c \in C$ and $d \in D$.

Definition 1.1.19 *The coalgebra structure just defined is the* tensor product coalgebra structure.

Exercise 1.1.5 Show that $\mathrm{C}_m(k) \otimes \mathrm{C}_n(k) \simeq \mathrm{C}_{mn}(k)$ as coalgebras for all $m, n \geq 1$, where the tensor product has the tensor product coalgebra structure.

Exercise 1.1.6 Show that $\mathrm{C}_n(k)$ is a simple coalgebra for all $n \geq 1$.

Exercise 1.1.7 Show that:

a) $\epsilon : C \longrightarrow k$ is a coalgebra map, and

b) $\Delta : C \longrightarrow C \otimes C$ is a coalgebra map if and only if C is cocommutative, where $C \otimes C$ is given the tensor product coalgebra structure.

Exercise 1.1.8 Show that all of the ways described in this section for computing $\Delta^{(n)}$ do in fact give equal results.

Exercise 1.1.9 Suppose that C is cocommutative. Show that for all $c \in C, n \geq 2$, and permutations $\sigma \in S_n$ that

$$c_{(1)} \otimes \cdots \otimes c_{(n)} = c_{(\sigma 1)} \otimes \cdots \otimes c_{(\sigma n)}.$$

Definition 1.1.20 *An element* $c \in C$ *is* cocommutative *if* $\Delta(c) = \Delta^{cop}(c)$.

Exercise 1.1.10 To say that C is cocommutative is to say that $\Delta(c) = \Delta^{cop}(c)$ for all $c \in C$. More generally,

a) Show that the set of cocommutative elements of C, denoted by coCom$C$, is a subspace of C.

b) Show that Dim coCom$\mathrm{C}_n(k) = 1$ for all $n \geq 1$. (Thus coCom$C$ is not necessarily a subcoalgebra of C.)

Exercise 1.1.11 Show that objects which are coalgebras over k and morphisms which are coalgebras maps with composition as multiplication form a category (the category of coalgebras over k, denoted by ${}_k$Coalg).

Exercise 1.1.12 Show that ${}_k$Coalg has direct sums.

1.2 The Algebra C*

Let C be a coalgebra over a field k. In this section we show that the space of functionals C^* on C has an algebra structure arising from the coalgebra structure on C. We refer to this algebra as the dual algebra of C. The main focus of the section is an examination of the relationship between the algebraic structures of C and those of the dual algebra C^*.

Suppose that C is a vector space over k and $\Delta : C \longrightarrow C \otimes C$ is a linear map. Define a linear map $m : C^* \otimes C^* \longrightarrow k$ by $m(c^* \otimes d^*) = c^*d^*$, where

$$<c^*d^*, c> = <c^*, c_{(1)}><d^*, c_{(2)}>$$

for all $c \in C$.

Proposition 1.2.1 *Suppose that* C *is a vector space over the field k. Let* $\Delta : C \longrightarrow C \otimes C$ *and* $\epsilon : C \longrightarrow k$ *be linear maps. Then* (C^*, m, η) *is an algebra over k if and only if* (C, Δ, ϵ) *is a coalgebra over k, where* m *is as defined above and* $\eta(1) = \epsilon$. ∎

We will leave the proof of the proposition as an exercise to the reader.

Definition 1.2.1 *If* (C, Δ, ϵ) *is a coalgebra over k, the algebra* (C^*, m, η) *is the* dual algebra *of* (C, Δ, ϵ).

We will frequently denote the dual algebra of C by C^*.

Transposes of coalgebra maps are algebra maps. The following is not hard to see.

Proposition 1.2.2 *Suppose that* C *and* D *are coalgebras over the field k. Then a linear map* $f : C \longrightarrow D$ *is a coalgebra map if and only if* $f^* : D^* \longrightarrow C^*$ *is an algebra map.* ∎

We now turn our attention to the algebraic connections between a coalgebra C and its dual algebra C^*. By the results of Section A.4.2 it follows that there is a one-one inclusion reversing correspondence $U \longmapsto U^\perp$ between the subspaces of C and the closed subspaces of C^*, where $U^\perp = \{c^* \in C^* \mid c^*(U) = (0)\}$. We will be interested in how algebraic structures relate via this correspondence.

First of all observe that C is a C^*-bimodule, where the actions are given by

$$c^* \rightharpoonup c = c_{(1)} <c^*, c_{(2)}> \quad \text{and} \quad c \leftharpoonup c^* = <c^*, c_{(1)}> c_{(2)}$$

for all $c^* \in C^*$ and $c \in C$. For $c^* \in C^*$ let $\ell_{c^*}, r_{c^*} \in \mathrm{End}(C^*)$ be defined by

$$\ell_{c^*}(d^*) = c^* d^* \quad \text{and} \quad r_{c^*}(d^*) = d^* c^*$$

for all $d^* \in C^*$.

Proposition 1.2.3 *Suppose that C is a coalgebra over the field k and C^* is the dual algebra of C. Then:*

a) *ℓ_{c^*} and r_{c^*} are continuous maps.*

b) *Finitely generated left or right ideals of C^* are closed.*

c) *Suppose that K, L, and M are subspaces of C^*. If $KL \subseteq M$ then $\overline{K}\,\overline{L} \subseteq \overline{M}$.*

Proof: For $c^* \in C^*$ let $L_{c^*}, R_{c^*} \in \mathrm{End}(C)$ be defined by $L_{c^*}(c) = c^* \rightharpoonup c$ and $R_{c^*}(c) = c \leftharpoonup c^*$ for all $c \in C$. Since $\ell_{c^*} = R^*_{c^*}$ and $r_{c^*} = L^*_{c^*}$ part a) follows. Now $\mathrm{Im}\ell_{c^*} = c^* C^*$ and $\mathrm{Im} r_{c^*} = C^* c^*$ are closed subspaces of C^* by part a) of Proposition A.4.5. Since a finite sum of closed subspaces of C^* is closed by part b) of Proposition A.4.4, part b) is established.

To show part c) we will use the fact that ℓ_{c^*} and r_{c^*} are continuous. Suppose that K, L, and M are subspaces of C^* and $KL \subseteq M$. Let $c^* \in K$. Then $\ell_{c^*}(L) \subseteq M$, so by part a) of Corollary A.4.2 we calculate $c^* \overline{L} = \ell_{c^*}(\overline{L}) = \overline{\ell_{c^*}(L)} \subseteq \overline{M}$. Therefore $c^* \overline{L} \subseteq \overline{M}$. Now suppose that $c^* \in \overline{L}$. Then same argument applied to $r_{c^*}(K)$ shows that $\overline{K} c^* \subseteq \overline{M}$. Thus $\overline{K}\,\overline{L} \subseteq \overline{M}$. This concludes the proof of part c). ∎

For vector spaces U and V over k we will think of $U^* \otimes V^*$ as a subspace of $(U \otimes V)^*$ by the identification $(u^* \otimes v^*)(u \otimes v) = u^*(u) v^*(v)$. See the discussion preceding Lemma A.4.3.

The multiplication map m for the dual algebra C^* is given by $m = \Delta^*|_{C^* \otimes C^*}$. Suppose that K, L and M are subspaces of C^*. Then $KL \subseteq M$ if and only if $\Delta^*(K \otimes M) \subseteq M$. The latter implies that $(\Delta^*(K \otimes M))^\perp \supseteq M^\perp$. By part b) of Corollary A.4.2 and Lemma A.4.3 we compute $(\Delta^*(K \otimes M))^\perp = \Delta^{-1}((K \otimes L)^\perp) = \Delta^{-1}(K^\perp \otimes C + C \otimes L^\perp)$. We have shown that

$$KL \subseteq M \quad \text{implies} \quad \Delta(M^\perp) \subseteq K^\perp \otimes C + C \otimes L^\perp. \tag{1.4}$$

Conversely, if U, V, and W are subspaces of C then by definition of multiplication in C^* it follows that

$$\Delta(U) \subseteq V \otimes C + C \otimes W \quad \text{implies} \quad V^{\perp}W^{\perp} \subseteq U^{\perp}. \tag{1.5}$$

Definition 1.2.2 *A subspace I of C is a* left coideal *if $\Delta(I) \subseteq C \otimes I$ and is a* right coideal *if $\Delta(I) \subseteq I \otimes C$.*

Proposition 1.2.4 *Suppose that C is a coalgebra over the field k. Let C^* be the dual algebra and suppose that A is a subalgebra of C^*. Then:*

a) *If A is a dense subalgebra C^* and I is a subalgebra (respectively ideal, left ideal, right ideal) of A then $I^{\perp}$ is a coideal (respectively subcoalgebra, left coideal, right coideal) of C.*

b) *If U is a coideal (respectively subcoalgebra, left coideal, right coideal) of C then $U^{\perp} \cap A$ is a subalgebra (respectively ideal, left ideal, right ideal) of A.*

c) *If I is a subalgebra (respectively ideal, left ideal, right ideal) of C^* then $\overline{I}$ is a subalgebra (respectively ideal, left ideal, right ideal) of C^*.*

d) *The one-one inclusion reversing correspondence $U \longmapsto U^{\perp}$ between the subspaces of C and the closed subspaces of C^* induces a one-one correspondence between the coideals (respectively subcoalgebras, left coideals, right coideals) of C and the closed subalgebras (respectively closed ideals, closed left ideals, closed right ideals) of C^*.*

Proof: Part a) follows by part c) of Proposition 1.2.3 and (1.4). Part b) follows by (1.5). Part c) follows by part c) of Proposition 1.2.3 or by parts a) and b) with $A = C^*$. Part d) follows from parts a) and b). This concludes the proof of the proposition. ∎

Let C be a finite-dimensional coalgebra over the field k. Then *all* subspaces of C^* are closed by Proposition A.4.2. Thus the correspondence of part d) of Proposition 1.2.4 relates structures of C and *all* subalgebras, ideals, left ideals and right ideals of C^*.

Definition 1.2.3 *The* wedge product *of subspaces U and V of C is*

$$U \wedge V = \Delta^{-1}(U \otimes C + C \otimes V).$$

The wedging operation in C corresponds to multiplication in C^*.

Lemma 1.2.1 *Suppose that C is a coalgebra over the field k and let K and L be subspaces of C^*. Then $(KL)^{\perp} = K^{\perp} \wedge L^{\perp}$.*

Proof: By (1.4) we see that $(KL)^{\perp} \subseteq K^{\perp} \wedge L^{\perp}$. By (1.5) it follows that $\overline{K}\,\overline{L} \subseteq (K^{\perp} \wedge L^{\perp})^{\perp}$. Therefore $K^{\perp} \wedge L^{\perp} \subseteq (\overline{K}\,\overline{L})^{\perp}$. By part c) of Proposition 1.2.3 it follows that $\overline{K}\,\overline{L}$ is a dense subspace of $\overline{KL}$. Consequently $(\overline{K}\,\overline{L})^{\perp} = (KL)^{\perp}$, and thus $K^{\perp} \wedge L^{\perp} \subseteq (KL)^{\perp}$ which is what we needed to establish that $K^{\perp} \wedge L^{\perp} = (KL)^{\perp}$. ∎

Corollary 1.2.1 *Suppose that C is a coalgebra over the field k. Then:*

a) $U \wedge V = (U^{\perp}V^{\perp})^{\perp}$ *for all subspaces U and V of C.*

b) $U \wedge (V \wedge W) = (U \wedge V) \wedge W$ *for all subspaces U, V, and W of C.*

Proof: Part a) follows by Lemma 1.2.1 with $K = U^{\perp}$ and $L = V^{\perp}$. We use part a) to calculate

$$(U \wedge V) \wedge W = ((U \wedge V)^{\perp}W^{\perp})^{\perp} = ((U^{\perp}V^{\perp})^{\perp\perp}W^{\perp})^{\perp} = ((U^{\perp}V^{\perp})W^{\perp})^{\perp}.$$

Likewise $U \wedge (V \wedge W) = (U^{\perp}(V^{\perp}W^{\perp}))^{\perp}$. This concludes our proof. ∎

Since wedging is associative, the wedge product of subspaces $U_1, \ldots, U_n$ of C defined inductively by $U_1 \wedge U_2 \wedge \cdots \wedge U_n = U_1 \wedge (U_2 \wedge \cdots \wedge U_n)$ can be evaluated according to any full parenthesization of the formal expression $U_1 \wedge U_2 \wedge \cdots \wedge U_n$.

We will let the reader explore implications of Proposition 1.2.4 and part a) of Corollary 1.2.1 for the wedge product. We do note the important fact that the wedge product $D \wedge E$ of two subcoalgebras D and E of C is a subcoalgebra of C and $D, E \subseteq D \wedge E$.

Let U be a subspace of C. Set $\wedge^0 U = U$ and $\wedge^n U = U \wedge \cdots \wedge U$ (n-wedges) for $n \geq 1$. If D is a subcoalgebra of C then $\wedge^n D$ is a subcoalgebra of C for all $n \geq 0$ and $D = \wedge^0 D \subseteq \wedge^1 D \subseteq \wedge^2 D \subseteq \cdots$.

Note that (0) is a coideal of C and that the sum of coideals of C is a coideal of C. Therefore there is a unique coideal of C which is maximal among all those contained in U. Notice that a subspace I of C is a coideal of C if and only if $\epsilon(I) = (0)$ and $I \subseteq I \wedge I$.

Corollary 1.2.2 *Suppose that C is a coalgebra over the field k.*

a) *Let $U_1, \ldots, U_n$ be subspaces of C and $I_1, \ldots, I_n$ be subspaces of C^* such that $U_i = I_i^{\perp}$ for $1 \leq i \leq n$. Then $U_1 \wedge \cdots \wedge U_n = (I_1 \cdots I_n)^{\perp}$.*

b) *Let U be a subspace of C and set $V = U \cap (\mathrm{Ker}\epsilon)$. Then the unique coideal of C maximal among those contained in U is $\mathcal{I} = \cap_{n=0}^{\infty}(\wedge^n V)$.*

Proof: We will prove part a) by induction on n. We may assume $n > 1$. Now $U_1 \wedge \cdots \wedge U_n = (U_1^{\perp}(U_2 \wedge \cdots \wedge U_n)^{\perp})^{\perp}$ by part a) of Corollary 1.2.1. Let $K = I_1$ and $L = I_2 \cdots I_n$. By assumption $U_1^{\perp} = \overline{K}$. By our induction hypothesis

$(U_2 \wedge \cdots \wedge U_n)^\perp = \overline{L}$. By part c) of Proposition 1.2.3 it follows that $\overline{K}\,\overline{L}$ is a dense subspace of $\overline{KL}$. Therefore $(KL)^\perp = (\overline{K}\,\overline{L})^\perp$ which completes the proof of part a).

To show part b) we let I be any coideal of C contained in U. Then $I \subseteq \mathrm{Ker}\epsilon$ which means that $I \subseteq V$. As $I \subseteq I \wedge I$ it follows by induction that $I \subseteq \wedge^n V$ for all $n \geq 0$. Therefore $I \subseteq \cap_{n=0}^{\infty}(\wedge^n V) = \mathcal{I}$. Set $J = V^\perp$. Then using part a) we compute

$$\mathcal{I} = \cap_{n=1}^{\infty}(J^n)^\perp = (\sum_{n=1}^{\infty} J^n)^\perp.$$

But $\mathcal{B} = \sum_{n=1}^{\infty} J^n$ is a subalgebra of C^* since $\epsilon \in J$. Therefore $\mathcal{I} = \mathcal{B}^\perp$ is a coideal of C by part a) of Proposition 1.2.4. Since $I \subseteq \mathcal{I} \subseteq U$ the proof of part b) is complete.
∎

Throughout the following exercises C and D are coalgebras over the field k.

Exercise 1.2.1 Let $n \geq 1$ and $\{E^i_j\}_{1\leq i,j\leq n}$ be the dual basis for $\mathrm{C}_n(k)^*$ of a standard basis $\{e^i_j\}_{1\leq i,j\leq n}$ for the coalgebra $\mathrm{C}_n(k)$ of Example 3 of Section 1.1. Show that

$$E^i_j E^k_\ell = \delta^k_j E^i_\ell$$

for all $1 \leq i,j,k,\ell \leq n$ and therefore $\mathrm{C}_n(k)^* \simeq \mathrm{M}_n(k)$, the algebra of $n \times n$ matrices over k.

Exercise 1.2.2 Suppose that A is an n-dimensional algebra over k. Then there is a one-one algebra map from A to the algebra of $n \times n$ matrices $\mathrm{M}_n(k)$ over k. What is the connection between this fact and Proposition 1.1.1?

Exercise 1.2.3 Let $C = k[S]$ be the coalgebra of Example 1 of Section 1.1. Show that C^* is isomorphic to the algebra of functions from S to k under point-wise multiplication.

Exercise 1.2.4 Let V be a vector space over k and g be a symbol. Endow $C = kg \oplus V$ with the coalgebra structure determined by

$$\Delta(g) = g \otimes g$$

and

$$\Delta(v) = g \otimes v + v \otimes g$$

for all $v \in V$. Show that $C^* = k\epsilon \oplus I$ as a vector space, where $I^2 = (0)$.

Exercise 1.2.5 Let C be the special case of Example 2 of Section 1.1 which has linear basis $\{c_0, c_1, \ldots\}$ and coalgebra structure determined by $\Delta(c_n) = \sum_{i=0}^{n} c_{n-i} \otimes c_i$ and $\epsilon(c_n) = \delta_{0,n}$ for all $n \geq 0$.

a) Show that $\pi : k[[X]] \longrightarrow C^*$ defined by $\pi(\sum_{n=0}^{\infty} \alpha_n X^n)(c_m) = \alpha_m$ is an algebra isomorphism of the algebra of formal power series in X over k and the dual algebra C^*.

b) Define $X \in C^*$ by $X(c_n) = \delta_{1,n}$ for all $n \geq 0$. Show that the subalgebra of C^* generated by X is dense and is the polynomial algebra in X over k.

Exercise 1.2.6 We will generalize the previous exercise. Let $S = N \times \cdots \times N$ be the n-fold Cartesian product of the additive semigroup of non-negative integers. Then $C = k[S]$ has basis $\{c_{\bar{n}} \mid \bar{n} \in S\}$ and

$$\Delta(c_{\bar{n}}) = \sum_{\bar{a}+\bar{b}=\bar{n}} c_{\bar{a}} \otimes c_{\bar{b}} \quad \text{and} \quad \epsilon(\bar{n}) = \delta_{\bar{0},\bar{n}}$$

for all $\bar{n} \in S$.

a) Show that $k[[X_1, \ldots, X_n]] \simeq C^*$ as algebras.

b) Show that $k[X_1, \ldots, X_n]$ can be realized as a dense subalgebra of C^*.

Exercise 1.2.7 Show that a maximal ideal of C^* is either a dense subspace of C^* or is a closed ideal of C^*.

Exercise 1.2.8 Let D be a subcoalgebra of C. Show that if $D^{\perp}$ is a maximal ideal of C^* then D is a simple subcoalgebra of C.

Exercise 1.2.9 Suppose that $C^* = I_1 \oplus \cdots \oplus I_r$ is the direct sum of left (or right) ideals of C^*. Show that $I_1, \ldots, I_r$ are closed.

Exercise 1.2.10 Redo Exercise 1.1.8 by describing an equivalent problem for the dual algebra C^* and solving it.

Exercise 1.2.11 Show that C is cocommutative if and only if the dual algebra C^* is commutative.

Exercise 1.2.12 Suppose that C is cocommutative. Redo Exercise 1.1.9 by describing an equivalent problem for the commutative algebra C^* and solving it.

Exercise 1.2.13 Suppose that A is an algebra over k.

a) Show that there is an algebra structure on $\mathrm{Hom}(C, A)$ where multiplication is defined by

$$(f{*}g)(c) = f(c_{(1)})g(c_{(2)})$$

for all $f, g \in \mathrm{Hom}(C, A)$ and $c \in C$ and the unit is $\eta\epsilon$.

Definition 1.2.4 *The algebra* $\mathrm{Hom}(C, A)$ *is the* convolution algebra *of C and A.*

b) Show that $\mathrm{Hom}(k, A) \simeq A$ and $\mathrm{Hom}(C, k) \simeq C^*$ as algebras.

Exercise 1.2.14 The tensor product $A \otimes B$ of two algebras A and B over k is an algebra over k where $(a \otimes b)(a' \otimes b') = aa' \otimes bb'$ and $1_A \otimes 1_B$ is the unit of $A \otimes B$.

Definition 1.2.5 *The algebra structure just described is the* tensor product algebra structure.

Show that the one-one map $\iota : C^* \otimes D^* \longrightarrow (C \otimes D)^*$ defined by $\iota(c^* \otimes d^*)(c \otimes d) = c^*(c)d^*(d)$ is an algebra map, where $C^* \otimes D^*$ has the tensor product algebra structure and $(C \otimes D)^*$ is the dual algebra of $C \otimes D$ with the tensor product coalgebra structure.

Exercise 1.2.15 In this exercise we relate coderivations and certain types of derivations.

Definition 1.2.6 *A linear endomorphism $f : C \longrightarrow C$ is a* coderivation *of C if*

$$\Delta f = (1_C \otimes f + f \otimes 1_C)\Delta.$$

Definition 1.2.7 *A linear endomorphism $F : A \longrightarrow A$ of an algebra A is a* derivation *of A if*

$$D(ab) = D(a)b + aD(b)$$

for all $a, b \in A$.

Show that f is a coderivation of C if and only if $F = f^*$ is a derivation of C^*. (Thus $f \longmapsto f^*$ gives a one-one correspondence between the coderivations of C and the *continuous* derivations of C^*.)

Exercise 1.2.16 Very important nonassociative algebras over k are Lie algebras.

Definition 1.2.8 *Suppose that k does not have characteristic 2. Then a* Lie algebra *over k is a pair (A, m), where $m : A \otimes A \longrightarrow A$ is a linear map satisfying*

$$m(a \otimes b) + m(b \otimes a) = 0$$

and

$$m(a \otimes m(b \otimes c)) + m(c \otimes m(a \otimes b)) + m(b \otimes m(c \otimes a)) = 0$$

for all $a, b, c \in A$.

(Writing $m(a \otimes b) = [a, b]$ the axioms translate to $[a, b] = -[b, a]$ and $[a, [b, c]] + [c, [a, b]] + [b, [c, a]] = 0$ for all $a, b, c \in A$. The usual axiom $[a, a] = 0$ is equivalent to $[a, b] = -[b, a]$ since the characteristic of k is not 2.)

Definition 1.2.9 *A* Lie coalgebra *over k is a pair (C, δ), where C is a vector space over k and $\delta : C \longrightarrow C \otimes C$ is a linear map satisfying*

$$c_{(1)} \otimes c_{(2)} + c_{(2)} \otimes c_{(1)} = 0$$

and

$$c_{(1)} \otimes c_{(2)(1)} \otimes c_{(2)(2)} + c_{(2)(2)} \otimes c_{(1)} \otimes c_{(2)(1)} + c_{(2)(1)} \otimes c_{(2)(2)} \otimes c_{(1)} = 0$$

for all $c \in C$, where we write $\delta(c) = c_{(1)} \otimes c_{(2)}$.

a) Suppose that (C, δ) is a Lie coalgebra over k. Show that (C^*, m) is a Lie algebra over k, where $m = \delta^*|_{C^* \otimes C^*}$.

b) Suppose that (A, m) is a finite-dimensional Lie algebra over k. Show that (A^*, m^*) is a Lie coalgebra over k.

For a treatment of Lie coalgebras see [Michaelis, 1980].

1.3 The Coalgebra A^o

Let C be a coalgebra over the field k. In Section 1.2 we used the transpose of the coproduct to define a product for C^*. Suppose that A is an algebra over the field k. We would like to use the transpose of multiplication to define a coproduct for A^*. This usually can not be done. However there is a unique subspace A^o of A^* maximal among those for which this procedure is meaningful.

Regard A^* as an A-bimodule where

$$(a{\cdot}f)(b) = f(ba) = f{\cdot}b(a)$$

for $a, b \in A$ and $f \in A^*$. We will think of $A^* \otimes A^*$ as a subspace of $(A \otimes A)^*$ where $(a^* \otimes b^*)(a \otimes b) = a^*(a)b^*(b)$. See the paragraph preceding the proof of Lemma A.4.3.

1.3.1 The Construction and Characterizations of A^o

Let A^o be the subset of A^* consisting of all $f \in A^*$ such that there exists an element $\Delta_f \in A^* \otimes A^*$ satisfying

$$f(ab) = \Delta_f(a \otimes b)$$

for all $a, b \in A$. It is easy to see that A^o is a subspace of A^*. Moreover:

Lemma 1.3.1 *Suppose that A is an algebra over the field k. Then A^o is a locally finite A-sub-bimodule of A^*.*

Proof: Let $f \in A^o$ and write $\Delta_f = \sum_{i=1}^r f_i \otimes g_i$. Then for $a, b, x, y \in A$ we have

$$\begin{aligned}(a{\cdot}f{\cdot}b)(xy) &= f(bxya)\\ &= \sum_{i=1}^r f_i(bx)g_i(ya)\\ &= \sum_{i=1}^r (f_i{\cdot}b)(x)(a{\cdot}g_i)(y).\end{aligned}$$

Therefore $\Delta_{a{\cdot}f{\cdot}b}$ exists and

$$\Delta_{a\cdot f\cdot b} = \sum_{i=1}^r (f_i{\cdot}b) \otimes (a{\cdot}g_i).$$

Since A^o is a subspace of A^*, our calculation shows that A^o is an A-sub-bimodule of A^*. By definition of A^o we have that $a{\cdot}f = \sum_{i=1}^r g_i(a)f_i$ and $f{\cdot}a = \sum_{i=1}^r f_i(a)g_i$. Thus A^o is a locally finite left and right A-submodule of A^*. ∎

We will continue with the calculation begun in the proof the lemma. Assume further that $f \neq 0$ and that $r = \text{Rank}\Delta_f$. Then the f_i's and the g_i's form linearly independent sets by virtue of Lemma A.4.1. Fix $1 \leq i \leq r$. Then there is an $a \in A$ such that $g_j(a) = \delta_{i,j}$ for $1 \leq j \leq r$ by Corollary A.4.1. Thus for $x \in A$ we calculate

$$f(xa) = \sum_{j=1}^{r} f_j(x)g_j(a) = f_i(x).$$

Therefore $f_i = a{\cdot}f$. Likewise choose $b \in A$ such that $f_j(b) = \delta_{i,j}$ for all $1 \leq j \leq r$. Then

$$f(bx) = \sum_{j=1}^{r} f_j(b)g_j(x) = g_i(x)$$

which means that $g_i = f{\cdot}b$. We have shown that

$$\Delta_f \in C \otimes C \quad \text{where} \quad C = A{\cdot}f{\cdot}A. \tag{1.6}$$

By virtue of Lemma 1.3.1 and (1.6) there exists a linear map $\Delta : A^o \longrightarrow A^o \otimes A^o$ defined by $\Delta(f) = \Delta_f$ for $f \in A^o$. The property $\Delta(f)(a \otimes b) = f(ab)$ for all $a, b \in A$ uniquely determines Δ. Let $\epsilon : A^o \longrightarrow k$ be the linear map defined by $\epsilon(f) = f(1)$ for $f \in A^o$.

Proposition 1.3.1 *Suppose that A is an algebra over the field k. Then:*

a) *(A^o, Δ, ϵ) is a coalgebra.*

b) *Suppose that B is an algebra over k and $f : A \longrightarrow B$ is an algebra map. Let $F : B^* \longrightarrow A^*$ be the transpose of f. Then $F(B^o) \subseteq A^o$ and the restriction $f^o = F|_{B^o}$ is a coalgebra map $f^o : B^o \longrightarrow A^o$.*

Proof: Part a) is left as an straightforward exercise for the reader. We will show part b).

Let $b^o \in B^o$ and write $\Delta(b^o) = b^o_{(1)} \otimes b^o_{(2)} \in B^o \otimes B^o$. Then for $a, b \in A$ we compute

$$\begin{aligned} F(b^o_{(1)})(a)F(b^o_{(2)})(b) &= b^o_{(1)}(f(a))b^o_{(2)}(f(b)) \\ &= b^o(f(a)f(b)) \\ &= b^o(f(ab)) \\ &= F(b^o)(ab). \end{aligned}$$

Therefore $\Delta_{F(b^o)}$ exists and $\Delta_{F(b^o)} = F(b^o_{(1)}) \otimes F(b^o_{(2)})$. By definition $F(b^o) \in A^o$. As $\epsilon f^o = \epsilon$ the remainder of part b) now follows. ∎

Definition 1.3.1 *Let A be an algebra over k. The coalgebra A^o described in the previous proposition is the* dual coalgebra *of A.*

Of fundamental importance to the study of A^o is the characterization of the subcoalgebras of A^o generated by a single element.

Theorem 1.3.1 *Suppose that A is an algebra over the field k and let $f \in A^o$. Let $C = A{\cdot}f{\cdot}A$ be the A-sub-bimodule of A^o generated by f. Then:*

a) *C is the subcoalgebra of A^o generated by f.*

b) *C is finite-dimensional.*

Let $I = \{a \in A \mid g(a) = 0 \quad \text{for all} \quad g \in C\}$. Then:

c) *I is a cofinite ideal of A.*

d) *Let $\pi : A \longrightarrow A/I$ be the projection. Then $\mathrm{Im}\pi^* = C$ and π^* induces an isomorphism of coalgebras $(A/I)^* \simeq C$. In particular $C = I^\perp$.*

Proof: We first show part a). By definition of the coproduct of A^o any subcoalgebra of A^o is a left and right A-submodule of A^o. By (1.6) we have that $\Delta(C) \subseteq C \otimes C$. Thus part a) follows. Part b) follows since A^o is a locally finite A-sub-bimodule of A^*.

We now show part c). Since C is a left and right A-submodule of A^o it follows that I is an ideal of A. Since C is finite-dimensional we conclude that I is a cofinite subspace of A (I is the kernel of the map $\iota : A \longrightarrow C^*$ given by $\iota(a)(c) = c(a)$). By part b) of Proposition 1.3.1 the transpose of π restricts to a coalgebra map $\pi^o : (A/I)^o \longrightarrow A^o$. Since A/I is finite-dimensional it follows that $(A/I)^* = (A/I)^o$. Now π onto means that π^o is one-one. Since $C \subseteq \mathrm{Im}\pi^o$ and is identified with a dense subspace of $(A/I)^*$, it follows that $\mathrm{Im}\pi^o = C$ since all subspaces of $(A/I)^*$ are closed. We have completed the proof of part d) and hence the proof of the theorem. ∎

As a consequence of Theorem 1.3.1 there are several different characterizations of A^o. For an ideal I of A let $\pi_I : A \longrightarrow A/I$ denote the projection of A onto A/I.

Corollary 1.3.1 *Suppose that A is an algebra over the field k. Then:*

a) *$A^o = \sum_I \mathrm{Im}\pi_I^o = \sum_I I^\perp$, where I runs over the cofinite ideals of A.*

b) *A^o is the unique maximal locally finite left A-submodule of A^*.*

c) *A^o is the unique maximal locally finite right A-submodule of A^*.*

Proof: In light of the preceding theorem and part b) of Proposition 1.3.1 we need only show parts b) and c). We only show part b) since the proof of part c) is similar.

Suppose that M is the unique maximal locally finite left A-submodule of A^*. Then $A^o \subseteq M$ by part a). Conversely, assume $f \in M$. We may suppose that $f \neq 0$ and let $\{f_1, \ldots, f_r\}$ be a basis for $A{\cdot}f$. Then there are $g_1, \ldots, g_r \in A^*$ such that

$$b{\cdot}f = \sum_{i=1}^{r} g_i(b) f_i$$

for all $b \in A$; that is $f(ab) = \sum_{i=1}^{r} f_i(a) g_i(b)$ for all $a, b \in A$. By definition $f \in A^o$. ∎

1.3.2 Double Duals

We explore the connections between original structures and "double duals" in the next two propositions.

Proposition 1.3.2 *Let A be an algebra over the field k and let $\iota : A \longrightarrow (A^o)^*$ be given by $\iota(a)(a^o) = a^o(a)$ for $a \in A$ and $a^o \in A^o$. Then:*

a) *ι is an algebra map.*

b) Imι *is a dense subalgebra of $(A^o)^*$.*

c) Kerι *is the intersection of the cofinite ideals of A.*

d) *If A is finite-dimensional then ι is an algebra isomorphism.*

Proof: Parts a) and b) are straightforward exercises based on definitions. Part c) follows by part a) of the preceding corollary. Part d) follows from part parts a) and c) and the fact that A is a finite-dimensional vector space. ∎

The coalgebra structure of A^o can be at least partially understood through the algebra structure of A by means of part a) of Propositions 1.3.2 and 1.2.4.

Proposition 1.3.3 *Let C be a coalgebra over the field k and let $\iota : C \longrightarrow (C^*)^*$ be the map defined by $\iota(c)(c^*) = c^*(c)$ for $c \in C$ and $c^* \in C^*$. Then:*

a) Im$\iota \subseteq (C^*)^o$ *and $\iota : C \longrightarrow (C^*)^o$ is a coalgebra map.*

b) Imι *consists of all $f \in (C^*)^*$ which vanish on a closed cofinite ideal of C^*.*

c) *ι is one-one.*

d) *If C is finite-dimensional then ι is a coalgebra isomorphism.*

Proof: To show part a) we let $c \in C$. Then it is easy to see that $\Delta_{\iota(c)}$ exists and that $\Delta_{\iota(c)} = \iota(c_{(1)}) \otimes \iota(c_{(2)})$. Thus $\iota(c) \in (C^*)^o$. As $\epsilon\iota = \epsilon$ the rest of part a) now follows. Part c) follows directly from definition. Part d) follows from parts a) and c) and the fact that C is a finite-dimensional vector space. It remains to show part b).

To show part b) we let $c \in C$. Then $\iota(c)$ generates a finite-dimensional subcoalgebra of A^o, where $A = C^*$, by parts a) and b) of Theorem 1.3.1. This subcoalgebra must be a subcoalgebra of $\mathrm{Im}\iota$ by part a). Therefore, since ι is one-one, we conclude that c generates a finite-dimensional subcoalgebra D of C. Now $I = D^\perp$ is a closed cofinite ideal of C^* and $\iota(c)$ vanishes on I by definition of ι. Conversely, suppose that I is a closed cofinite ideal of C^* and $f \in A^*$ vanishes on I. Then $f \in \mathrm{Im}\iota$ by Proposition A.4.6. ∎

1.3.3 The Fundamental Theorem of Coalgebras

In establishing part b) of the previous proposition we proved the fundamental theorem of coalgebras.

Theorem 1.3.2 *Suppose that C is a coalgebra over a field k and let $c \in C$. Then c generates a finite-dimensional subcoalgebra of C. Consequently every finite-dimensional subspace of C is contained in a finite-dimensional subcoalgebra of C.* ∎

By virtue of the fundamental theorem many questions about coalgebras over a field can be reduced to questions about finite-dimensional coalgebras, and therefore reduced to questions about finite-dimensional algebras.

Throughout the following exercises A and B are algebras over the field k.

Exercise 1.3.1 Show that $G(A^o) = \mathrm{Alg}(A, k)$ is the set of algebra homomorphisms $\eta : A \longrightarrow k$. (Thus by Lemma 1.1.1 the set $\mathrm{Alg}(A, k)$ is linearly independent.)

Exercise 1.3.2 First we make the following definition.

Definition 1.3.2 *Let A be an algebra over k and $\eta, \rho \in \mathrm{Alg}(A, k)$. An* $\eta{:}\rho$-derivation *of A is a linear map $f : A \longrightarrow k$ which satisfies*

$$f(ab) = \eta(a)f(b) + f(a)\rho(b)$$

for all $a, b \in A$.

Show that $P_{\eta,\rho}(A^o)$ is the set of $\eta{:}\rho$-derivations of A. (See Exercise 1.1.3).

Exercise 1.3.3 Let $A = A_1 \oplus \cdots \oplus A_n$ be the direct sum of algebras over k. Show that $A^o \simeq A_1^o \oplus \cdots \oplus A_n^o$ as coalgebras.

Exercise 1.3.4 Show that $(A \otimes B)^o = A^o \otimes B^o$, where we think of $A^* \otimes B^*$ as a subspace of $(A \otimes B)^*$ by $(a^* \otimes b^*)(a \otimes b) = a^*(a)b^*(b)$, and where $A \otimes B$ and $A^o \otimes B^o$ have the tensor product algebra and coalgebra structures respectively.

Exercise 1.3.5 Suppose that A is the n-dimensional algebra over k generated by x where $x^n = 0$. Show that $C = A^*$ has basis $\{c_0, \ldots, c_{n-1}\}$ such that

$$\Delta(c_m) = \sum_{i=0}^{m} c_{m-i} \otimes c_i \quad \text{and} \quad \epsilon(c_m) = \delta_{0,m} \tag{1.7}$$

for all $0 \leq m \leq n-1$.

Exercise 1.3.6 Assume that the field k is algebraically closed and $A = k[X]$ is the algebra of polynomials in indeterminant X over k.

a) Show that A^o is the sum of subcoalgebras with bases $\{c_0, \ldots, c_{n-1}\}$ such that (1.7) is satisfied for all $0 \leq m \leq n-1$.

b) Show that there is an infinite-dimensional subcoalgebra C of A^o with basis $\{c_0, c_1, c_2, \ldots\}$ such that (1.7) is satisfied for all $m \geq 0$ and C is a dense subspace of A^*.

Exercise 1.3.7 Find an infinite-dimensional algebra A over k such that $A^* = A^o$. Find an infinite-dimensional algebra A over k such that $A^o = (0)$.

Exercise 1.3.8 Let $F = k[\alpha]$ be a degree 2 field extension of k such that $\alpha^2 = a \in k$. Show that $C = F^*$ has a basis $\{s, c\}$ such that

$$\begin{aligned} \Delta(s) &= s \otimes c + c \otimes s, & \epsilon(s) &= 0, \\ \Delta(c) &= c \otimes c + a(s \otimes s), & \epsilon(c) &= 1. \end{aligned}$$

Exercise 1.3.9 Suppose that k is algebraically closed. Find all algebras and coalgebras (up to isomorphism) over k of dimension 1 or 2.

Exercise 1.3.10 Let C be a coalgebra over k. Show that the linear map

$$C^* \otimes A \xrightarrow{\ \iota\ } \mathrm{Hom}(C, A)$$

defined by

$$\iota(c^* \otimes a)(c) = c^*(c)a$$

for all $c^* \in C^*$, $a \in A$, and $c \in C$ is a one-one algebra map, where $C^* \otimes A$ has the tensor product algebra structure and $\mathrm{Hom}(C, A)$ has the convolution algebra structure of Exercise 1.2.13.

Exercise 1.3.11 Let C be a coalgebra over k.

Definition 1.3.3 *The coalgebra C is* reflexive *if the one-one coalgebra map $\iota : C \longrightarrow (C^*)^o$ of Proposition 1.3.3 is an isomorphism.*

a) Show that C is reflexive if and only if all cofinite ideals of C^* are closed.

b) If C^* contains a dense subalgebra A such that all cofinite left ideals of A are finitely generated, show that C is reflexive.

The reader is referred to [Heyneman and Radford, 1974], [Radford, 1973], [Taft, 1972].

1.4 Rational Modules and Comodules

Let C be a coalgebra over the field k and let $A = C^*$ be the dual algebra. Rational A-modules are those locally finite A-modules whose elements are annihilated by a *closed* cofinite ideal of A. Such a (left) module structure $\mu : A \otimes M \longrightarrow M$ can be accounted for by a linear map $\rho : M \longrightarrow M \otimes C$ which satisfies properties reflecting the module axioms.

1.4.1 Rational Modules

Suppose that M is a left C^*-module. We will let M_r be the subset of all $m \in M$ such that there exists an element $\rho_m \in M \otimes C$ satisfying the condition

$$c^* \cdot m = (1_M \otimes c^*)(\rho_m) \tag{1.8}$$

for all $c^* \in C^*$. For $m \in M$ observe that there is at most one such element $\rho_m \in M \otimes C$ which satisfies (1.8).

Proposition 1.4.1 *Let C be a coalgebra over the field k and suppose that M is a left C^*-module.*

a) *Assume that $m \in M_r$ is not zero and write $\rho_m = \sum_{i=1}^{r} m_i \otimes c_i$, where $r =$ Rankρ_m. Then $\{m_1, \ldots, m_r\}$ is a basis for $C^* \cdot m$. In particular* Dim$C^* \cdot m =$ Rankρ_m.

b) *M_r is a locally finite C^*-submodule of M.*

c) *Let N be a submodule of M. Then $N_r = M_r \cap N$. In particular $N = N_r$ when $N \subseteq M_r$.*

d) *Suppose that $f : M \longrightarrow N$ is a map of left C^*-modules. Then $f(M_r) \subseteq N_r$. Thus the restriction $f_r = f|_{M_r}$ is a module map $f_r : M_r \longrightarrow N_r$.*

Proof: The fact that M_r is a subspace of M is a straightforward exercise. Let $m \in M_r$ and assume that $m \neq 0$. Write $\rho_m = \sum_{i=1}^{r} m_i \otimes c_i$ where $r =$ Rankρ_m. Then the m_i's and the c_i's form linearly independent sets by Lemma A.4.1. As $c^* \cdot m = \sum_{i=1}^{r} c^*(c_i) m_i$ for $c^* \in C^*$, by definition it follows that $C^* \cdot m$ is contained in the span of $\{m_1, \ldots, m_r\}$. Since $\{c_1, \ldots, c_r\}$ is linearly independent, for given $1 \leq i \leq r$ there is a $c^* \in C^*$ such that $c^*(c_j) = \delta_{i,j}$ for all $1 \leq j \leq r$. For such a functional c^* we have $c^* \cdot m = m_i$. Therefore $C^* \cdot m$ is the span of $\{m_1, \ldots, m_r\}$. Since this set is independent part a) now follows. To complete the proof of part b) we need only show that $d^* \cdot m \in M_r$ whenever $d^* \in C^*$. For $c^* \in C^*$ the calculation

$$c^* \cdot (d^* \cdot m) \quad = \quad (c^* d^*) \cdot m$$

$$
\begin{aligned}
&= \sum_{i=1}^{r}(c^*d^*)(c_i)m_i \\
&= \sum_{i=1}^{r} c^*(c_{i(1)})d^*(c_{i(2)})m_i \\
&= (1_M \otimes c^*)(\sum_{i=1}^{r} m_i \otimes (d^* \rightharpoonup c_i))
\end{aligned}
$$

shows that $d^* \cdot m \in M_r$ for all $d^* \in C^*$. Therefore M_r is a submodule of M. Part c) follows directly from part a). Part d) follows with the observation that $\rho_{f(m)}$ exists for $m \in M$ and $\rho_{f(m)} = (f \otimes 1_C)(\rho_m)$. ∎

Definition 1.4.1 *A left C^*-module M is* rational *if $M = M_r$.*

In this case the map $\rho : M \longrightarrow M \otimes C$ defined by $\rho(m) = \rho_m$ for $m \in M$ is linear and is uniquely determined by the condition

$$c^* \cdot m = (1_M \otimes c^*)(\rho(m)) \tag{1.9}$$

for all $c^* \in C^*$ and $m \in M$. Since $\epsilon \cdot m = m$ for all $m \in M$ it follows that

$$(1_M \otimes \epsilon)\rho = 1_M. \tag{1.10}$$

Observe (1.10) implies that ρ is one-one. Since $c^* \cdot (d^* \cdot m) = (c^* d^*) \cdot m$ for all $c^*, d^* \in C^*$ and $m \in M$ we conclude that

$$(\rho \otimes 1_C)\rho = (1_M \otimes \Delta)\rho. \tag{1.11}$$

1.4.2 Comodules

Definition 1.4.2 *A right C-*comodule *is a pair (M, ρ) where M is a vector space over k and $\rho : M \longrightarrow M \otimes C$ is a linear map satisfying (1.10) and (1.11).*

We will frequently denote (M, ρ) by M. Notice that C is a right C-comodule with $\rho = \Delta$.

When M is a rational left C^*-module the map $\rho : M \longrightarrow M \otimes C$ determined by (1.9) affords M a right C-comodule structure. This comodule structure is referred to as the *underlying comodule structure* for the (rational) module action. Conversely, if (M, ρ) is a right C-comodule then (1.9) defines a module action of C^* on M which is referred to as the *associated rational module* action. We will use the symbolic summation notation

$$\rho(m) = m^{<1>} \otimes m^{(2)} \in M \otimes C$$

to denote the image of $m \in M$ under ρ. It is customary to use an arrow to designate rational module action as follows:

$$c^* \rightharpoonup m = m^{<1>} <c^*, m^{(2)}>$$

where $c^* \in C^*$ and $m \in M$.

Definition 1.4.3 *Let (M, ρ_M) and (N, ρ_N) be right C-comodules. A linear map $f : M \longrightarrow N$ is a* map of right C-comodules *if it is a map of rational left C^*-modules.*

Thus $f : M \longrightarrow N$ is a comodule map if and only if

$$(f \otimes 1_C)\rho_M = \rho_N f.$$

Definition 1.4.4 *Suppose that (M, ρ) is a right C-comodule. Then a* sub-comodule *of M is a subspace N of M such that $\rho(N) \subseteq N \otimes C$.*

Suppose that N is a sub-comodule of M. Then $(N, \rho|_N)$ is a right C-comodule in its own right. Observe that the quotient space M/N has a unique right C-comodule structure such that the projection $\pi : M \longrightarrow M/N$ is a comodule map. Observe that the kernel and image of a map of right C-comodules $f : M \longrightarrow N$ are sub-comodules of M and N respectively. We leave the formulation and proof a fundamental homomorphism theorem for comodules as an exercise to the reader.

By part c) of Proposition 1.4.1 the sub-comodules of M are the submodules of associated rational C^*-module structure on M. Let V be a subspace of M. Then among all sub-comodules of M which contain V there is a unique minimal one which is the submodule of M generated by V. Since rational modules are locally finite we have shown:

Proposition 1.4.2 *Let M be a right C-comodule where C is a coalgebra over the field k. Then a finite-dimensional subspace V of C is contained in a finite-dimensional sub-comodule of M.* ∎

The comatrix identities arise in a very natural way in connection with (right) comodules.

Proposition 1.4.3 *Suppose that (M, ρ) is a right C-comodule where C is a coalgebra over the field k. Then:*

a) *There is a unique subspace $C(\rho)$ of C minimal with respect to the property that $\rho(M) \subseteq M \otimes C(\rho)$.*

b) *$C(\rho)$ is a subcoalgebra of C.*

c) *Suppose that $M \neq (0)$ and is finite-dimensional. Let $\{m_1, \ldots, m_r\}$ be a basis for M and write $\rho(m_j) = m_i \otimes e^i_j$ for $1 \leq j \leq r$ where $e^i_j \in C$. Then $\{e^i_j\}_{1 \leq i,j \leq r}$ satisfies the comatrix identities and spans $C(\rho)$.*

Proof: Suppose the proposition holds whenever M is finite-dimensional. An arbitrary right C-comodule (M,ρ) is the sum of its finite-dimensional sub-comodules by Proposition 1.4.2. The subspace $C(\rho) = \sum_N C(\rho|_N)$, where N runs over the finite-dimensional sub-comodules of M, meets the requirements of part a). Now $C(\rho)$ is a subcoalgebra of C since each summand $C(\rho|_N)$ is a subcoalgebra of C. We have reduced the proof of the proposition to the finite-dimensional case.

Now assume that $M \neq (0)$ and is finite-dimensional. Let $\{m_1,\ldots,m_r\}$ be any basis for M and write $\rho(m_j) = m_i \otimes e^i_j$. The e^i_j's must be contained in any subspace V of C such that $\rho(M) \subseteq M \otimes V$ since $\{m_1,\ldots,m_r\}$ is linearly independent. Thus $C(\rho)$ does exist and is the span of the e^i_j's. Applying (1.10) to m_j we deduce that $m_j = m_i\epsilon(e^i_j)$. Therefore $\epsilon(e^i_j) = \delta^i_j$. Applying (1.11) to m_j we obtain

$$m_i \otimes e^i_\ell \otimes e^\ell_j = m_\ell \otimes e^\ell_i \otimes e^i_j = m_i \otimes \Delta(e^i_j).$$

Comparing terms on both sides of the equations we see that $\Delta(e^i_j) = e^i_\ell \otimes e^\ell_j$ for $1 \leq i,j \leq r$. This concludes our proof. ∎

The theory of left C^*-modules and the theory of right C-comodules coincides when C is finite-dimensional.

Theorem 1.4.1 *Suppose that C is a finite-dimensional coalgebra over the field k. Then all left C^*-modules are rational.*

Proof: Let M be a left C^*-module. Since C^* is finite-dimensional it follows that M is locally finite. As the sum of rational submodules of M is rational by Proposition 1.4.1, to prove the theorem we may assume that M is finite-dimensional.

Suppose that $M \neq (0)$ and is finite-dimensional. Let $\{m_1,\ldots,m_r\}$ be any basis for M and let $\{m^1,\ldots,m^r\}$ be the corresponding dual basis for M^* (thus $m^i(m_j) = \delta^i_j$ for $1 \leq i,j \leq r$). Fix $1 \leq j \leq r$. It suffices to show that ρ_{m_j} exists by part b) of Proposition 1.4.1. Let $\nu = m_i \otimes e^i_j \in M \otimes C$. Then $\rho_{m_j} = \nu$ if and only if $c^*{\cdot}m_j = c^*(e^i_j)m_i$ for all $c^* \in C^*$. But the latter is the case if and only if $m^\ell(c^*{\cdot}m_j) = c^*(e^\ell_j)$ for all $1 \leq \ell \leq r$ and $c^* \in C^*$. Now for fixed $1 \leq \ell, j \leq n$ the map $f : C^* \longrightarrow k$ defined by $f(c^*) = m^\ell(c^*{\cdot}m_j)$ is linear. Since C is finite-dimensional there exists a unique $e^\ell_j \in C$ such that $f(c^*) = c^*(e^\ell_j)$ for all $c^* \in C^*$. Thus ρ_{m_j} exists and hence M is rational. ∎

The notions of rational right C^*-module and left C-comodule are defined in the expected ways. Replacing C with C^{cop}, and hence C^* with $(C^*)^{op} = (C^{cop})^*$, the results of this section gives analogous results for right C^*-modules and left C-comodules.

1.4.3 M_r *and* M^r

Let A be an algebra over a field k and suppose that M is a left A-module. Then M^* is a right A-module according to the rule

$$<\alpha{\cdot}a, m> = <\alpha, a{\cdot}m>$$

for all $\alpha \in M^*, a \in A$, and $m \in M$. This action on M^* is referred to as the *transpose action*. The transpose action arises implicitly in the proof of Theorem 1.4.1. See Exercise 1.4.5 at the end of this section. When $\mu : A \otimes M \longrightarrow M$ denotes the left module action on M we write $\mu^T : M^* \otimes A \longrightarrow M^*$ for the transpose action.

Now let C be a coalgebra over k and $A = C^*$. Set $M^r = (M^*)_r$. By definition

$$\alpha{\cdot}c^* = (c^* \otimes 1_{M^r})(\rho_\alpha) \quad \text{and} \quad c^*{\cdot}m = (1_M \otimes c^*)(\rho_m)$$

for all $\alpha \in M^r, c^* \in C^*$, and $m \in M$. We let (M^r, ρ^T) denote the underlying left C-comodule structure for the right C^*-module (M^r, μ^T).

Regard M as a subspace of M^{**} in the usual way by $m(\alpha) = \alpha(m)$ for $m \in M$ and $\alpha \in M^*$. Therefore

$$(\alpha{\cdot}c^*)(m) = c^*((1_C \otimes m)(\rho_\alpha)) \quad \text{and} \quad \alpha(c^*{\cdot}m) = c^*((\alpha \otimes 1_C)(\rho_m)).$$

Consequently

$$(1_C \otimes m)(\rho_\alpha) = (\alpha \otimes 1_C)(\rho_m) \tag{1.12}$$

for $m \in M_r$ and $\alpha \in M^r$. In terms of the H-S notation for comodule action (1.12) is expressed as

$$\alpha^{(1)}<\alpha^{<2>}, m> = <\alpha, m^{<1>}>m^{(2)} \tag{1.13}$$

for $\alpha \in M^r$ and $m \in M_r$.

1.4.4 M_r *Characterized in Terms of Annihilators*

Let A be an algebra over the field k and suppose that M is a left A-module. For a subspace V of M we let

$$\mathrm{ann}_A(V) = \{a \in A \mid a{\cdot}V = (0)\}$$

be the annihilator of V. Observe that $\mathrm{ann}_A(V)$ is a left ideal of A and is an ideal of A when V is a submodule of M. For $m \in M$ let $\mathrm{ann}_A(m) = \mathrm{ann}_A(km)$.

Now fix $m \in M$. The linear map $A \longrightarrow A{\cdot}m$ defined by $a \longmapsto a{\cdot}m$ induces an isomorphism of left A-modules $A/\mathrm{ann}_A(m) \simeq A{\cdot}m$. Consequently $A{\cdot}m$ is finite-dimensional if and only if $\mathrm{ann}_A(m)$ is a cofinite left ideal of A.

Proposition 1.4.4 *Suppose that C is a coalgebra over the field k and let M be a left C^*-module. For $m \in M$ the following are equivalent:*

a) $\mathrm{ann}_{C^*}(C^*{\cdot}m)$ *contains a cofinite closed subspace of* C^*.

b) $\mathrm{ann}_{C^*}(C^*{\cdot}m)$ *is a cofinite closed ideal of* C^*.

c) $\mathrm{ann}_{C^*}(m)$ *contains a cofinite closed subspace of* C^*.

d) $\mathrm{ann}_{C^*}(m)$ *is a cofinite closed cofinite left ideal of* C^*.

e) $m \in M_r$.

Proof: Suppose that I and J are subspaces of C^* and $I \subseteq J$. If I is cofinite and closed then J is as well by Lemma A.4.4. Therefore part a) implies part b) which in turn implies part c) which in turn implies part d).

We now show that part d) implies part e). Suppose that $L = \mathrm{ann}_{C^*}(m)$ is a cofinite closed left ideal of C^*. Since $\mathrm{ann}_{C^*}(m)$ is cofinite it follows that $C^*{\cdot}m$ is finite-dimensional. We may as well assume that $m \neq 0$. Choose a basis $\{m_1, \ldots, m_r\}$ for $C^*{\cdot}m$. Then there are $f_1, \ldots, f_r \in C^*$ such that

$$c^*{\cdot}m = f_1(c^*)m_1 + \cdots + f_r(c^*)m_r$$

for all $c^* \in C^*$. Since $L{\cdot}m = (0)$ it follows that $f_1(L) = \cdots = f_r(L) = (0)$. Since L is a cofinite closed subspace of C^* we conclude by Proposition A.4.6 that there are $c_1, \ldots, c_r \in C$ such that $f_i(c^*) = c^*(c_i)$ for all $1 \leq i \leq r$ and $c^* \in C^*$. Therefore ρ_m exists and $\rho_m = \sum_{i=1}^{r} m_i \otimes c_i$. By definition $m \in M_r$. We have shown that part d) implies part e).

Lastly, we show that part e) implies part a). Let $m \in M_r$. Then $N = C^*{\cdot}m$ is a finite dimensional submodule of M_r by part b) of Proposition 1.4.1. Let (N, ρ) be the underlying right C-comodule structure. Then $\rho(N) \subseteq N \otimes V$ for some finite-dimensional subspace V of C since N is finite-dimensional. Observe that $V^\perp \subseteq \mathrm{ann}_{C^*}(C^*{\cdot}m)$ and is a closed cofinite subspace of C^*. Therefore part e) implies part a) and the proposition is proved. ∎

We conclude this section with a discussion the concept of strongly rational module. See [Heyneman and Radford, 1974] for a discussion of the related concept of strongly reflexive coalgebra. Let C be a coalgebra over k and suppose that M is a left C^*-module. We will let $M_{(r)}$ be the set of all $m \in M$ such that $\mathrm{ann}_{C^*}(m)$ contains a cofinite ideal I which is finitely generated as a left ideal. The latter property means that I is closed by part b) of Proposition 1.2.3. Suppose that I and J are two ideals of C^* which are finitely generated as left ideals. It is easy to see that the product IJ is finitely generated as a left ideal. By Exercise 1.4.2 it follows that IJ is a cofinite subspace of C^*. It thus follows that $M_{(r)}$ is a submodule of M_r.

Definition 1.4.5 *A comodule M is* strongly rational *if $M = M_{(r)}$.*

Observe that if $f : M \longrightarrow N$ is a map of left C^*-modules then $f(M_{(r)}) \subseteq N_{(r)}$. Submodules and quotients of strongly rational modules are strongly rational. Conversely:

Proposition 1.4.5 *Suppose that C is a coalgebra over the field k and N is a submodule of a left C^*-module M. If N and M/N are strongly rational then M is strongly rational.*

Proof: Let $\pi : M \longrightarrow M/N$ be the projection and assume that N and M/N are strongly rational. Let $m \in M$. Then $\pi(I{\cdot}m) = I{\cdot}\pi(m) = (0)$ for some cofinite ideal I of C^* which is finitely generated as a left ideal. Let $f_1, \ldots, f_r$ generate I as a left ideal of C^*. Since $f_1{\cdot}m, \ldots, f_r{\cdot}m \in \operatorname{Ker}\pi = N$ it follows that there are cofinite ideals $I_1, \ldots, I_r$ of C^* which are finitely generated as left ideals such that $I_i{\cdot}(f_i{\cdot}m) = (0)$ for all $1 \leq i \leq r$. Therefore the product $J = I_1 \cdots I_r I$ is a cofinite ideal of C^* which is finitely generated as a left ideal. Since $J{\cdot}m = (0)$ we conclude that $m \in M_{(r)}$. We have shown that $M = M_{(r)}$. ∎

1.4.5 Another Proof of the Fundamental Theorem of Coalgebras

Here is a very short proof of Theorem 1.3.2 using elementary aspects of the theory of comodules developed thus far. Suppose that C is a coalgebra over the field k and c is a non-zero element of C. Regard C as a right C-comodule under the coproduct Δ. Then $N = C^* \rightharpoonup c$ is a finite-dimensional sub-comodule of C which contains c. By Proposition 1.4.3 there is a finite-dimensional subcoalgebra $\mathcal{C}$ of C such that $\Delta(N) \subseteq N \otimes \mathcal{C}$. Therefore

$$c \in N = (\epsilon \otimes I)\Delta(N) \subseteq \mathcal{C}.$$

Throughout the following exercises C is a coalgebra over the field k.

Exercise 1.4.1 Formulate and prove a fundamental homomorphism theorem for comodules. (See Theorem 1.1.1.)

Exercise 1.4.2 Let A be an algebra over the field k and suppose that M is a finitely-generated left A-module. Show that if V is a cofinite subspace of A then $V{\cdot}M$ is a cofinite subspace of M. [Hint: Suppose that $\{m_1, \ldots, m_r\}$ generates M as a left A-module. Show that the composite $A \oplus \cdots \oplus A \longrightarrow M \longrightarrow M/(V{\cdot}M)$ of the map defined by $(a_1, \ldots, a_r) \longmapsto a_1{\cdot}m_1 + \cdots + a_r{\cdot}m_r$ followed by the projection factors through $A/V \oplus \cdots \oplus A/V$.]

Exercise 1.4.3 Left and right comodules are formally related as are left and right modules.

a) Suppose that A is an algebra over the field k and (M, μ) is a left A-module. Show that (M, μ^{op}) is a right A^{op}-module, where $\mu^{op} = \mu\tau_{A,M}$. (We write

$$m{\cdot}^{op} a = \mu^{op}(m \otimes a) = \mu(a \otimes m) = a{\cdot}m$$

for $m \in M$ and $a \in A$.)

b) Suppose that (M, ρ) is a right C-comodule. Show that (M, ρ^{cop}) is a left C^{cop}-comodule, where $\rho^{cop} = \tau_{M,C}\rho$. (We write

$$m^{(1)^{cop}} \otimes m^{<2>^{cop}} = \rho^{cop}(m) = (\tau_{M,C}\rho)(m) = m^{(2)} \otimes m^{<1>}$$

for $m \in M$.)

c) Reformulate part a) for right A-modules and reformulate part b) for left C-comodules.

Exercise 1.4.4 Suppose that M is a right C-comodule.

a) Show that $C(\rho)^{\perp} = \mathrm{ann}_{C^*}(C^* \cdot m)$.

Definition 1.4.6 *A right C-comodule M is* simple *if $M \neq (0)$ and M has no proper subcomodules.*

b) Show that if M is a simple right C-comodule then $C(\rho)$ is a simple subcoalgebra of C.

Exercise 1.4.5 Let (M, μ) be a left C^*-module. Suppose that both C and M are finite-dimensional. There is a natural basis free argument that M is rational. Let

$$M^* \otimes C^* \xrightarrow{\mu^T} M^*$$

denote the right transpose action on the dual space M^* of M. Thus

$$\mu^T(m^* \otimes c^*)(m) = (m^* \cdot c^*)(m) = m^*(c^* \cdot m) = m^*(\mu(c^* \otimes m)).$$

Show that M is rational and the composite

$$M \simeq M^{**} \xrightarrow{(\mu^T)^*} (M^* \otimes C^*)^* \simeq M \otimes C$$

is the underlying comodule structure for the module (M, μ).

Exercise 1.4.6 Show that Proposition 1.4.5 does not hold when "rational" replaces "strongly rational".

Exercise 1.4.7 Show that there exists a coalgebra C over k and left C^*-modules M and N such that

a) $M_r \neq 0$ and $M^r = (0)$,

b) $N_r = (0)$ and $N^r \neq (0)$.

Exercise 1.4.8 Suppose that A is a dense subalgebra of C^* and $\rho : M \longrightarrow M \otimes C$ is linear. Define $\mu : A \otimes M \longrightarrow M$ by $\mu(a \otimes m) = (1_M \otimes a)(\rho(m))$.

a) Prove that the following are equivalent:

i) (M, μ) is a left A-module.

ii) (M, ρ) is a right C-comodule.

b) If either i) or ii) holds, show that the A-submodules of M are the C-sub-comodules of M.

Exercise 1.4.9 Let A be an algebra over k. Recall from Proposition 1.3.2 that

$$\begin{array}{ccc} A & \xrightarrow{\iota} & (A^o)^* \\ a & \longmapsto & (a^o \longmapsto a^o(a)) \end{array}$$

is an algebra map and that $\mathrm{Im}\iota$ is a dense subalgebra of $(A^o)^*$.

a) Suppose that (M, ρ) is a right A^o-comodule. Show that the rule

$$a{\cdot}m = m^{<1>} <m^{(2)}, a>$$

for $a \in A$ and $m \in M$ turns M into a locally finite left A-module.

b) Suppose that (M, μ) is a locally finite left A-module.

i) Show that there is a unique right A^o-comodule structure (M, ρ_μ) on M such that

$$a{\cdot}m = m^{<1>^o} <m^{(2)^o}, a>$$

holds for all $a \in A$ and $m \in M$, where we write $\rho_\mu(m) = m^{<1>^o} <m^{(2)^o}, a>$.

ii) Suppose that M is finite-dimensional. Show that $A^o(\rho_\mu) = \mathrm{ann}_A(M)^\perp$.

Exercise 1.4.10 Let (M, ρ) be a right C-comodule and suppose that D is a subcoalgebra of C. If $N = \rho^{-1}(M \otimes D)$ show that N is a sub-comodule of M and $\rho(N) \subseteq N \otimes D$. (Thus $(N, \rho|_N)$ is a right D-comodule as well.)

Exercise 1.4.11 Suppose that $C = \oplus_{i \in I} C_i$ is the direct sum of coalgebras.

a) Let M be a right C-comodule. If $M_i = \rho^{-1}(M \otimes C_i)$ show that M_i is a C-sub-comodule of M and that $M = \oplus_{i \in I} M_i$ as C-comodules. (By Exercise 1.4.10 we have that $(M_i, \rho|_{M_i})$ is a right C_i-comodule.)

b) Suppose that (M_i, ρ_i) is a right C_i-comodule for each $i \in I$. Show that $M = \oplus_{i \in I} M_i$ has a unique right C-comodule structure (M, ρ) such that $\rho|_{M_i} = \rho_i$ for each $i \in I$.

Exercise 1.4.12 Show that the direct sum of left rational C^*-modules is rational. Is the product of left rational C^*-modules always rational? Give a proof or counterexample.

Exercise 1.4.13 Let M be a rational left C^*-module and regard M^* as a right C^* module under the transpose action. Show that finitely generated submodules of M^* are closed subspaces of M^*.

Exercise 1.4.14 Suppose C is the coalgebra over k with basis $\{c_0, c_1, c_2, \ldots\}$ and whose coalgebra structure is determined by $\epsilon(c_n) = \delta_{n,0}$ and $\Delta(c_n) = \sum_{i=0}^{n} c_{n-i} \otimes c_i$ for all $n \geq 0$.

Show that the finite-dimensional right C-comodules (M, ρ) are described in the following manner: there exists a nilpotent endomorphism $T \in \mathrm{End}(M)$ such that

$$\rho(m) = \sum_{i=0}^{r} T^i(m) \otimes c_i$$

for $m \in M$, where $r = \mathrm{Dim}M$. [Hint: We may regard the polynomial algebra $k[X]$ in one indeterminant as a dense subalgebra of C^* where $X(c_n) = \delta_{n,1}$. See Exercise 1.4.8.]

Exercise 1.4.15 Generalize Exercise 1.4.14 to the coalgebras of Exercise 1.2.6.

Exercise 1.4.16 Let $M = C^*$ be regarded a left C^*-module under left multiplication. Show that:

a) M_r is the sum of the finite-dimensional left ideals of C^*.

b) M_r consists of the functionals of C^* which vanish on a cofinite left coideal of C.

Exercise 1.4.17 Let A be an algebra over the field k.

Definition 1.4.7 *The algebra A is* left almost Noetherian *if all cofinite left ideals of A are finitely generated.*

Suppose that C^* contains a dense subalgebra which is left almost Noetherian. Show that all finite-dimensional left C^*-modules are strongly rational. [Hint: See Exercise 1.3.11. The reader is also referred to [Heyneman and Radford, 1974].]

1.5 Bialgebras

Bialgebras are vector spaces with compatible coalgebra and algebra structures.

Lemma 1.5.1 *Suppose that A is a vector space over the field k with an algebra structure (A, m, η) and a coalgebra structure (A, Δ, ϵ). Endow $A \otimes A$ with the tensor product algebra and coalgebra structures. Then the following are equivalent:*

a) *m and η are coalgebra maps.*

b) *Δ and ϵ are algebra maps.*

Proof: First of all observe that m is a coalgebra map if and only if $\Delta m = (m \otimes m)\Delta_{A\otimes A}$ and $\epsilon m = \epsilon_{A\otimes A}$. Therefore m is a coalgebra map if and only if

$$\Delta(ab) = a_{(1)}b_{(1)} \otimes a_{(2)}b_{(2)} = (\Delta(a))(\Delta(b))$$

and

$$\epsilon(ab) = \epsilon(a)\epsilon(b)$$

for all $a, b \in A$. Now η is a coalgebra map if and only if $\Delta\eta = (\eta\otimes\eta)\Delta_k$ and $\epsilon\eta = \epsilon_k$. Thus η is a coalgebra map if and only if

$$\Delta(1) = 1 \otimes 1 \quad \text{and} \quad \epsilon(1) = 1.$$

The lemma follows from these calculations. ∎

Definition 1.5.1 *A* bialgebra *over k is a quintuple* $(A, m, \eta, \Delta, \epsilon)$ *where* (A, m, η) *is an algebra over k and* (A, Δ, ϵ) *is a coalgebra over k such that either of the equivalent conditions of Lemma 1.5.1 are satisfied.*

We will frequently denote a bialgebra $(A, m, \eta, \Delta, \epsilon)$ simply by A.

Definition 1.5.2 *A bialgebra A over k is* commutative *if the underlying algebra structure of A is commutative and A is* cocommutative *if the underlying coalgebra structure of A is cocommutative.*

Bialgebras have an advantage over algebras and coalgebras in representation theory. Let A be a bialgebra over k. If M and N are left A-modules then the tensor product $M \otimes N$ is a left A-module where

$$a\cdot(m \otimes n) = a_{(1)}\cdot m \otimes a_{(2)}\cdot n \tag{1.14}$$

for all $a \in A, m \in M$ and $n \in N$.

Definition 1.5.3 *Let A be a bialgebra over k. The module structure on the tensor product of two left A-modules M and N defined by (1.14) is the* tensor product A-module structure.

If M and N are right A-comodules then $M \otimes N$ is a right A-comodule where

$$\rho(m \otimes n) = (m^{<1>} \otimes n^{<1>}) \otimes m^{(2)}n^{(2)} \tag{1.15}$$

for all $m \in M$ and $n \in N$.

Definition 1.5.4 *Let A be a bialgebra over k. The comodule structure on the tensor product of two right A-comodules M and N defined by (1.15) is the* tensor product A-comodule structure.

If M and N are right A-modules then the tensor product A-module structure on $M \otimes N$ is given by $(m \otimes n)\cdot a = m\cdot a_{(1)} \otimes n\cdot a_{(2)}$ for all $a \in A$, $m \in M$, and $n \in N$. Likewise if M and N are left A-comodules then the tensor product A-comodule structure on $M \otimes N$ is given by $\rho(m \otimes n) = m^{(1)}n^{(1)} \otimes (m^{<2>} \otimes n^{<2>})$ for all $m \in M$ and $n \in N$.

Definition 1.5.5 *A* sub-bialgebra *of A is a subspace B of A which is a simultaneously a subalgebra and a subcoalgebra of A.*

Thus the intersection of a family of sub-bialgebras of A is a sub-bialgebra of A. For a subspace V of A we conclude there is a unique sub-bialgebra B of A which is minimal with respect to the property that B contains V.

Definition 1.5.6 *The sub-bialgebra B of A just described is the* sub-bialgebra *of A* generated by V.

Now let B also be a bialgebra over k. Then the tensor product $A \otimes B$ is a bialgebra with the tensor product algebra and coalgebra structures. We shall refer to this bialgebra structure as the *tensor product bialgebra structure*.

Definition 1.5.7 *Let A and B be bialgebras over k. A linear map $f : A \longrightarrow B$ is a* bialgebra map *if f is both an algebra map and a coalgebra map.*

Definition 1.5.8 *A* bi-ideal *of a bialgebra A over k is a subspace of A which is both an ideal and a coideal of A.*

Observe the kernel of a bialgebra map is a bi-ideal. If I is a bi-ideal of A then the quotient space A/I has a unique bialgebra structure such that the projection $\pi : A \longrightarrow A/I$ is a bialgebra map. We leave it to the reader at this point to formulate and prove a fundamental homomorphism theorem for bialgebras.

Recall that an associative algebra A has an associated Lie algebra structure defined by $[a, b] = ab - ba$ for all $a, b \in A$. Recall from Exercise 1.1.3 that $P_1(A) = P_{1,1}(A)$, which is the set of all $a \in A$ satisfying $\Delta(a) = 1 \otimes a + a \otimes 1$, is a coideal of A.

Definition 1.5.9 *Let A be a bialgebra over the field k. An element $a \in A$ which satisfies $\Delta(a) = 1 \otimes a + a \otimes 1$ is* primitive.

Proposition 1.5.1 *Suppose that A is a bialgebra over the field k. Then:*

a) *$G(A)$ is a multiplicative semigroup of A. If $g \in G(A)$ has a multiplicative inverse in A then $g^{-1} \in G(A)$.*

b) *Let $g \in G(A)$. Then $\ell_g, r_g : A \longrightarrow A$ defined by $\ell_g(a) = ga$ and $r_g(a) = ag$ for $a \in A$ are coalgebra maps.*

c) *Let $g, h, \ell \in G(A)$. Then*

$$\ell(P_{g,h}(A)) \subseteq P_{\ell g,\ell h}(A)$$

and

$$(P_{g,h}(A))\ell \subseteq P_{g\ell,h\ell}(A).$$

d) *Let $L = P_1(A)$ be the coideal of primitive elements of A. Then $[L, L] \subseteq L$. Thus L is a Lie subalgebra of the Lie algebra associated to A.*

e) *Suppose that $x \in A$ is primitive. Then ℓ_x and r_x are coderivations of A.*

∎

The proof is a elementary exercise for the reader. Coderivations are discussed in Exercise 1.2.15. We do point out that if $g \in G(A)$ has an inverse in A then $g^{-1} \in G(A)$. For if g has an inverse in A then

$$\Delta(g^{-1}) = (\Delta(g))^{-1} = (g \otimes g)^{-1} = g^{-1} \otimes g^{-1}$$

since Δ is an algebra map and

$$\epsilon(g^{-1}) = (\epsilon(g))^{-1} = 1$$

since ϵ is an algebra map.

The previous proposition relates aspects of the algebra and coalgebra structures of a bialgebra A. Along the same lines, note that the left ideal (respectively right ideal, ideal) of A generated by a left coideal, right coideal or coideal of A is also a left coideal (respectively right coideal, coideal) of A since Δ and ϵ are multiplicative.

There are two very basic examples of bialgebras. The first is the semigroup algebra $k[S]$ over k of a semigroup S. We have given $k[S]$ a coalgebra structure determined by $\Delta(s) = s \otimes s$ for $s \in S$. With this coalgebra structure $k[S]$ is a bialgebra over k. The second basic example is the universal enveloping algebra $(\iota, U(L))$ of a Lie algebra L over k [Jacobson, 1962]. The pair satisfies the following universal mapping property:

a) $U(L)$ is an associative algebra over k and $\iota : L \longrightarrow U(L)$ is a Lie algebra map of L to the Lie algebra associated to $U(L)$, and

b) if A is an associative algebra over k and $f : L \longrightarrow A$ is a map of the Lie algebra L to the Lie algebra associated to A, then there is a map of associative algebras $F : U(L) \longrightarrow A$ uniquely determined by $F\iota = f$.

Observe that the universal mapping property implies that $\mathrm{Im}\iota$ generates $U(L)$ as an associative algebra. To determine a coalgebra structure for $U(L)$ we note that $0 : L \longrightarrow k$ and $\delta : L \longrightarrow U(L) \otimes U(L)$ defined by $\delta(\ell) = \iota(\ell) \otimes 1 + 1 \otimes \iota(\ell)$ are Lie algebra maps. Therefore there are associative algebra maps $\epsilon : U(L) \longrightarrow k$ and $\Delta : U(L) \longrightarrow U(L) \otimes U(L)$ such that

$$\epsilon\iota = 0 \quad \text{and} \quad \Delta\iota = \delta.$$

We need only to show that $(U(L), \Delta, \epsilon)$ is a coalgebra. To this end we note that $(1_{U(L)} \otimes \Delta)\Delta$ and $(\Delta \otimes 1_{U(L)})\Delta$ are algebra maps. An easy calculation shows that both agree on $\mathrm{Im}\iota$ which generates $U(L)$ as an algebra. Therefore $(1_{U(L)} \otimes \Delta)\Delta = (\Delta \otimes 1_{U(L)})\Delta$. Likewise $(1_{U(L)} \otimes \epsilon)\Delta$ and $(\epsilon \otimes 1_{U(L)})\Delta$ are associative algebra maps which agree with $1_{U(L)}$ on $\mathrm{Im}\iota$. Thus $(1_{U(L)} \otimes \epsilon)\Delta = 1_{U(L)} = (\epsilon \otimes 1_{U(L)})\Delta$.

Notice that $\iota(L)$ consists of primitive elements by virtue of the way we constructed the bialgebra structure on $U(L)$.

The commutative polynomial algebra $A = k[X_1, \ldots, X_r]$ over k is in fact $U(L)$ where L is the r-dimensional abelian Lie algebra. Thus A is a bialgebra where $\Delta(X_i) = X_i \otimes 1 + 1 \otimes X_i$ for all $1 \leq i \leq r$. One could circumvent the Lie algebra connection and use the universal mapping property of the polynomial algebra to establish this coalgebra structure on A.

Suppose that A is a bialgebra over k and that C is a subcoalgebra of A. Let B be the sub-bialgebra of A generated by C. Since Δ is multiplicative the product of subcoalgebras of A is again a subcoalgebra of A. Therefore the subalgebra $k1 + C + C^2 + C^3 + \cdots$ of A generated by C is also a subcoalgebra of A. We have shown that $B = k1 + C + C^2 + C^3 + \cdots$. Thus C generates A as a bialgebra if and only if $A = k1 + C + C^2 + C^3 + \cdots$.

We now construct the free bialgebra on a coalgebra. Let (C, δ, e) be a coalgebra over the field k, and let $(\iota, T(C))$ be the tensor algebra on the vector space C. Generally the tensor algebra on a vector space V over k is a pair $(\iota, T(V))$ which satisfies the following universal property:

a) $T(V)$ is an algebra over k and $\iota : V \longrightarrow T(V)$ is linear map, and

b) if A is an algebra over k and $f : V \longrightarrow A$ is a linear map then there is an algebra map $F : T(V) \longrightarrow A$ uniquely determined by $F\iota = f$.

Now the linear maps $e : C \longrightarrow k$ and $(\iota \otimes \iota)\delta : C \longrightarrow T(C) \otimes T(C)$ determine algebra maps $\epsilon : T(C) \longrightarrow k$ and $\Delta : T(C) \longrightarrow T(C) \otimes T(C)$ satisfying $\epsilon\iota = e$ and $\Delta\iota = (\iota \otimes \iota)\delta$ by the universal mapping property of the tensor algebra. To show that $T(C)$ is a bialgebra with these structures is a matter of showing that $(T(C), \Delta, \epsilon)$ is a coalgebra. Again it is an easy exercise to show that the algebra maps $(1_{T(C)} \otimes \Delta)\Delta$ and $(\Delta \otimes 1_{T(C)})\Delta$ agree on $\iota(C)$. Since $\iota(C)$ generates $T(C)$ as an algebra we conclude that $(1_{T(C)} \otimes \Delta)\Delta = (\Delta \otimes 1_{T(C)})\Delta$. Likewise $(\epsilon \otimes 1_{T(C)})\Delta = 1_{T(C)} = (1_{T(C)} \otimes \epsilon)\Delta$.

Definition 1.5.10 *Let C be a coalgebra over the field k. The pair $(\iota, T(C))$ described above is the* free bialgebra on the coalgebra C, *or the* tensor bialgebra on the coalgebra C.

The free bialgebra on a coalgebra is characterized as follows.

Theorem 1.5.1 *Suppose that C is a coalgebra over the field k. Then there exists a pair $(\iota, T(C))$ such that*

a) *$T(C)$ is a bialgebra over k and $\iota : C \longrightarrow T(C)$ is a coalgebra map, and*

b) *if A is a bialgebra over k and $f : C \longrightarrow A$ is a coalgebra map then there exists a bialgebra map $F : T(C) \longrightarrow A$ uniquely determined by $F\iota = f$.*

Proof: We need only add to our discussion preceding the statement of the theorem by justifying part b). Suppose that A is a bialgebra over k and $f : C \longrightarrow A$ is a coalgebra map. By the universal mapping property of the tensor algebra of the vector space C there is a algebra map $F : T(C) \longrightarrow A$ uniquely determined by $F\iota = f$. It remains to show that F is a coalgebra map. It is easy to see that algebra maps $(F \otimes F)\Delta$ and ΔF agree on $\iota(C)$. Thus $(F \otimes F)\Delta = \Delta F$. Likewise $\epsilon F = \epsilon$. ∎

Suppose that A is a finitely generated bialgebra. Let V be a finite generating set. Then V is contained in a finite-dimensional subcoalgebra C of A by Theorem 1.3.2. By Proposition1.1.1 for some $n > 0$ there is coalgebra map $\pi : \mathrm{C}_n(k) \longrightarrow C$ which is onto. Regard π as a coalgebra map to A. Since the image of a bialgebra map $f : A \longrightarrow B$ is a sub-bialgebra of B, it now follows by the previous theorem that

Corollary 1.5.1 *Suppose that A is a finitely generated bialgebra over the field k. Then A is the homomorphic image of $T(\mathrm{C}_n(k))$ for some $n > 0$.* ∎

Suppose that A is a bialgebra over k. Then the underlying algebra structure on A accounts for the dual coalgebra A^o. The underlying coalgebra structure on A accounts for the dual algebra structure on A^*. A very natural question to ask in this context is whether or not A^o is a subalgebra of A^*.

Proposition 1.5.2 *Suppose that A is a bialgebra over the field k. Then:*

a) *The coalgebra A^o is a subalgebra of the dual algebra A^*.*

b) *A^o is a bialgebra with the dual coalgebra structure and the subalgebra structure of the dual algebra A^*.*

c) *Let $\iota : A \longrightarrow (A^o)^*$ be the algebra map defined by $\iota(a)(a^o) = a^o(a)$. Then $\mathrm{Im}\iota \subseteq (A^o)^o$ and $\iota : A \longrightarrow (A^o)^o$ is a bialgebra map.*

Proof: We will show part a). Parts b) and c) are left as straightforward exercises to the reader. First of all notice that Δ_ϵ exists and $\Delta_\epsilon = \epsilon \otimes \epsilon$ since ϵ is an algebra map. Now suppose that $a^o, b^o \in A^o$. Then the calculation

$$\begin{aligned}(a^o_{(1)}b^o_{(1)})(a)(a^o_{(2)}b^o_{(2)})(b) &= a^o_{(1)}(a_{(1)})b^o_{(1)}(a_{(2)})a^o_{(2)}(b_{(1)})b^o_{(2)}(b_{(2)}) \\ &= a^o(a_{(1)}b_{(1)})b^o(a_{(2)}b_{(2)}) \\ &= a^o((ab)_{(1)})b^o((ab)_{(2)}) \\ &= (a^ob^o)(ab)\end{aligned}$$

for all $a, b \in A$ shows that $\Delta_{a^ob^o}$ does exist and $\Delta_{a^ob^o} = a^o_{(1)}b^o_{(1)} \otimes a^o_{(2)}b^o_{(2)}$. Therefore $a^ob^o \in A^o$ by definition, and part a) follows. This concludes our proof. ∎

Let A be a bialgebra over k. Then twisting the product and the coproduct separately give rise to bialgebras A^{op} and A^{cop} respectively. In particular $A^{op\ cop}$ is a bialgebra.

Now A^o is a bialgebra by Proposition 1.5.2. Therefore $(A^o)^{op} = (A^{cop})^o$, $(A^o)^{cop} = (A^{op})^o$, and $(A^o)^{op\ cop} = (A^{cop\ op})^o$ are bialgebras. Hence a bialgebra formally gives rise to eight, including itself.

Definition 1.5.11 *Let A be a bialgebra over the field k. Then $A^{op\ cop}$ is the* opposite bialgebra.

Throughout the following exercises A and B are bialgebras over the field k.

Exercise 1.5.1 Formulate and prove a fundamental homomorphism theorem for bialgebras. (See Theorem 1.1.1.)

Exercise 1.5.2 Show that $G(A^o) = \mathrm{Alg}(A, k)$ is multiplicative semigroup of A.

Exercise 1.5.3 Show that the subspace $\mathrm{coCom}A$ of cocommutative elements of A is a subalgebra of A.

Exercise 1.5.4 Show that $(A \otimes B)^o \simeq A^o \otimes B^o$ as bialgebras, where the tensor products have the tensor product bialgebra structures.

Exercise 1.5.5 We will make the following definition of simple bialgebra.

Definition 1.5.12 *A bialgebra A over a field k is* simple *if the only bi-ideals of A are* (0) *and* $\mathrm{Ker}\epsilon$.

Suppose that A is finite-dimensional. Show that A is simple if and only if the only subbialgebras of A^* are $k\epsilon$ and A^*.

Exercise 1.5.6 Suppose that the characteristic of k is 0 and $x \in A$ is a non-zero primitive element. Show that $\{1, x, x^2, \ldots\}$ is linearly independent. (Thus the subalgebra of A generated by x is the polynomial algebra $k[x]$ over k.)

Exercise 1.5.7 Suppose that the characteristic of k is 0 and the polynomial algebra $A = k[x]$ has the bialgebra structure determined by making x primitive. Show that the only bi-ideals of A are (0) and $\mathrm{Ker}\epsilon$; thus A is simple.

Exercise 1.5.8 This exercise relates the associated Lie algebra structure of A to the coalgebra structure of A.

a) For $a, b \in A$ show that

$$\epsilon([a, b]) = 0$$

and

$$\Delta([a, b]) = [a_{(1)}, b_{(1)}] \otimes a_{(2)}b_{(2)} + b_{(1)}a_{(1)} \otimes [a_{(2)}, b_{(2)}].$$

b) Suppose that C and D are subcoalgebras of A. Show that $[C, D]$ is a coideal of A.

Exercise 1.5.9 Let L be a Lie algebra over k. Construct the bialgebra $U(L)$ directly from the tensor algebra $(\iota, T(L))$ of the vector space L directly as follows:

a) Show that the algebra maps $\Delta : T(L) \longrightarrow T(L) \otimes T(L)$ and $\epsilon : T(L) \longrightarrow k$ which arise from the linear maps $\delta : L \longrightarrow T(L) \otimes T(L)$ and $0 : L \longrightarrow k$ respectively, where $\delta(\ell) = \iota(\ell) \otimes 1 + 1 \otimes \iota(\ell)$, give the tensor algebra $T(L)$ a bialgebra structure.

b) For $\ell, \ell' \in L$ show that $d(\ell, \ell') = [\ell, \ell']_a - \iota([\ell, \ell'])$ spans a coideal, where $[\ell, \ell']_a = \iota(\ell)\iota(\ell') - \iota(\ell')\iota(\ell)$ is the Lie product of $\iota(\ell)$ and $\iota(\ell')$ arising from the associative algebra structure on $T(L)$.

c) Let I be the ideal of $T(L)$ generated by the differences $d(\ell, \ell')$, where $\ell, \ell' \in L$. Show that $U(L) = T(L)/I$ with the induced quotient bialgebra structure is the universal enveloping algebra of L with the bialgebra structure described in the preceding section.

Exercise 1.5.10 Prove that there is a pair (π, A_c) which satisfies the following universal mapping property:

a) A_c is a commutative bialgebra over k and $\pi : A \longrightarrow A_c$ is a bialgebra map, and

b) if B is a commutative bialgebra over k and $f : A \longrightarrow B$ is a bialgebra map, then there exists a map of (commutative) bialgebras $F : A_c \longrightarrow B$ uniquely determined by $F\pi = f$.

Exercise 1.5.11 Suppose that C is a coalgebra over the field k. Show that there is a pair $(\pi, S(C))$ which satisfies the following universal mapping property:

a) $S(C)$ is a commutative bialgebra over k and $\pi : C \longrightarrow S(C)$ is a coalgebra map, and

b) if B is a commutative bialgebra over k and $f : C \longrightarrow B$ is a coalgebra map then there exists a map of (commutative) bialgebras $F : S(C) \longrightarrow B$ uniquely determined by $F\pi = f$.

Definition 1.5.13 *The pair* $(\pi, S(C))$ *is the* free commutative bialgebra *on* C.

Exercise 1.5.12 Show that every finitely generated commutative bialgebra over the field k is the homomorphic image of $S(\mathrm{C}_n(k))$ for some $n \geq 1$.

Exercise 1.5.13 Part d) of Proposition 1.5.1 can be generalized. Suppose that $g, h, g', h' \in G(A)$ commute with each other and centralize $P_{g,h}(A)$ and $P_{g',h'}(A)$. Show that

$$[P_{g,h}(A), P_{g',h'}(A)] \subseteq P_{gg',hh'}(A).$$

1.6 Hopf Algebras

A Hopf algebra over the field k is a bialgebra over k with a linear endomorphism which can be thought of as analogous to the inverse map for groups. It may be useful to think of the relationship of bialgebras to Hopf algebras as similar to the relationship of semigroups to groups. There is a richness in the theory for groups which is lacking for semigroups.

Before introducing Hopf algebras we briefly discuss the convolution algebra. The definition of Hopf algebra is made in terms of this structure.

1.6.1 The Convolution Algebra

Let (C, Δ, ϵ) be a coalgebra and (A, m, η) be an algebra over the field k. The convolution algebra of C and A, introduced in Exercise 1.2.13, is the linear space $\mathrm{Hom}(C, A)$ with product defined by

$$f*g = m(f \otimes g)\Delta \tag{1.16}$$

for all $f, g \in \mathrm{Hom}(C, A)$. The map $\eta\epsilon$ is the multiplicative identity for this algebra. In terms of the H-S notation for the coproduct, (1.16) is expressed by the equations

$$(f*g)(c) = f(c_{(1)})g(c_{(2)})$$

for all $f, g \in \mathrm{Hom}(C, A)$ and $c \in C$. When $A = k$ the convolution algebra

$$\mathrm{Hom}(C, k) = C^*$$

is the dual algebra of C.

Let $\pi : C \longrightarrow D$ be a coalgebra map and $\iota : A \longrightarrow B$ be an algebra map. Then an easy exercise in definitions shows that

$$\mathrm{Hom}(D, A) \xrightarrow{\pi^*} \mathrm{Hom}(C, A) \tag{1.17}$$

defined by $\pi^*(f) = f\pi$ and

$$\mathrm{Hom}(C, A) \xrightarrow{\iota_*} \mathrm{Hom}(C, B) \tag{1.18}$$

defined by $\iota_*(f) = \iota f$ for $f \in \mathrm{Hom}(C, A)$ are algebra maps. By Exercise A.4.2 the linear map

$$C^* \otimes A \xrightarrow{\iota} \mathrm{Hom}(C, A)$$

defined by $\iota(c^* \otimes a)(c) = c^*(c)a$ is one-one and $\mathrm{Im}\iota = \mathrm{Hom}_f(C, A)$ is the subspace of all linear maps $f : C \longrightarrow A$ with finite rank. It is easy to see that ι is an algebra map where $C^* \otimes A$ is given the tensor product algebra structure. In particular $\mathrm{Hom}_f(C, A)$ is a subalgebra of $\mathrm{Hom}(C, A)$. Notice that ι is an isomorphism if and only if either C or A is finite-dimensional.

Throughout the following exercises C is a coalgebra over the field k and A is an algebra over k.

Exercise 1.6.1 Suppose that C is a bialgebra over k and $f \in \mathrm{Hom}(C, A)$ has a convolution inverse f^{-1}. Show that:

a) If $f : C \longrightarrow A$ is an algebra map then $f^{-1} : C \longrightarrow A^{op}$ is an algebra map.

b) If $f : C \longrightarrow A^{op}$ is an algebra map then $f^{-1} : C \longrightarrow A$ is an algebra map.

[Hint: Let $\mathcal{C} = C\otimes C$ have the tensor product coalgebra structure and let m be the multiplication of C. Then $m^*(f)$ has an inverse $m^*(f^{-1})$ in the convolution algebra $\mathrm{Hom}(\mathcal{C}, A)$. Show that $\ell : C \otimes C \longrightarrow A$ defined by $\ell(c \otimes d) = f^{-1}(d)f^{-1}(c)$ is also a left convolution inverse for $m^*(f)$. Show that part a) is equivalent to part b).]

Exercise 1.6.2 Suppose that C is a bialgebra and $f : C \longrightarrow A$ and $g : C \longrightarrow A^{op}$ are algebras maps. Let V be a subspace of A which generates C as an algebra. Show that $f{*}g = \eta\epsilon$ if and only if $(f{*}g)|_V = \eta\epsilon|_V$.

Exercise 1.6.3 Suppose that A is a bialgebra over k and $f \in \mathrm{Hom}(C, A)$ has a convolution inverse f^{-1}. Show that:

a) If $f : C \longrightarrow A$ is a coalgebra map then $f^{-1} : C \longrightarrow A^{cop}$ is a coalgebra map.

b) If $f : C \longrightarrow A^{cop}$ is a coalgebra map then $f^{-1} : C \longrightarrow A$ is a coalgebra map.

Exercise 1.6.4 Suppose that $(\iota, T(C))$ is the free bialgebra on C and $f : T(C) \longrightarrow A$ is an algebra map. Show that f has a left (respectively right, two sided) inverse in the convolution algebra $\mathrm{Hom}(T(C), A)$ if and only if $f\iota$ has a left (respectively right, two sided) inverse in the convolution algebra $\mathrm{Hom}(C, A)$.

1.6.2 The Definition of Hopf Algebra and Basic Properties of the Antipode

Let $(A, m, \eta, \Delta, \epsilon)$ be a bialgebra over a field k. Let A_c and A_m denote the underlying coalgebra and algebra structures on A respectively. Give $\mathrm{End}(A) = \mathrm{Hom}(A_c, A_m)$ the convolution algebra structure.

Definition 1.6.1 *A bialgebra A over the field k is a* Hopf algebra *if the identity map $1_A \in \mathrm{End}(A)$ has an inverse s in the convolution algebra* $\mathrm{End}(A)$. *In this case s is an* antipode *of A.*

Since inverses are unique when they exist, the bialgebra A has at most one antipode. In terms of the H-S notation for the coproduct, to say that $s \in \mathrm{End}(A)$ is an antipode for A is equivalent to the equations

$$s(a_{(1)})a_{(2)} = \epsilon(a)1 = a_{(1)}s(a_{(2)})$$

holding for all $a \in A$. From these equations it is clear that if A is a Hopf algebra with antipode s then the bialgebra $A^{cop\ op}$ is a Hopf algebra with antipode s as well. We will generally denote Hopf algebras by H or K instead of A or B.

Lemma 1.6.1 *Suppose that H is a Hopf algebra with antipode s_H over the field k.*

a) *If B is an algebra over k and $\iota : H \longrightarrow B$ is an algebra map then ι has an inverse in the convolution algebra* $\mathrm{Hom}(H, B)$ *which is ιs_H.*

b) *If C is a coalgebra over k and $\pi : C \longrightarrow H$ is a coalgebra map then π has an inverse in the convolution algebra* $\mathrm{Hom}(C, H)$ *which is* $s_H\pi$.

c) *If K is a Hopf algebra over k with antipode s_K and $f : H \longrightarrow K$ is a bialgebra map then $fs_H = s_Kf$.*

Proof: Let B be an algebra over k and C be a coalgebra over k. Now 1_H and s_H are inverses in the convolution algebra $\mathrm{End}(H)$. Thus by (1.18) it follows that $\iota_*(1_H) = \iota$ and $\iota_*(s_H) = \iota s_H$ are inverses in the convolution algebra $\mathrm{Hom}(H, B)$ and by (1.17) it follows that $\pi^*(1_H) = \pi$ and $\pi^*(s_H) = s_H\pi$ are inverses in the convolution algebra $\mathrm{Hom}(C, H)$. Now let $f : H \longrightarrow K$ be a map of bialgebras. We have shown that both s_Kf and fs_H are inverses for f in the convolution algebra $\mathrm{Hom}(H, K)$. Consequently $s_Kf = fs_H$ and the lemma is proved. ∎

Definition 1.6.2 *Let H and K be Hopf algebras over k with antipodes s_H and s_K respectively. A linear map $f : H \longrightarrow K$ is a* Hopf algebra map *if f is a bialgebra map and $fs_H = s_Kf$.*

By part c) of the preceding lemma bialgebra maps between Hopf algebras are automatically Hopf algebra maps.

Definition 1.6.3 *A* sub-Hopf algebra *of a Hopf algebra H with antipode s over k is a sub-bialgebra B of H which satisfies $s(B) \subseteq B$.*

A sub-Hopf algebra B of a Hopf algebra H is a Hopf algebra in its own right with antipode $s|_B$. We have noted that the intersection of sub-bialgebras of H is a sub-bialgebra. Thus it follows that the intersection of sub-Hopf algebras of H is again a sub-Hopf algebra of H. Therefore for every subspace V of H there is a unique sub-Hopf algebra K of H which is minimal with respect to the property that K contains V.

Definition 1.6.4 *This sub-Hopf algebra K of H just described is the* sub-Hopf algebra *of H* generated by *V.*

Definition 1.6.5 *Suppose that H is a Hopf algebra with antipode s over k. A* Hopf ideal *of H is a bi-ideal I of H such that $s(I) \subseteq I$.*

Suppose that $f : H \longrightarrow K$ is a Hopf algebra map. Then $\mathrm{Ker} f$ is a Hopf ideal of H. Let I be any Hopf ideal of H. Then there is a unique Hopf algebra structure on the quotient space H/I such that the projection $\pi : H \longrightarrow H/I$ is a Hopf algebra map.

Fundamental properties of the antipode are described in our next proposition.

Proposition 1.6.1 *Suppose that H is a Hopf algebra with antipode s over the field k. Then:*

a) $s(1) = 1$ *and* $s(ab) = s(b)s(a)$ *for all* $a, b \in H$.

b) $\epsilon(s(a)) = \epsilon(a)$ *and* $\Delta(s(a)) = s(a_{(2)}) \otimes s(a_{(1)})$ *for all* $a \in H$.

c) $s : H \longrightarrow H^{op\ cop}$ *is a Hopf algebra map and* Ker$s$ *is a Hopf ideal of* H.

d) *Suppose that* B *is a finite-dimensional sub-bialgebra of* H. *Then* $s(B) \subseteq B$; *thus* B *is a sub-Hopf algebra of* H.

e) $G(H)$ *is a multiplicative group of* H *and* $g^{-1} = s(g)$ *for* $g \in G(H)$.

f) $s(P_{g,h}(H)) = P_{h^{-1},g^{-1}}(H)$ *for* $g, h \in G(H)$. *In particular* $s(x) = -h^{-1}xg^{-1}$ *for all* $x \in P_{g,h}(H)$ *and* s *restricts to a linear isomorphism*

$$P_{g,h}(H) \simeq P_{h^{-1},g^{-1}}(H).$$

Proof: Since $1_H : H \longrightarrow H$ is both an algebra map and a coalgebra map, parts a) and b) follows by Exercises 1.6.1 and 1.6.3 respectively.

Part c) follows from parts a) and b). To show part d) let $\iota : B \longrightarrow H$ be the inclusion. Since ι is a coalgebra map we conclude that $\iota^*(1_H) = \iota$ has an inverse $\iota^*(s_H) = s|_B$ in the convolution algebra $\mathrm{Hom}(B, H)$ by part b) of Lemma 1.6.1. Now since ι is an algebra map we conclude that $\iota_* : \mathrm{Hom}(B, B) \longrightarrow \mathrm{Hom}(B, H)$ is a (one-one) algebra map by (1.18). As $\iota_*(1_B) = \iota$ and $\mathrm{Im}\iota_*$ is a finite-dimensional subalgebra of $\mathrm{Hom}(B, H)$, it follows by Exercise 1.6.6 that the inverse of ι is in $\mathrm{Im}\iota_*$. Therefore 1_B has an inverse s_B in $\mathrm{Hom}(B, B)$. Since both ιs_B and $s|_B$ are inverses of ι in the convolution algebra $\mathrm{Hom}(B, H)$ we conclude that $\iota s_B = s|_B$. Therefore $s(B) \subseteq B$ and part d) follows.

To show part e), note that for $g \in G(H)$, $s(g)g = \epsilon(g)1 = gs(g)$ by definition. As $\epsilon(g) = 1$ we conclude that g has an inverse and that $g^{-1} = s(g)$. By part a) of Proposition 1.5.1 it follows that $G(H)$ is a multiplicative semigroup of H and that $g^{-1} \in G(H)$.

It remains to show part f). Suppose that $x \in P_{g,h}(H)$ where $g, h \in G(H)$. Then $\epsilon(x) = 0$ and $\Delta(x) = g \otimes x + x \otimes h$ by definition. Therefore $0 = \epsilon(x)1 = gs(x) + xs(h) = gs(x) + xh^{-1}$, the last equation following by part e). Thus we calculate $s(x) = -g^{-1}xh^{-1}$ which is in $P_{h^{-1},g^{-1}}(H)$ since Δ is multiplicative. The remainder of the proof of part f) is straightforward. ∎

We have observed that if H is a Hopf algebra with antipode s then $H^{op\ cop}$ is a Hopf algebra also with antipode s. Whether or not H^{op} and H^{cop} are Hopf algebras depends on whether or not s is a linear automorphism of H. The following result is attributed to [Heyneman, 1966].

Proposition 1.6.2 *Suppose that* H *is a Hopf algebra with antipode* s *over the field* k. *Then the following are equivalent:*

a) *H^{op} is a Hopf algebra.*

b) *H^{cop} is a Hopf algebra.*

c) *s is a linear automorphism of H.*

If s is a linear automorphism of H then s^{-1} is the antipode of H^{op} and H^{cop}.

Proof: Since $H^{cop} = (H^{op})^{op\ cop}$ and $H^{op} = (H^{cop})^{op\ cop}$ parts a) and b) are equivalent. Suppose that H^{op} is a Hopf algebra with antipode t. Then for $a \in H$ we have $a_{(2)}t(a_{(1)}) = \epsilon(a)1 = t(a_{(2)})a_{(1)}$. Applying s to the left hand equation we obtain

$$(st)(a_{(1)})s(a_{(2)}) = \epsilon(a)1.$$

by part a) of Proposition 1.6.1. Replacing a with $s(a)$ in the right hand equation we obtain by part b) of the same

$$(ts)(a_{(1)})s(a_{(2)}) = \epsilon(a)1.$$

Therefore st and ts are both the inverse of s in the convolution algebra $\mathrm{End}(H)$. We have shown that $st = 1_H = ts$. Thus part a) implies part c).

Conversely, suppose that s is a linear automorphism and s^{-1} is its linear inverse. Since $s : H \longrightarrow H^{op}$ is an algebra map it follows that $s^{-1} : H^{op} \longrightarrow H$ is as well. Thus applying s^{-1} to the two equations $s(a_{(1)})a_{(2)} = \epsilon(a)1 = a_{(1)}s(a_{(2)})$, and noting that $\epsilon s = \epsilon$, we conclude that $s^{-1}(a_{(2)})a_{(1)} = \epsilon(a)1 = a_{(2)}s^{-1}(a_{(1)})$. This means that s^{-1} is an antipode for H^{op}. Thus part c) implies part a) and the proof of the proposition is complete. ▮

Corollary 1.6.1 *Suppose that H is a commutative or cocommutative Hopf algebra over the field k with antipode s. Then $s^2 = 1_H$.* ▮

When H is a Hopf algebra then the bialgebra H^o is a Hopf algebra.

Theorem 1.6.1 *Suppose that H is a Hopf algebra with antipode s over the field k. Then $s^*(H^o) \subseteq H^o$ and the restriction $s^o = s^*|_{H^o}$ is an antipode for H^o.*

Proof: Let $a^o \in H^o$. The calculation

$$\begin{aligned} s^*(a^o_{(2)})(a)s^*(a^o_{(1)})(b) &= a^o_{(2)}(s(a))a^o_{(1)}(s(b)) \\ &= a^o(s(b)s(a)) \\ &= a^o(s(ab)) \\ &= s^*(a^o)(ab) \end{aligned}$$

shows that $\Delta_{s^*(a^o)}$ exists and that $\Delta_{s^*(a^o)} = s^*(a^o_{(2)}) \otimes s^*(a^o_{(1)})$. Therefore $s^*(a^o) \in H^o$. We leave the calculation that the restriction $s^o = s^*|_{H^o}$ is an antipode for H^o as an exercise for the reader. ▮

Theorem 1.6.2 *Suppose that H is a finite-dimensional Hopf algebra over the field k with antipode s. Then s is a linear automorphism of H. Consequently H^{op} and H^{cop} are Hopf algebras with antipode s^{-1}.*

Proof: By Proposition 1.6.2 it suffices to show that s is a linear automorphism of H. We will prove this by induction on $\mathrm{Dim}H$. The case $\mathrm{Dim}H = 1$ is trivial since $H = k1$ and $s = 1_H$ in this case.

Assume that $\mathrm{Dim}H > 1$ and that antipodes of finite-dimensional Hopf algebras of dimension less that $\mathrm{Dim}H$ are linear automorphisms. By part c) of Proposition 1.6.1 we have that $K = s(H)$ is a sub-Hopf algebra of H and $\mathrm{Ker}s$ is a Hopf ideal of H. If $K = H$ then $\mathrm{Ker}s = (0)$ and we are done. If $K \neq H$ then by assumption the antipode $s|_K$ of K is a linear automorphism of K. Therefore $H = \mathrm{Ker}s \oplus K$ and $s|_K$ is one-one in any event.

Let $\pi : H \longrightarrow K$ be the projection onto K. Since $\mathrm{Ker}s$ is a Hopf ideal of H and $\mathrm{Ker}\pi = \mathrm{Ker}s$ it follows that $\epsilon(\mathrm{Ker}s) = (0)$ and $\Delta(\mathrm{Ker}s) \subseteq \mathrm{Ker}s \otimes H + H \otimes \mathrm{Ker}s = \mathrm{Ker}\pi \otimes H + H \otimes \mathrm{Ker}s$. Therefore $(\pi{*}s)(a) = 0 = \epsilon(a)1$ for $a \in \mathrm{Ker}s$. For $a \in K$ we compute $(\pi{*}s)(a) = \pi(a_{(1)})s(a_{(2)}) = a_{(1)}s(a_{(2)}) = \epsilon(a)1$. Therefore $\pi{*}s = \eta\epsilon$ This means that $\pi = 1_H$. Thus $K = H$ and consequently $s = s|_K$ is a linear automorphism of H. ∎

The fact that a finite-dimensional Hopf algebra over a field has a bijective antipode is established in [Sweedler, 1969] by a very different argument. Our proof of Theorem 1.6.2 uses ideas found in [Radford, 1977, Section 3].

Generally the antipode of a Hopf algebra is not a linear automorphism. For an example see [Takeuchi, 1971].

Suppose that H is a finite-dimensional Hopf algebra over k. Then H formally gives rise to eight Hopf algebras. We have shown that $H^{op\ cop}$, H^{op} and H^{cop} are Hopf algebras. We have noted that $H \simeq H^{op\ cop}$, and $H^{op} \simeq H^{cop}$ as Hopf algebras. Now H^* is a Hopf algebra. Thus a finite-dimensional Hopf algebra gives rise to eight, including itself, by twisting structures on H and H^*. Among these eight there are four potentially different isomorphism types. Observe that $(H^*)^{op\ cop} = (H^{cop\ op})^*$, $(H^*)^{op} = (H^{cop})^*$, and $(H^*)^{cop} = (H^{op})^*$.

Exercise 1.6.5 Formulate and prove a fundamental homomorphism theorem for Hopf algebras. (See Theorem 1.1.1.)

Exercise 1.6.6 Suppose that B is a finite-dimensional subalgebra of an algebra A over the field k. Assume that $a \in B$ has a multiplicative inverse a^{-1} in A. Show that $a^{-1} \in B$ in two different ways:

a) Show that $\ell_a, r_a : B \longrightarrow B$ defined by $\ell_a(b) = ab$ and $r_a(b) = ba$ for all $b \in B$ are one-one, and hence onto.

b) Consider the sequence $\{1, a, a^2, \ldots\}$. Since B is finite-dimensional, show that there must is a non-trivial dependency relation in this set. Show that one of minimal length must have the form $\alpha_n a^n + \cdots + \alpha_0 1 = 0$, where $\alpha_n \neq 0 \neq \alpha_0$, and thus a^{-1} is a linear combination of powers of a.

Exercise 1.6.7 Let L be a Lie algebra over k. Show that $U(L)$ is a Hopf algebra, where $(\iota, U(L))$ is the universal enveloping algebra of L. [Hint: First show that the bialgebra $T(L)$ of Exercise 1.5.9 is a Hopf algebra with antipode which is the algebra map $S : T(L) \longrightarrow T(L)^{op}$ arising from the linear map $\varsigma : L \longrightarrow T(L)^{op}$ defined by $\varsigma(\ell) = \iota(-\ell)$. See Exercise 1.6.2. Then show that S determines an antipode for $U(L)$.]

1.7 The Coradical and the Coradical Filtration

Throughout this section C is a coalgebra over the field k.

Definition 1.7.1 *C_0 is the sum of the simple subcoalgebras of C and is the* coradical *of C.*

By the fundamental theorem of coalgebras (Theorem 1.3.2) all simple subcoalgebras of C are finite-dimensional.

Lemma 1.7.1 *Let C be coalgebra over the field k, let $\{C_i\}_{i\in I}$ be a family of subcoalgebras of C, and let D be a subcoalgebra of C. Then:*

a) *If D is simple and $D \subseteq \sum_{i\in I} C_i$ then $D \subseteq C_i$ for some $i \in I$.*

b) *If $\sum_{i\in I} C_i$ is direct then $D \cap (\oplus_{i\in I} C_i) = \oplus_{i\in I}(D \cap C_i)$.*

c) *C_0 is the direct sum of the simple subcoalgebras of C.*

d) *$D_0 = C_0 \cap D$.*

Proof: We will first show part a). We have noted that D is finite-dimensional since it is simple. Therefore $D \subseteq \sum_{i\in S} C_i$ for some finite subset S of I. Fix $i_0 \in S$. If $D \not\subseteq C_{i_0}$ then $D \cap C_{i_0} = (0)$ since D is simple. Choose $f \in C^*$ such that $f|_D = \epsilon|_D$ and $f|_{C_{i_0}} = 0$. Then

$$D = D \leftharpoonup \epsilon = D \leftharpoonup f \subseteq \sum_{i\in S}(C_i \leftharpoonup f) \subseteq \sum_{i\in S\setminus\{i_0\}} C_i$$

since $C_i \leftharpoonup f \subseteq C_i$ and $C_{i_0} \leftharpoonup f = (0)$. Thus by induction on $|S|$ we conclude that $D \subseteq C_i$ for some $i \in S$.

To show part b) we first note $\oplus_{i\in I}(D \cap C_i) \subseteq D \cap (\oplus_{i\in I} C_i)$. To establish the other inclusion let $d \in D \cap (\oplus_{i\in I} C_i)$. Then $d \in D \cap (\oplus_{i\in S} C_i)$ for some finite subset $S \subseteq I$. For the proof we may assume that $C = \oplus_{i\in S} C_i$ and $D = D \cap C$. For each $i \in S$ define $\epsilon_i \in C^*$ by $\epsilon_i|_{C_i} = \epsilon|_{C_i}$ and $\epsilon_i|_{C_j} = 0$ when $j \neq i$. Then $\epsilon = \sum_{i\in S} \epsilon_i$ and

$\epsilon_i \rightharpoonup C = C_i$. As $\epsilon_i \rightharpoonup d \in D$ for all $i \in S$ we have that $d = \epsilon \rightharpoonup d = \sum_{i \in S}(\epsilon_i \rightharpoonup d) \in \sum_{i \in S} D \cap C_i$. Thus part b) follows.

Part c) comes down to noting that if $C_1, \ldots, C_r$ are distinct simple subcoalgebras of C then $C_1 + \cdots + C_r$ is direct. Suppose that $c_1 + \cdots + c_r = 0$, where $c_i \in C_i$, and that $c_r \neq 0$. Then $r > 1$ and $C_r \cap (C_1 + \cdots + C_{r-1}) \neq (0)$. Since C_r is simple $C_r \subseteq C_1 + \cdots + C_{r-1}$. Thus by part a) we have $C_r \subseteq C_i$ for some $1 \leq i < r$. Since C_i is simple $C_r = C_i$, a contradiction. Therefore $c_r = 0$. By induction on r we conclude that $c_1 = \ldots = c_r = 0$. Therefore $C_1 + \cdots + C_r$ is direct.

To show part d) we first note that $C_0 = \oplus_{i \in I} C_i$ is the direct sum of the simple subcoalgebras C_i of C by part c). By part b) we have that $D \cap C_0 = \oplus_{i \in I}(D \cap C_i)$. Now each $D \cap C_i$ is (0) or C_i since C_i is simple. Therefore $D \cap C_0$ is the sum of simple subcoalgebras of D. This means that $D \cap C_0 \subseteq D_0$. As simple subcoalgebras of D are simple subcoalgebras of C it follows that $D_0 \subseteq C_0$. Thus $D_0 \subseteq D \cap C_0$ and part d) is established. ∎

Definition 1.7.2 *A coalgebra C over k is* pointed *if all simple subcoalgebras of C are one-dimensional.*

Thus C is pointed if and only if C_0 is the span of $G(C)$.

Definition 1.7.3 *A coalgebra C over k is* irreducible *if C_0 is simple.*

Thus an irreducible coalgebra C has a unique simple subcoalgebra which must therefore be contained in all non-zero subcoalgebras of C.

Definition 1.7.4 *A coalgebra C over k is* pointed irreducible *if C_0 is one-dimensional.*

By the preceding lemma the properties of pointed, irreducible and pointed irreducible are hereditary.

Define $C_n = C_0 \wedge \cdots \wedge C_0$ ($n-1$ wedges) for $n \geq 1$. The reader is referred to Section 1.2 for the definition and basic properties of the wedge product. Since C_0 is a subcoalgebra of C, it follows that C_n is a subcoalgebra of C for $n \geq 1$.

Proposition 1.7.1 *Let C be a coalgebra over the field k and let C_n be defined for $n \geq 0$ as above. Then:*

a) $C_0 \subseteq C_1 \subseteq C_2 \subseteq \cdots \subseteq \cup_{n=0}^{\infty} C_n = C$.

b) $\Delta(C_n) \subseteq \sum_{i=0}^{n} C_{n-i} \otimes C_i$ *for* $n \geq 0$.

Proof: Let D be a subcoalgebra of C. Since $D_0 \subseteq C_0$ by part c) of Lemma 1.7.1 it follows by induction that $D_n \subseteq C_n$ for all $n \geq 0$. Now let $c \in C$. Then $c \in D$ for some finite-dimensional subcoalgebra D of C by Theorem 1.3.2. Thus to prove the proposition we may assume that $C \neq (0)$ and is finite dimensional.

Suppose that $C \neq (0)$ and is finite-dimensional. By Lemma 1.7.1 we see that C has finitely many simple subcoalgebras, which we label $D_1, \ldots, D_n$, and $C_0 = D_1 + \cdots + D_n$. Set $J = C_0^\perp$. Then $C_0^\perp = \cap_{i=1}^n D_i^\perp$. Since all subspaces of D^* are closed by Proposition A.4.2, the $D_i^\perp$'s are the maximal ideals of C by part d) of Proposition 1.2.4. Therefore J is the Jacobson radical of C^*.

Now $C_\ell = (J^{\ell+1})^\perp$ for all $\ell \geq 0$ by part a) of Corollary 1.2.2. From this fact we can deduce $C_0 \subseteq C_1 \subseteq C_2 \subseteq \ldots$. Since J is nilpotent $J^{N+1} = (0)$ for some $N \geq 0$. Therefore $C_N = (J^{N+1})^\perp = C$. Part a) is established.

Now we show part b). Since C is finite-dimensional, we conclude from part a) that there is a basis $\{c_1, \ldots, c_r\}$ for C and a sequence $1 \leq r(0) \leq r(1) \leq r(2) \leq \ldots \leq r$ such that $\{c_1, \ldots, c_{r(n)}\}$ is a basis for C_n for all $n \geq 0$. Now let $c \in C_n$. Since C_n is a subcoalgebra of C we can write

$$\Delta(c) = \sum_{i=1}^{r(n)} c_i \otimes d_i$$

where $d_i \in C_n$ for each $1 \leq i \leq r(n)$. Now fix an index i and let m be the smallest integer such that $c_i \in C_m$. We will show that $d_i \in C_{n-m}$. For this purpose we may assume that $m > 0$. Thus $r(m-1) < i \leq r(m)$. Consider the functional $f \in C^*$ which vanishes on all basis elements except c_i on which f takes the value 1. Then f vanishes on C_{m-1} which means $f \in J^m$. Now chose any $g \in J^{n-m+1}$. Then applying $f \otimes g$ to both sides of the equation above we calculate

$$0 = (f*g)(c) = (f \otimes g)(\Delta(c)) = g(d_i)$$

since $f*g \in J^m J^{n-m+1} = J^{n+1}$. Therefore $d_i \in C_{n-m}$. Part b) is now established and the proposition is proved. ∎

Definition 1.7.5 *A sequence of subspaces $V_0, V_1, V_2, \ldots$ of C is a* filtration *of C if $V_0 \subseteq V_1 \subseteq V_2 \subseteq \ldots \subseteq \cup_{n=0}^\infty V_n = C$ and $\Delta(V_n) \subseteq \sum_{i=0}^n V_{n-i} \otimes V_i$ for all $n \geq 0$.*

Notice that the terms of a filtration of C are subcoalgebras of C. By the previous proposition $C_0, C_1, C_2, \ldots$, where $C_n = C_0 \wedge \cdots \wedge C_0$ is defined above, is a filtration of C. This filtration is of fundamental importance.

Definition 1.7.6 *The filtration $C_0, C_1, C_2, \ldots$ of C is the* coradical filtration *of C.*

The first term of *any* filtration of C contains the first term of the coradical filtration C_0.

Proposition 1.7.2 *Let C be a coalgebra over the field k and let $V_0, V_1, V_2, \ldots$ be a filtration of C. If D is a simple subcoalgebra of C then $D \subseteq V_0$.*

Proof: Let D be a simple subcoalgebra of C. Since D is finite-dimensional it follows that $D \subseteq V_n$ for some $n \geq 0$. We will assume that n is the smallest integer satisfying this property.

Suppose that $D \not\subseteq V_0$. Then $D \cap V_0 = (0)$ since this intersection is a subcoalgebra of D and D is simple. Choose $f \in C^*$ such that $f|_D = \epsilon|_D$ and $f|_{V_0} = 0$. Then

$$D = D \leftharpoonup \epsilon = D \leftharpoonup f \subseteq \sum_{i=0}^{n} f(V_{n-i})V_i$$

means that $D \subseteq V_0 + \cdots + V_{n-1} \subseteq V_{n-1}$. Therefore $D \subseteq V_{n-1}$ which contradicts the minimality of n. Thus $D \subseteq V_0$ after all. ∎

If $f : C \longrightarrow D$ is a coalgebra map which is onto and $V_0, V_1, V_2, \ldots$ is a filtration of C then $f(V_0), f(V_1), f(V_2) \ldots$ is a filtration of D. Therefore by the previous proposition:

Corollary 1.7.1 *Suppose that $f : C \longrightarrow D$ is a coalgebra map which is onto. Then $f(C_0) \supseteq D_0$.* ∎

Corollary 1.7.2 *Suppose that A is a bialgebra over the field k and C is a subcoalgebra of A which generates A as an algebra. Then the coradical A_0 of A is contained in the subalgebra of A generated by C_0.*

Proof: For $n \geq 0$ define

$$V_n = k1 + \sum_{0 \leq i_1 + \cdots + i_m \leq n} C_{i_1} \cdot \cdots \cdot C_{i_m}.$$

Let B be the subalgebra of A generated by C. Then B is a sub-bialgebra of A and $V_0, V_1, V_2, \ldots$ defines a filtration of the coalgebra B. Now $B = A$ by assumption. The corollary follows from the previous proposition at this point. ∎

Important bialgebras which arise in the theory of quantum groups, and indeed in this text, are generated as an algebra by their grouplike and skew primitive elements. Let A be a bialgebra over the field k and suppose $S \subseteq G(A)$. Consider the subcoalgebra $C = k[S] + \sum_{g,h \in S} P_{g,h}(A)$ of A. Recall $P_{g,h}(A) = \{a \in A \mid \Delta(a) = g \otimes a + a \otimes h\}$ is the subspace of $g{:}h$ skew primitives of A. Then $C_{(0)} = k[S]$ and $C_{(1)} = C_{(2)} = \cdots = C$ defines a filtration of C. Therefore $C_0 \subseteq C_{(0)}$ by Proposition 1.7.2. Thus $C_0 = C_{(0)}$. Suppose further that C generates A as an algebra. Since $G(A)$ is linearly independent by Lemma 1.1.1 it follows by the preceding corollary that $A_0 = k[\mathcal{S}]$, where $\mathcal{S}$ is the sub-semigroup of $G(A)$ generated by S. Consequently:

Corollary 1.7.3 *Suppose that A is a bialgebra over the field k generated as a bialgebra by skew primitive elements. Then $A_0 = k[G(A)]$. More specifically, if $S \subseteq G(A)$ and*

$C = k[S] + \sum_{g,h\in S} P_{g,h}(A)$ generates A as an algebra then $A_0 = k[S]$ where $\mathcal{S}$ is the sub-semigroup of $G(A)$ generated by S.

∎

We note Corollary 1.7.3 is basically [Radford, 1993a, Lemma 1]. Let A be an algebra over k and suppose that $f : C \longrightarrow A$ vanishes on C_0. Then formal linear combinations of all powers of f in the convolution algebra $\mathrm{Hom}(C, A)$ are meaningful. For f^m vanishes on C_n whenever $n < m$. This is the case when $m = 1$ by assumption. If $m > 1$ then

$$f^m(C_n) = (f^{m-1}{*}f)(C_n) \subseteq \sum_{i=0}^{n} f^{m-1}(C_{n-i})f(C_i) = (0)$$

as $n - i < m - 1$ when $0 < i \leq n$. Let $h(X) = \sum_{n=0}^{\infty} \alpha_n X^n \in k[[X]]$ be a formal power series with coefficients in k and set $h(f) = \sum_{n=0}^{\infty} \alpha_n f^n$. Let $c \in C$. Then $c \in C_m$ for some $m \geq 0$. Therefore $f^{m+1}(c) = f^{m+2}(c) = \cdots = 0$ which means that

$$h(f)(c) = \sum_{n=0}^{\infty} \alpha_n f^n(c) = \sum_{n=0}^{m} \alpha_n f^n(c)$$

is a well-defined element of A.

Lemma 1.7.2 *Suppose that C is a coalgebra and A is an algebra over k. Let $f : C \longrightarrow A$ be a linear map.*

a) *Suppose that f vanishes on C_0. Then the substitution map*

$$k[[X]] \xrightarrow{\ \pi_f\ } \mathrm{Hom}(C, A)$$

defined by $\pi_f(h(X)) = h(f)$ is a well-defined algebra homomorphism. In particular $\eta\epsilon - f$ has an inverse in the convolution algebra $\mathrm{Hom}(C, A)$.

b) *The linear map f has a left inverse (respectively right inverse, inverse) in the convolution algebra* $\mathrm{Hom}(C, A)$ *if and only if the restriction $f|_{C_0}$ has an left inverse (respectively right inverse, inverse) in the convolution algebra* $\mathrm{Hom}(C_0, A)$.

Proof: We will leave the straightforward details of part a) to the reader with the remark that $1 - X$ has an inverse in $k[[X]]$.

To show part b) we first observe that it suffices to show that f has a left inverse in the convolution algebra $\mathrm{Hom}(C, A)$ if and only if $f|_{C_0}$ has a left inverse in the convolution algebra $\mathrm{Hom}(C_0, A)$. For the right invertibility statement follows from the left invertibility statement when C^{cop} replaces C and A^{op} replaces A. The invertibility statement follows from these two.

If f has a left inverse in $\mathrm{Hom}(C, A)$ then clearly $f|_{C_0}$ has a left inverse in $\mathrm{Hom}(C_0, A)$. Conversely, suppose that $f|_{C_0}$ has a left inverse $G \in \mathrm{Hom}(C_0, A)$. Extend G to $g \in \mathrm{Hom}(C, A)$ in any manner whatsoever. Then $(g*f)|_{C_0} = (\eta\epsilon)|_{C_0}$ which means that $g*f - \eta\epsilon$ vanishes on C_0. Therefore $g*f = \eta\epsilon - (\eta\epsilon - g*f)$ has an inverse in $\mathrm{Hom}(C, A)$ by part a). In particular $g*f$ has a left inverse in $\mathrm{Hom}(C, A)$. Therefore f has a left inverse in $\mathrm{Hom}(C, A)$. ▮

The reader should note that the proof of the preceding lemma is basically found in the details of the proof of [Takeuchi, 1971, Lemma 14]. There are immediate and interesting corollaries to Lemma 1.7.2.

Corollary 1.7.4 *Suppose that A is a bialgebra over the field k. Then A is a Hopf algebra if and only if the inclusion $i : C_0 \longrightarrow A$ has an inverse in the convolution algebra* $\mathrm{Hom}(C_0, A)$. ▮

Corollary 1.7.5 *Suppose that A is a pointed bialgebra over the field k. Then A is a Hopf algebra if and only if all $g \in G(A)$ have an inverse in A. In particular a pointed irreducible bialgebra is a Hopf algebra.* ▮

In the proof of Proposition 1.7.1 we showed that $\mathrm{Rad}(C^*) = C_0^\perp$ when C is finite-dimensional. We conclude this section by showing that finite dimensionality is not necessary for this result for the sake of completeness. We note that the characterization of $\mathrm{Rad}(C^*)$ in part a) of the following theorem is found in [Heyneman and Radford, 1974, Proposition 2.1.4].

Theorem 1.7.1 *Let C be a coalgebra over the field k. Then* $\mathrm{Rad}(C^*)$ *is:*

a) $C_0^\perp = \cap_D D^\perp$, *where D runs over the simple subcoalgebras of C;*

b) *the intersection of the closed cofinite maximal deals of C^*;*

c) *the intersection of the closed cofinite maximal left ideals of C^*.*

Proof: Since C_0 is the sum of the simple subcoalgebras D of C it follows that $C_0^\perp = \cap_D D^\perp$ where D runs over the simple subcoalgebras of C. Since simple subcoalgebras of C are finite-dimensional, by part d) of Proposition 1.2.4 and Lemma A.4.4 the association $D \longmapsto D^\perp$ gives a one-one correspondence between the set of simple subcoalgebras of C and the set of cofinite closed maximal ideals of C^*. Therefore $\mathrm{Rad}(C^*) \subseteq C_0^\perp$. Now a cofinite maximal ideal in any algebra over k is the intersection of the maximal left ideals containing it. By Lemma A.4.4 subspaces of C^* which contain a closed cofinite subspace are themselves cofinite and closed. Thus to conclude the proof we need only show that $C_0^\perp \subseteq \mathrm{Rad}(C^*)$.

Let L be a maximal left ideal of C^*. If $C_0^\perp \not\subseteq L$ then $L + C_0^\perp = C^*$ since L is maximal. Assume this is the case. Then $\epsilon = \ell + a$ for some $\ell \in L$ and $a \in C_0^\perp$. Since

$a = \epsilon - \ell$ vanishes on C_0 we conclude that $\ell = \epsilon - a$ has an inverse in C^* by part a) of Lemma 1.7.2. This contradiction shows that $C_0^\perp \subseteq L$ after all, and the theorem is proved. ∎

Throughout the following exercises C and D are coalgebras over the field k.

Exercise 1.7.1 Suppose that A is an algebra over k and let $\mathrm{Hom}(C, A)$ be the convolution algebra of C and A. Show that

$$\mathrm{Hom}(C, A)^{op} = \mathrm{Hom}(C^{cop}, A^{op})$$

as algebras. (Thus $\mathrm{Hom}(C, A)^{op}$ is a convolution algebra.)

Exercise 1.7.2 Let C be pointed irreducible and denote its unique grouplike element by g.

a) If $c \in C_n$ show that $\Delta(c) = c \otimes g + g \otimes c + u$, where $u \in C_{n-1} \otimes C_{n-1}$, when $n \geq 1$.

b) Show that $C_1 = kg \oplus P_g(C)$.

Exercise 1.7.3 Suppose that C is pointed. Show that

$$C_1 = k[G(C)] + \sum_{g,h \in G(C)} P_{g,h}(C).$$

See [Taft and Wilson, 1974].

Exercise 1.7.4 Give an example of an onto coalgebra map $f : C \longrightarrow D$ such that $f(C_0) \neq f(C)_0$. (Generally $f(C_0) \supseteq f(C)_0$ for onto coalgebra maps by Corollary 1.7.1.)

Exercise 1.7.5 Give $C \otimes D$ the tensor product coalgebra structure.

a) Show that $(C \otimes D)_0 \subseteq C_0 \otimes D_0$.

b) Suppose that $\mathcal{E}$ is a simple subcoalgebra of $C \otimes D$. Show that $\mathcal{E} \subseteq \mathcal{C} \otimes \mathcal{D}$ for some simple subcoalgebras $\mathcal{C}$ and $\mathcal{D}$ of C and D respectively.

Exercise 1.7.6 Set $\mathrm{pi}(C) = C/C_0^+$.

a) Show that $\mathrm{pi}(C)$ is a pointed irreducible coalgebra. Also show that $\pi(C_0) = \mathrm{pi}(C)_0$, where $\pi : C \longrightarrow \mathrm{pi}(C)$ is the projection.

b) Show that the pair $(\pi, \mathrm{pi}(C))$ satisfies the following universal mapping property:

   i) $\mathrm{pi}(C)$ is a pointed irreducible coalgebra over k and $\pi : C \longrightarrow \mathrm{pi}(C)$ is a coalgebra map such that $\pi(C_0) = \mathrm{pi}(C)_0$, and

   ii) if D is a pointed irreducible coalgebra over k and $f : C \longrightarrow D$ is a coalgebra map which satisfies $f(C_0) = D_0$ then there exists a coalgebra map $F : \mathrm{pi}(C) \longrightarrow D$ uniquely determined by $F\pi = f$.

c) Show that $\pi(C_n) = \mathrm{pi}(C)_n$ for $n \geq 0$.

The reader is referred to [Heyneman and Radford, 1974].

Exercise 1.7.7 Let C be pointed irreducible and denote the unique grouplike element of C by g. If I is a coideal of C, show that the following are equivalent:

a) $I \cap C_1 \neq (0)$.

b) $I \cap P_g(C) \neq (0)$.

c) $I \neq (0)$.

Exercise 1.7.8 Let $f : C \longrightarrow D$ be a map of coalgebras. Show that f is one-one if and only if the restriction $f|_{C_1}$ is one-one. [Hint: Note that $\mathrm{Ker} f$ is a coideal of C and consider the pointed irreducible case first.]

Exercise 1.7.9 Let k be a field of characteristic 0 and let $A = k[x_1, \ldots, x_r]$ be the polynomial algebra over k in commuting indeterminants $x_1, \ldots, x_r$ with bialgebra structure determined by letting $x_1, \ldots, x_r$ be primitive. Show that A_n is the span of all monomials $x_1^{n_1} \cdots x_r^{n_r}$ such that $0 \leq n_1 + \cdots + n_r \leq n$. [Hint: There is a certain (dense) subalgebra of A^* which can be identified with the polynomial algebra $k[X_1, \ldots, X_r]$ and $J^\perp = k1$, where $J = (X_1, \ldots, X_r)$. See Exercise 1.2.6.]

1.8 Pointed Hopf Algebras

As we stated in the Forward, pointed bialgebras play an important role for us in the algebra related to the quantum Yang–Baxter equation which is developed in this book. Many important Hopf algebras which arise in quantum groups are pointed, for example the quantized enveloping algebras and their quotients. The quantized enveloping algebras are generated as algebras by the first term of their coradical filtrations.

Suppose that K is a pointed Hopf algebra over the field k and suppose that $a \in G(K)$. We show that the condition

$$\Delta(x) = x \otimes a + 1 \otimes x \quad \text{and} \quad xa = qax, \tag{1.19}$$

where $x \in K\backslash 0$, $a \in G(K)$, and $q \in k\backslash 0$ is very natural. The defining relations for the quantized enveloping algebras can be described in terms of elements satisfying (1.19).

The sub-Hopf algebra H of K generated by $a \in G(K)$ and x satisfying (1.19) is a very basic pointed Hopf algebra. Let s be the antipode of K. Since $s(x) = -xa^{-1}$ we conclude by parts a) and e) of Proposition 1.6.1 that $s^2(x) = q^{-1}x$. Consequently $s^{2n}(x) = q^{-n}x$ for all $n \geq 0$ and ς^2 is a diagonalizable operator, where $\varsigma = s|_H$ is the antipode of H.

Let K be a pointed Hopf algebra over the field k. Recall from Exercise 1.7.3 that $K_1 = k[G(K)] + \sum_{g,h \in G(K)} P_{g,h}(K)$. Let $g, h \in G(K)$ and $u \in P_{g,h}(K)$. Then by definition $\Delta(u) = g \otimes u + u \otimes h$. Since $G(K)$ is a group under multiplication in K

and Δ is an algebra map it follows that

$$g^{-1}(P_{g,h}(K)) = P_{1,g^{-1}h}(K).$$

Fix $a \in G(K)$ and let $V_a = P_{1,a}(K)$. Suppose that T is the algebra automorphism of K defined by $T(v) = a^{-1}va$ for all $v \in K$. Then $T(V_a) = V_a$. Therefore $T_a = T|_{V_a}$ is a linear automorphism of V_a. If $x \in K\backslash 0$ notice that (1.19) holds for x if and only if $x \in V_a$ and x is an eigenvector of T_a belonging to q. In this case $q \neq 0$ since T_a is one-one.

Suppose further that k is algebraically closed and V_a is finite-dimensional and is not (0). Then T_a has an eigenvalue. Thus there is an $x \in V_a\backslash 0$ such that $xa = qax$ for some $q \in k\backslash 0$. Assume that K is finite-dimensional. Then $G(K)$ is a finite group since $G(K)$ is linearly independent. Therefore $a^n = 1$ for some $n \geq 1$. Consequently $T_a^n = I$. When the characteristic of k is 0 it follows that T_a is a diagonalizable operator. In this case V_a has a basis $\{x_1, \ldots, x_r\}$ such that $x_i a = q_i a x_i$, where q_i is an nth root of unity, for all $1 \leq i \leq r$.

Exercise 1.8.1 Suppose that H is a finite-dimensional Hopf algebra over k. Assume that k is an algebraically closed field of characteristic 0 and $\mathcal{G}$ is a commutative subgroup of $G(H)$. For $a \in \mathcal{G}$ show that

$$V_a = \oplus_{\omega \in \mathcal{W}} V_{a,\omega},$$

where $\mathcal{W} = G(k[\mathcal{G}]^*) = \text{Alg}(k[\mathcal{G}], k)$ and

$$V_{a,\omega} = \{x \in V_a \mid xg = \omega(g)gx \quad \text{for all} \quad g \in \mathcal{G}\}.$$

Exercise 1.8.2 Show that there exists a 4-dimensional Hopf algebra H over k determined by the following conditions [Sweedler, 1966]: as a k-algebra H is generated by a and x which satisfy the relations

$$a^2 = 1, \quad x^2 = 0, \quad \text{and} \quad xa = -ax,$$

and the coalgebra structure of H is determined by

$$\Delta(a) = a \otimes a \quad \text{and} \quad \Delta(x) = x \otimes a + 1 \otimes x.$$

Note that when the characteristic of k is not 2 the antipode of H has order 4 as a linear operator.

1.9 (Co)Module (Co)Algebras

Up to this point our discussion of algebras, coalgebras, bialgebras, and related structures has taken place for the most part in ${}_k\text{Vec}$, the category of whose objects are vector spaces over the field k and whose morphisms are linear maps. Definitions of algebras, coalgebras, bialgebras, and related structures can be made over a commutative ring. However, the ensuing structure theory is not nearly as rich and more often than not requires technical assumptions on the base ring or the structures themselves.

In many applications algebras and coalgebras over a field are also modules or comodules over a bialgebra or Hopf algebra. In this section we will introduce these structures and give a few examples. Morphisms of the categories we discuss in this section are certain types of functions and their multiplication is function composition.

1.9.1 H-Module Algebras and Coalgebras

Suppose that H is a bialgebra over the field k. Let ${}_H\mathrm{Mod}$ denote the category whose objects are left H-modules and whose morphisms are maps of left H-modules. We regard k as an object of ${}_H\mathrm{Mod}$ according to

$$h{\cdot}1_k = \epsilon(h)1_k$$

for all $h \in H$. For objects M and N of ${}_H\mathrm{Mod}$ we regard $M \otimes N$ as an object of ${}_H\mathrm{Mod}$ by

$$h{\cdot}(m \otimes n) = h_{(1)}{\cdot}m \otimes h_{(2)}{\cdot}n$$

for all $h \in H, m \in M$ and $n \in N$. Let $f : M \longrightarrow M'$ and $g : N \longrightarrow N'$ be morphisms of ${}_H\mathrm{Mod}$, that is maps of left H-modules. Then the tensor product of linear maps $f \otimes g : M \otimes N \longrightarrow M' \otimes N'$ is a morphism of ${}_H\mathrm{Mod}$.

Observe that ${}_k\mathrm{Vec}$ can be thought of as sub-category of ${}_H\mathrm{Mod}$ where vector spaces V over k are given the left H-module structure defined by

$$h{\cdot}v = \epsilon(h)v \tag{1.20}$$

for all $h \in H$ and $v \in V$.

Definition 1.9.1 *Let H be a bialgebra over the field k and V be a vector space over k. The left H-module structure on V defined by (1.20) is the* trivial left H-module structure *on V.*

For an object M of ${}_H\mathrm{Mod}$ notice that the linear isomorphisms

$$k \otimes M \simeq M \quad \text{and} \quad M \otimes k \simeq M \tag{1.21}$$

defined by $\alpha \otimes m \longmapsto \alpha m$ and $m \otimes \alpha \longmapsto \alpha m$ respectively are morphisms of ${}_H\mathrm{Mod}$. For objects L, M, and N of ${}_H\mathrm{Mod}$ observe that the linear isomorphism

$$(L \otimes M) \otimes N \simeq L \otimes (M \otimes N) \tag{1.22}$$

defined by $(\ell \otimes m) \otimes n \longmapsto \ell \otimes (m \otimes n)$ is also a morphism of ${}_H\mathrm{Mod}$. We are now in a position to describe algebras and coalgebras in the category ${}_H\mathrm{Mod}$.

An algebra in ${}_H\mathrm{Mod}$ is a triple (A, m, η), where A is an object of ${}_H\mathrm{Mod}$ and $m : A \otimes A \longrightarrow A, \eta : k \longrightarrow A$ are morphisms such that

$$m(m \otimes 1_A) = m(1_A \otimes m) \quad \text{and} \quad m(\eta \otimes 1_A) = 1_A = m(1_A \otimes \eta). \tag{1.23}$$

Implicit in these equations are the identifications $(A \otimes A) \otimes A \simeq A \otimes (A \otimes A)$ and $k \otimes A \simeq A \simeq A \otimes k$ described above.

Definition 1.9.2 *An algebra of* ${}_H\mathrm{Mod}$ *is a* left H-module algebra.

Thus a left H-module algebra can be thought of as a quadruple (A, μ, m, η), where (A, μ) is a left H-module and (A, m, η) in an algebra over k such that

$$h{\cdot}1_A = \epsilon(h)1_A \quad \text{and} \quad h{\cdot}(ab) = (h_{(1)}{\cdot}a)(h_{(2)}{\cdot}b) \tag{1.24}$$

for all $h \in H$ and $a, b \in A$. A morphism $f : A \longrightarrow B$ of algebras of ${}_H\mathrm{Mod}$ is a morphism of ${}_H\mathrm{Mod}$ which is a map of k-algebras.

What it means to say that an algebra A of ${}_H\mathrm{Mod}$ is commutative is problematic. We would like to say that A is commutative if $m\tau_{A,A} = m$, where $\tau_{A,A} : A \otimes A \longrightarrow A \otimes A$ is the twist map defined by $\tau_{A,A}(a \otimes b) = b \otimes a$ for $a, b \in A$. Generally $\tau_{A,A}$ is not a morphism. The existence of suitable "twist" morphisms $\sigma_{M,N} : M \otimes N \longrightarrow N \otimes M$ for pairs of objects (M, N) of certain categories is a matter of great interest and importance in the theory of quantum groups.

We give two basic examples of left H-module algebras.

Example 1: The dual algebra $A = H^*$ is a left H-module algebra where $(h{\cdot}a)(h') = a(h'h)$ for $h, h' \in H$ and $a \in A$.

Example 2: Suppose that H has antipode s. Then $A = H$ is a left H-module algebra where $h{\bullet}a = h_{(1)}a(s(h_{(2)}))$ for $h, a \in H$.

Definition 1.9.3 *The left module action of H on itself of Example 2 is the* left adjoint action.

It is an interesting exercise to see what constructions involving algebras over k go through in the category ${}_H\mathrm{Mod}$. In the process some very interesting categorical issues arise which we leave the reader to ponder.

The tensor product of two algebras A and B of ${}_H\mathrm{Mod}$ may *not* be an algebra of ${}_H\mathrm{Mod}$ with the *usual* tensor product algebra structure of algebras of ${}_k\mathrm{Vec}$. If $A \otimes B$ is an algebra of ${}_H\mathrm{Mod}$ with the tensor product algebra structure, then

$$h{\cdot}(a \otimes b) = h_{(1)}{\cdot}a \otimes h_{(2)}{\cdot}b$$

on the one hand and

$$\begin{aligned} h{\cdot}(a \otimes b) &= h{\cdot}((1 \otimes b)(a \otimes 1)) \\ &= (h_{(1)}{\cdot}(1 \otimes b))(h_{(2)}{\cdot}(a \otimes 1)) \\ &= (1 \otimes h_{(1)}{\cdot}b)(h_{(2)}{\cdot}a \otimes 1)) \\ &= h_{(2)}{\cdot}a \otimes h_{(1)}{\cdot}b \end{aligned}$$

on the other. Thus necessarily

$$h_{(1)}\cdot a \otimes h_{(2)}\cdot b = h_{(2)}\cdot a \otimes h_{(1)}\cdot b$$

for all $h \in H, a \in A$, and $b \in B$ if $A \otimes B$ is an algebra in ${}_H\text{Mod}$ with the tensor product algebra structure of ${}_k\text{Vec}$. Examples where this condition does not hold are found in the exercises.

The free algebra in ${}_H\text{Mod}$ on an object M of ${}_H\text{Mod}$ does exist. There exists a pair $(\iota, T(M))$ such that

i) $T(M)$ is an algebra of ${}_H\text{Mod}$ and $\iota : M \longrightarrow T(M)$ is a morphism,

ii) if A is an algebra of ${}_H\text{Mod}$ and $f : M \longrightarrow A$ is a morphism then there exists a morphism of algebras $F : T(M) \longrightarrow A$ uniquely determined by $F\iota = f$.

The underlying k-algebra structure of the free algebra on M is the tensor algebra $T(M)$ of the vector space M.

A coalgebra of ${}_H\text{Mod}$ is a triple (C, Δ, ϵ) where C is an object of ${}_H\text{Mod}$ and $\Delta : C \longrightarrow C \otimes C$, $\epsilon : C \longrightarrow k$ are morphisms such that

$$(\Delta \otimes 1_C)\Delta = (1_C \otimes \Delta)\Delta \quad \text{and} \quad (\epsilon \otimes 1_C)\Delta = 1_C = (1_C \otimes \epsilon)\Delta. \tag{1.25}$$

Implicit in these equations are the identifications $(C \otimes C) \otimes C \simeq C \otimes (C \otimes C)$ and $k \otimes C \simeq C \simeq C \otimes k$ of (1.22) and (1.21) respectively.

Definition 1.9.4 *A coalgebra of* ${}_H\text{Mod}$ *is a* left H-module coalgebra.

Thus a left H-module coalgebra can be thought of as a quadruple $(C, \mu, \Delta, \epsilon)$, where (C, μ) is a left H-module and (C, Δ, ϵ) is a coalgebra over k which satisfies

$$\epsilon(h\cdot c) = \epsilon(h)\epsilon(c) \quad \text{and} \quad \Delta(h\cdot c) = h_{(1)}\cdot c_{(1)} \otimes h_{(2)}\cdot c_{(2)} \tag{1.26}$$

for all $h \in H$ and $c \in C$. A morphism of coalgebras $f : C \longrightarrow D$ in ${}_H\text{Mod}$ is a morphism of ${}_H\text{Mod}$ which is a map of coalgebras over k.

We give two basic examples of left H-module coalgebras.

Example 3: $C = H$ is a left H-module coalgebra where $h\cdot c = hc$ for $h, c \in H$.

Example 4: Suppose that H has an antipode s. Then the dual coalgebra $C = H^o$ is a left H-module coalgebra where $h\bullet c = h_{(2)}\cdot c\cdot s(h_{(1)})$ for $h \in H$ and $c \in H^o$.

We regard $C = H^o$ as an H-bimodule in the usual way by $(h\cdot c\cdot h')(h'') = c(h'h''h)$ for all $c \in H^o$ and $h, h', h'' \in H$. See Section 1.3.

Right H-module algebras and right H-module coalgebras are defined to be algebras and coalgebras respectively in the category Mod_H whose objects are right H-modules and whose morphisms are maps of right H-modules in the same way left H-module algebras and left H-module coalgebras are defined to be algebras and coalgebras in the category ${}_H\mathrm{Mod}$. We regard k as an object of Mod_H by

$$1_k{\cdot}h = \epsilon(h)1_k$$

for $h \in H$. For objects M and N of Mod_H we regard $M \otimes N$ as an object of Mod_H by

$$(m \otimes n){\cdot}h = m{\cdot}h_{(1)} \otimes n{\cdot}h_{(2)}$$

for all $m \in M, n \in N$ and $h \in H$. The reader is left with the exercise of working out analogs of (1.24) and (1.26).

A left module for an algebra A of the category ${}_H\mathrm{Mod}$ is a pair (M, μ), where M is an object of ${}_H\mathrm{Mod}$ and $\mu : A \otimes M \longrightarrow M$ is a morphism such that the usual axioms for modules of ${}_k\mathrm{Vec}$ hold. In the obvious manner we define the notion of comodule for a coalgebra of ${}_H\mathrm{Mod}$.

Throughout the following exercises H is a bialgebra over the field k.

Exercise 1.9.1 Let ${}_k\mathrm{Alg}$ denote the category whose objects are algebras over the field k and whose morphisms are algebra maps. Notice that k is an object of ${}_k\mathrm{Alg}$, and if A and B are objects of ${}_k\mathrm{Alg}$ then the tensor product $A \otimes B$ over k is an object of ${}_k\mathrm{Alg}$ with the tensor product algebra structure.

a) Show that the linear maps defined by (1.21) and (1.22) are morphisms for objects L, M and N of ${}_k\mathrm{Alg}$.

b) Let A be an algebra over k. Show that A is an object of ${}_k\mathrm{Alg}$ if and only if A is commutative.

c) Show that the (coassociative) coalgebras of ${}_k\mathrm{Alg}$ are the bialgebras over k.

Exercise 1.9.2 For vector spaces M and N over k recall the twist map $\tau_{M,N} : M \otimes N \longrightarrow N \otimes M$ is defined by $\tau_{M,N}(m \otimes n) = n \otimes m$ for all $m \in M$ and $n \in N$. Show that the following are equivalent:

a) H is cocommutative.

b) $\tau_{M,N}$ is a morphism for all objects M, N of ${}_H\mathrm{Mod}$.

Exercise 1.9.3 Suppose that H has an antipode s. Regard $A = H$ as a left H-module under the adjoint action of Example 2 defined by

$$h{\bullet}a = h_{(1)}a(s(h_{(2)}))$$

for all $h, a \in H$.

a) Show that $ha = (h_{(1)}{\bullet}a)h_{(2)}$ for all $h, a \in H$.

b) Let $a \in H$. Show that a is in the center of H if and only if $h{\bullet}a = \epsilon(h)a$ for all $h \in H$.

(In particular non-zero central elements generate one-dimensional submodules of H under the left adjoint action.)

Exercise 1.9.4 Let $A = H^*$ be the left H-module algebra of Example 1. Show that $A \otimes A$, with the tensor product algebra structure over k, is an algebra of ${}_H\mathrm{Mod}$ if and only if H is cocommutative.

Exercise 1.9.5 Let M be an object of ${}_H\mathrm{Mod}$. Carry out the details of the construction of the free algebra of ${}_H\mathrm{Mod}$ on M.

Exercise 1.9.6 Let A be an algebra of ${}_H\mathrm{Mod}$. Show that the tensor product $A \otimes H$ over k is an algebra over k where

$$1_{A\otimes H} = 1_A \otimes 1_H \quad \text{and} \quad (a \otimes h)(a' \otimes h') = a(h_{(1)}\cdot a') \otimes h_{(2)}h'$$

for all $a, a' \in A$ and $h, h' \in H$.

Definition 1.9.5 *The algebra just described is the* smash product *of A and H.*

The smash product can be motivated by analyzing the structure of the group algebra of a semidirect product of groups. Notice that when A has the trivial left H-module structure then the smash product is the tensor product algebra over k.

Exercise 1.9.7 Suppose that H has an antipode s and let M and N be objects of ${}_H\mathrm{Mod}$.

a) Show that $(\mathrm{Hom}(M, N), \mu)$ is an object of ${}_H\mathrm{Mod}$, where μ is defined by

$$(h\bullet f)(m) = h_{(1)}\cdot(f(s(h_{(2)})\cdot m))$$

for all $h \in H$ and $m \in M$.

Let $f \in \mathrm{Hom}(M, N)$.

b) Show that

$$h\cdot(f(m)) = (h_{(1)}\bullet f)(h_{(2)}\cdot m)$$

for all $h \in H$ and $m \in M$.

c) Show that f is a module map (morphism) if and only if $h\bullet f = \epsilon(h)f$ for all $h \in H$. (In particular the non-zero module maps generate one-dimensional H-submodules of $\mathrm{Hom}(M, N)$.)

d) Show that the space of module maps $\mathrm{Mor}(M, N) = \mathrm{Hom}_H(M, N)$ is a sub-object of $(\mathrm{Hom}(M, N), \mu)$.

1.9.2 H-Comodule Algebras and Coalgebras

Let Comod^H denote the category whose objects are right H-comodules and whose morphisms are maps of right H-comodules. We regard k as an object of Comod^H by

$$\rho(1_k) = 1_k \otimes 1_H.$$

For objects M and N of Comod^H we regard the tensor product of vector spaces $M \otimes N$ as an object of Comod^H by

$$\rho(m \otimes n) = (m^{<1>} \otimes n^{<1>}) \otimes m^{(2)}n^{(2)}$$

for all $m \in M$ and $n \in N$. Let $f : M \longrightarrow M'$ and $g : N \longrightarrow N'$ be morphisms of Comod^H, that is maps of right H-comodules. Then the tensor product of linear maps $f \otimes g : M \otimes N \longrightarrow M' \otimes N'$ is a morphism of Comod^H.

Observe that ${}_k\text{Vec}$ can be thought of as a sub-category of Comod^H where vector spaces V over k are given the right H-comodule structure defined by

$$\rho(v) = v \otimes 1$$

for all $v \in V$.

Definition 1.9.6 *The comodule structure just defined is the* trivial right H-comodule structure *on* V.

Note that the linear maps defined by (1.21) and (1.22) are morphisms for objects L, M, and N of Comod^H.

An algebra in Comod^H is a triple (A, m, η), where A is an object of Comod^H and $m : A \otimes A \longrightarrow A$, $\eta : k \longrightarrow A$ are morphisms which satisfy (1.23). Again, implicit in these equations are the identifications $(A \otimes A) \otimes A \simeq A \otimes (A \otimes A)$ and $k \otimes A \simeq A \simeq A \otimes k$ of (1.22) and (1.21) respectively.

Definition 1.9.7 *An algebra of* Comod^H *is a* right A-comodule algebra.

Thus we may think of a right H-comodule algebra as a quadruple (A, ρ, m, η), where (A, ρ) is a right H-comodule and (A, m, η) in an algebra over k such that

$$\rho(1_A) = 1_A \otimes 1_H \quad \text{and} \quad \rho(ab) = a^{<1>}b^{<1>} \otimes a^{(2)}b^{(2)} \tag{1.27}$$

for all $a, b \in A$. A morphism of algebras $f : A \longrightarrow B$ of Comod^H is a morphism of Comod^{H} which is a map of k-algebras.

We give two elementary examples of right H-comodule algebras.

Example 5: Let $A = H$. Then (A, ρ) is a right H-comodule algebra where $\rho = \Delta$.

Example 6: Suppose that H is finite-dimensional and has antipode s. Then the dual algebra $A = H^*$ is a left H^*-module under the module action of Example 2. Let (A, ρ) be the underlying right H-comodule structure for this action. Observe that

$$b_{(1)}aS(b_{(2)}) = a^{<1>}b(a^{(2)})$$

for all $a, b \in A$, where $S = s^*$ is the antipode of H^*. The algebra A with the comodule structure (A, ρ) is a right H-comodule algebra.

A coalgebra of Comod^H is a triple (C, Δ, ϵ) where C is an object of Comod^H and $\Delta : C \longrightarrow C \otimes C$, $\epsilon : C \longrightarrow k$ are morphisms such that (1.25) hold. Again, implicit in these equations are the identifications $(C \otimes C) \otimes C \simeq C \otimes (C \otimes C)$ and $k \otimes C \simeq C \simeq C \otimes k$ of (1.22) and (1.21) respectively.

Definition 1.9.8 *A coalgebra of* Comod^H *is a* right H-comodule coalgebra.

Thus we may think of a right H-comodule coalgebra as a quadruple $(C, \rho, \Delta, \epsilon)$, where (C, ρ) is a right H-comodule and (C, Δ, ϵ) is a coalgebra over k which satisfies

$$\epsilon(c^{<1>})c^{(2)} = \epsilon(c)1_H \tag{1.28}$$

and

$$\Delta(c^{<1>}) \otimes c^{(2)} = ({c_{(1)}}^{<1>} \otimes {c_{(2)}}^{<1>}) \otimes {c_{(1)}}^{(2)} {c_{(2)}}^{(2)} \tag{1.29}$$

for all $c \in C$. A morphism $f : C \longrightarrow D$ of coalgebras of Comod^H is a morphism of Comod^H which is a map of coalgebras over k.

We give two basic examples of right H-comodule coalgebras.

Example 7: Suppose that H is finite-dimensional. Then $C = H^*$ is a left H^*-module under left multiplication. Let (C, ρ) be the underlying right H-comodule structure of this module action. Thus

$$ba = a^{<1>}b(a^{(2)})$$

for all $a, b \in C$. The coalgebra C together with the right H-comodule structure (C, ρ) is a right H-comodule coalgebra.

Example 8: Suppose that H has an antipode s. Then the $C = H$ is a right H-comodule coalgebra where (C, ρ) is the right H-comodule structure defined by

$$\rho(c) = c_{(2)} \otimes s(c_{(1)})c_{(3)}$$

for all $c \in H$.

Definition 1.9.9 *The right comodule action of H on itself of Example 8 is the* right coadjoint action *of H on itself.*

Left H-comodule algebras and left H-comodule coalgebras are defined to be algebras and coalgebras respectively in the category ^{H}Comod whose objects are left H-comodules and morphisms are maps of left H-comodules in the same way right H-comodule algebras and right H-comodule coalgebras are defined to be algebras and coalgebras in the category Comod^H. We regard k to be an object of ^{H}Comod by

$$\rho(1_k) = 1_H \otimes 1_k$$

for all $h \in H$. For objects M and N of ^{H}Comod we regard $M \otimes N$ as an object of ^{H}Comod by

$$\rho(m \otimes n) = m^{(1)}n^{(1)} \otimes (m^{<2>} \otimes n^{<2>})$$

for all $m \in M$ and $n \in N$. We leave the reader with the exercise of working out analogs of (1.27)–(1.29).

Throughout the following exercises H is a bialgebra over the field k.

Exercise 1.9.8 Recall that ${}_k\text{Coalg}$ is the category whose objects are coalgebras over the field k and whose morphisms are coalgebra maps. Notice that k is an object of ${}_k\text{Coalg}$, and if C and D are objects of ${}_k\text{Coalg}$ then the tensor product $C \otimes D$ over k is an object of ${}_k\text{Coalg}$ with the tensor product coalgebra structure of ${}_k\text{Vec}$.

a) Show that the linear maps defined by (1.21) and (1.22) are morphisms for objects L, M and N of ${}_k\text{Coalg}$.

b) Let C be a coalgebra over k. Show that C is an object of ${}_k\text{Coalg}$ if and only if C is cocommutative.

c) Show that the (associative) algebras of ${}_k\text{Coalg}$ are the bialgebras over k.

Exercise 1.9.9 Show that the twist map $\tau_{M,N}$ of Exercise 1.9.2 is a morphism for all objects M and N of Comod^H if and only if H is commutative.

Exercise 1.9.10 Suppose that H has an antipode s. Let $C = H$ be the right H-comodule coalgebra with the right coadjoint action (C, ρ) given by

$$\rho(h) = h_{(2)} \otimes s(h_{(1)})h_{(3)}$$

for all $h \in H$ in Example 8.

a) Writing $\rho(h) = h^{<1>} \otimes h^{(2)}$, show that

$$h_{(1)} \otimes h_{(2)} = h^{<1>}{}_{(2)} \otimes s^2(h^{<1>}{}_{(1)})h^{(2)}$$

for all $h \in H$.

b) Show that $\rho(h) = h \otimes 1$ if and only if $h_{(2)} \otimes h_{(1)} = s^2(h_{(1)}) \otimes h_{(2)}$.

Exercise 1.9.11 Let C and D be coalgebras of Coalg^H.

a) Show that $C \otimes D$ with the tensor product coalgebra structure of ${}_k\text{Vec}$ is a coalgebra of Coalg^H if and only if $(c^{<1>} \otimes d^{<1>}) \otimes c^{(2)}d^{(2)} = (c^{<1>} \otimes d^{<1>}) \otimes d^{(2)}c^{(2)}$ for all $c \in C$ and $d \in D$

b) Let $C = H^*$ be the right H-comodule coalgebra of Example 7. Show that $C \otimes C$, with the usual tensor product coalgebra structure over k, is a coalgebra of Coalg^H if and only if H is commutative.

Exercise 1.9.12 Determine whether or not the free algebra of Comod^H on an object M of Comod^H exists.

Exercise 1.9.13 Let C be a coalgebra of Comod^H. Show that the tensor product $C \otimes H$ over k is a coalgebra over k, where

$$\epsilon(c \otimes h) = \epsilon(c)\epsilon(h)$$

and

$$\Delta(c \otimes h) = ({c_{(1)}}^{<1>} \otimes h_{(1)}) \otimes (c_{(2)} \otimes h_{(2)}{c_{(1)}}^{(2)})$$

for all $c \in C$ and $h \in H$. See Exercise 1.9.6.

Definition 1.9.10 *The coalgebra just described is a* cosmash product.

Notice that the cosmash product is the tensor product coalgebra over k when C has the trivial right H-comodule structure.

Exercise 1.9.14 Suppose that H has antipode s and that M and N are objects of Comod^H, where M is finite-dimensional.

a) Show that there is a right H-comodule structure $(\text{Hom}(M, N), \rho)$ on $\text{Hom}(M, N)$ such that

$$f^{<1>}(m) \otimes f^{(2)} = f(m^{<1>})^{<1>} \otimes f(m^{<1>})^{(2)} s(m^{(2)})$$

for all $f \in \text{Hom}(M, N)$ and $m \in M$. (Thus $(\text{Hom}(M, N), \rho)$ is an object of Comod^H.)

b) Show that $f \in \text{Hom}(M, N)$ is a map of right H-comodules if and only $\rho(f) = f \otimes 1$.

c) Show that the space of comodule maps $\text{Mor}(M, N) = \text{Hom}^H(M, N)$ is a sub-object of $(\text{Hom}(M, N), \rho)$.

2 THE QUANTUM YANG-BAXTER EQUATION (QYBE)

As mentioned in the Introduction, the quantum Yang-Baxter equation arises in variety of contexts and has a number of forms. In this chapter we consider three fundamental forms of the equation: the constant, the one-parameter, and the two-parameter. The constant and one-parameter forms are connected to bialgebras through the FRT construction. We introduce the FRT construction in this chapter. The constant form of the quantum Yang–Baxter equation is very closely related to the braid equation, an equation of considerable importance for invariants of knots and links. Our treatment of compatibility conditions in the constant case is based on [Lambe, 1994] and [Lambe and Radford, 1993]. Our treatment of symmetries in Section 2.3 is based on [Hietarinta, 1993b, Section 2]. From this point on in the text we will refer to the quantum Yang–Baxter equation as the QYBE.

2.1 The Constant Form of the Quantum Yang–Baxter Equation

Let M be a vector space over k and suppose that $R : M \otimes M \longrightarrow M \otimes M$ is a linear map. Define $R_{(i,j)} \in \mathrm{End}(M \otimes M \otimes M)$ for $1 \leq i < j \leq 3$ by

$$R_{(1,2)} = R \otimes 1_M, \quad R_{(2,3)} = 1_M \otimes R,$$

and

$$R_{(1,3)} = (1_M \otimes \tau_{M,M})(R \otimes 1_M)(1_M \otimes \tau_{M,M}),$$

where $\tau_{M,M} : M \otimes M \longrightarrow M \otimes M$ is the twist map defined by $\tau_{M,M}(m \otimes n) = n \otimes m$ for all $m, n \in M$.

Definition 2.1.1 *The* constant form *of the QYBE is*

$$R_{(1,2)}R_{(1,3)}R_{(2,3)} = R_{(2,3)}R_{(1,3)}R_{(1,2)}.$$

We will refer to operators $R : M \otimes M \longrightarrow M \otimes M$ which satisfy the constant form of the QYBE as *constant solutions*, or simply as solutions to the QYBE.

Definition 2.1.2 *Let M be a vector space over the field k. Then* QYB(M) *denotes the* set of solutions $R : M \otimes M \longrightarrow M \otimes M$ to the constant *QYBE.*

In the exercises below M is a vector space over the field k.

Exercise 2.1.1 There are some very simple solutions to the QYBE.

a) Show that the twist map $\tau_{M,M} : M \otimes M \longrightarrow M \otimes M$ is a solution to the QYBE.

b) Suppose that B is a basis for M and $\omega_{m,n} \in k$ for all $m, n \in B$. Show that $R : M \otimes M \longrightarrow M \otimes M$ determined by $R(m \otimes n) = \omega_{m,n} m \otimes n$ for all $m, n \in B$ is a solution to the QYBE. (In particular $R = 1_{M \otimes M}$ is solution to the QYBE.)

Exercise 2.1.2 Suppose that $R : M \otimes M \longrightarrow M \otimes M$ is a solution to the QYBE.

a) If R is an invertible operator, show that R^{-1} is a solution to the QYBE.

b) Let $u : M \longrightarrow M$ be a linear automorphism of M. Show that $R_u = (u \otimes u)R(u^{-1} \otimes u^{-1})$ is a solution to the QYBE.

c) Suppose that M is finite-dimensional. Show that the transpose $R^* : M^* \otimes M^* \longrightarrow M^* \otimes M^*$ of R is a solution to the QYBE. (We identify $M^* \otimes M^* = (M \otimes M)^*$ in the usual way where $(m^* \otimes n^*)(m \otimes n) = m^*(m)n^*(n)$ for $m^*, n^* \in M^*$ and $m, n \in M$.)

Exercise 2.1.3 Suppose that $R : M \otimes M \longrightarrow M \otimes M$ is a linear map. Define $R^\tau : M \otimes M \longrightarrow M \otimes M$ by

$$R^\tau = \tau_{M,M} R \tau_{M,M}.$$

Show that R is a solution to the QYBE if and only if R^τ is a solution to the QYBE.

Exercise 2.1.4 Show that QYB(M) is closed under scalar multiplication but is not necessarily a subspace of End$(M \otimes M)$.

2.1.1 The Constant Form of the Quantum Yang–Baxter Equation in H-S Notation

Suppose that $R : M \otimes M \longrightarrow M \otimes M$ is linear. We adapt the H-S notation to express $R(m \otimes n)$ succinctly by

$$R(m \otimes n) = m_{[1]} \otimes n_{[2]}$$

for all $m, n \in M$. We will use the notation $m = m_{[0]}$ to denote the result of applying 1_M to m for clarity if necessary. Thus we write

$$R_{(2,3)}(\ell \otimes m \otimes n) = \ell_{[0]} \otimes m_{[1]} \otimes n_{[2]}$$

for example. Observe that

$$R_{(1,2)}R_{(1,3)}R_{(2,3)}(\ell \otimes m \otimes n) = \ell_{[0][1][1]} \otimes m_{[1][0][2]} \otimes n_{[2][2][0]}$$

and

$$R_{(2,3)}R_{(1,3)}R_{(1,2)}(\ell \otimes m \otimes n) = \ell_{[1][1][0]} \otimes m_{[2][0][1]} \otimes n_{[0][2][2]}.$$

Thus R satisfies the QYBE if and only if

$$\ell_{[0][1][1]} \otimes m_{[1][0][2]} \otimes n_{[2][2][0]} = \ell_{[1][1][0]} \otimes m_{[2][0][1]} \otimes n_{[0][2][2]} \tag{2.1}$$

holds for all $\ell, m, n \in M$. We will see that (2.1) is a very useful way of expressing the constant form of the QYBE.

2.1.2 The Constant Form of the Quantum Yang–Baxter Equation in Coordinates

Suppose now that M is finite dimensional and $B = \{m_1, \ldots, m_n\}$ is a basis for M. We will frequently write B more informally as $\{m_i\}$. Let $R : M \otimes M \longrightarrow M \otimes M$ be a linear map. Write

$$R(m_i \otimes m_j) = R^{k,l}_{i,j} m_k \otimes m_l. \tag{2.2}$$

Thus $\{R^{k,l}_{i,j}\}$ is the set of B-coordinates of R. We formulate what it means for R to be a solution to the QYBE in terms of B-coordinates, which we will informally refer to as coordinates.

Consider the left hand side of the QYBE applied to $m_i \otimes m_j \otimes m_k$. Observe that

$$\xi_1 = R_{(2,3)}(m_i \otimes m_j \otimes m_k) = R^{s_2,s_3}_{j,k} m_i \otimes m_{s_2} \otimes m_{s_3},$$

$$\xi_2 = R_{(1,3)}(\xi_1) = R^{s_2,s_3}_{j,k} R^{s_1,c}_{i,s_3} m_{s_1} \otimes m_{s_2} \otimes m_c,$$

and

$$R_{(1,2)}(\xi_2) = R^{s_2,s_3}_{j,k} R^{s_1,c}_{i,s_3} R^{a,b}_{s_1,s_2} m_a \otimes m_b \otimes m_c.$$

Therefore

$$R_{(1,2)}R_{(1,3)}R_{(2,3)}(m_i \otimes m_j \otimes m_k) = R_{j,k}^{s_2,s_3}R_{i,s_3}^{s_1,c}R_{s_1,s_2}^{a,b}m_a \otimes m_b \otimes m_c.$$

To evaluate the right hand side of the QYBE on $m_i \otimes m_j \otimes m_k$, we compute in a similar manner

$$\chi_1 = R_{(1,2)}(m_i \otimes m_j \otimes m_k) = R_{i,j}^{r_1,r_2}m_{r_1} \otimes m_{r_2} \otimes m_k,$$

$$\chi_2 = R_{(1,3)}(\chi_1) = R_{i,j}^{r_1,r_2}R_{r_1,k}^{a,r_3}m_a \otimes m_{r_2} \otimes m_{r_3},$$

and

$$R_{(2,3)}(\chi_2) = R_{i,j}^{r_1,r_2}R_{r_1,k}^{a,r_3}R_{r_2,r_3}^{b,c}m_a \otimes m_b \otimes m_c.$$

Therefore

$$R_{(2,3)}R_{(1,3)}R_{(1,2)}(m_i \otimes m_j \otimes m_k) = R_{i,j}^{r_1,r_2}R_{r_1,k}^{a,r_3}R_{r_2,r_3}^{b,c}m_a \otimes m_b \otimes m_c.$$

By comparing coefficients, we see that R satisfies the QYBE if and only if

$$R_{j,k}^{s_2,s_3}R_{i,s_3}^{s_1,c}R_{s_1,s_2}^{a,b} = R_{i,j}^{r_1,r_2}R_{r_1,k}^{a,r_3}R_{r_2,r_3}^{b,c}. \tag{2.3}$$

The coordinate form of the QYBE is probably the most common expression of the equation.

Exercise 2.1.5 Suppose that M is a vector space over k and $f, g : M \longrightarrow M$ are invertible linear maps. Define $R : M \otimes M \longrightarrow M \otimes M$ by $R = f \otimes g$ (thus $R(m \otimes n) = f(m) \otimes g(n)$ for $m, n \in M$). Find necessary and sufficient conditions in terms of f and g for R to be a solution to the QYBE.

2.2 The Braid Equation

Suppose that M is a vector space over the field k and $R : M \otimes M \longrightarrow M \otimes M$ is linear.

Definition 2.2.1 *The* braid equation *is*

$$R_{(1,2)}R_{(2,3)}R_{(1,2)} = R_{(2,3)}R_{(1,2)}R_{(2,3)}.$$

Solving the braid equation is equivalent to solving the constant form of the QYBE.

Proposition 2.2.1 *Suppose that M is a vector space over the field k and $R : M \otimes M \longrightarrow M \otimes M$ is linear. Let $\tau = \tau_{M,M}$. Then the following are equivalent:*

a) *R satisfies the QYBE.*

b) *$\tau R\tau$ satisfies the QYBE.*

c) *τR satisfies the braid equation.*

d) *$R\tau$ satisfies the braid equation.*

Proof: By Exercise 2.1.3 parts a) and b) are equivalent. As $R\tau = \tau(\tau R\tau)$, to complete the proof of the proposition we need only show that parts a) and c) are equivalent.

Let $B = \tau R$. Writing $R(m \otimes n) = m_{[1]} \otimes n_{[2]}$ in the H-S notation we have $B(m \otimes n) = n_{[2]} \otimes m_{[1]}$ for all $m, n \in M$. Thus

$$B_{(1,2)}B_{(2,3)}B_{(1,2)}(\ell \otimes m \otimes n) = n_{[0][2][2]} \otimes m_{[2][0][1]} \otimes \ell_{[1][1][0]}$$

and

$$B_{(2,3)}B_{(1,2)}B_{(2,3)}(\ell \otimes m \otimes n) = n_{[2][2][0]} \otimes m_{[1][0][2]} \otimes \ell_{[0][1][1]}$$

for all $\ell, m, n \in M$. Therefore B satisfies the braid equation if and only if R satisfies (2.1). ∎

Observe that $B : M \otimes M \longrightarrow M \otimes M$ satisfies the braid equation if and only if

$$(B \otimes 1_M)(1_M \otimes B)(B \otimes 1_M) = (1_M \otimes B)(B \otimes 1_M)(1_M \otimes B). \tag{2.4}$$

The braid equation is related to the Artin braid group as we will now explain.

Let n be a fixed positive integer, M be a vector space over the field k, and let $M^n = \otimes^n M$ be the n-fold tensor product of M with itself. For $1 \leq i < n$ define $\sigma_{i,B} \in \mathrm{End}(M^n)$ by

$$\sigma_{i,B} = 1_M \otimes \cdots \otimes B \otimes \cdots \otimes 1_M,$$

where B is the ith tensorand. Write $\sigma_i = \sigma_{i,B}$. Then by (2.4) we have the relations

$$\sigma_i\sigma_{i+1}\sigma_i = \sigma_{i+1}\sigma_i\sigma_{i+1} \quad \text{for} \quad i = 1, 2, \ldots, n-1 \tag{2.5}$$

and

$$\sigma_i\sigma_j = \sigma_j\sigma_i \quad \text{for} \quad | i - j | \leq 2. \tag{2.6}$$

The relations above are well-known and have been studied extensively [Artin, 1947], [Birman, 1975].

Definition 2.2.2 *The group generated by formal symbols $\sigma_1, \ldots, \sigma_n$ modulo the relations described in (2.5) and (2.6) is called the* Artin braid group *and is denoted by* $\mathbb{B}_n$.

Now note that $\sigma_{i,B}$ is an invertible endomorphism of M^n if and only if B is an invertible endomorphism of $M \otimes M$. The latter is the case if and only if $R : M \otimes M \longrightarrow M \otimes M$ is an invertible solution to the QYBE, where $R = \tau_{M,M}B$. In summary, we have

Theorem 2.2.1 *Suppose that M is a vector space over the field k and $R : M \otimes M \longrightarrow M \otimes M$ is an invertible solution to the QYBE. Let $B = \tau R$. Then B is a solution to the braid equation and the map*

$$\mathbb{B}_n \xrightarrow{\pi} \mathrm{End}(M^n)$$

defined by

$$\pi(\sigma_i) = \sigma_{i,B}$$

for $1 \leq i < n$ is a representation of the braid group. ▮

For more information about the relationship between invertible QYBE solutions, braid group representations, and knot and link invariants see the collections [Yang and Ge, 1989], [Yang and Ge, 1994] and the book [Kauffman, 1991] along with the references therein.

2.3 Symmetries

We begin this section with a slight generalization of part b) of Exercise 2.1.2.

Lemma 2.3.1 *Suppose that M and M' are vector spaces over the field k and $R : M \otimes M \longrightarrow M \otimes M$ is linear. Suppose that $u : M \longrightarrow M'$ is any linear isomorphism and set $R_u = (u \otimes u)R(u^{-1} \otimes u^{-1})$. Then R is a solution to the QYBE if and only if R_u is a solution to the QYBE.*

Proof: Let $\mathcal{R} = R_u$. Then observe that $\mathcal{R}_{(i,j)} = (u \otimes u \otimes u)R(u^{-1} \otimes u^{-1} \otimes u^{-1})$ for $1 \leq i < j \leq 3$. The proof is now quickly concluded. ▮

Definition 2.3.1 *Suppose that M and M' are vector spaces over the field k and $R : M \otimes M \longrightarrow M \otimes M$ and $R' : M' \otimes M' \longrightarrow M' \otimes M'$ are linear. Then R and R' are* congruent *if there exists a linear isomorphism $u : M \longrightarrow M'$ such that $R' = R_u$.*

Observe that this notion of congruence defines an equivalence relation on the class of all QYBE solutions. We will write $R' \approx R$, or $R' \approx_u R$, when $R' = R_u$. By virtue of Lemma 2.3.1 if R is solution to the QYBE and $R' \approx R$ then R' is also a solution to the QYBE.

Now suppose that M is any vector space over the field k. Recall that $\mathrm{QYB}(M)$ denotes the set of solutions $R : M \otimes M \longrightarrow M \otimes M$ to the QYBE. Observe that

$$\mathrm{Aut}(M) \xrightarrow{\pi} \mathrm{End}(\mathrm{QYB}(M))$$

defined by $\pi(u)(R) = R_u$ describes a representation of the group of linear automorphisms of M on the set of solutions $R : M \otimes M \longrightarrow M \otimes M$ to the QYBE.

Now suppose that M is finite-dimensional and has basis $B = \{m_1, \ldots, m_n\}$. Identifying $R \in \mathrm{End}(M \otimes M)$ with its matrix of B-coordinates, the representation of $\mathrm{Aut}(M)$ is realized more concretely as a representation

$$\mathrm{GL}(n, k) \longrightarrow \mathrm{M}_{n^2}(k).$$

Observe that the symmetric group $\mathbb{S}_n$ can be considered a subgroup of $\mathrm{GL}(n, k)$ by $\sigma(m_i) = m_{\sigma(i)}$ for all $\sigma \in \mathbb{S}_n$ and $1 \le i \le n$. Thus for $u = \sigma$ we have

$$(R_{\sigma^{-1}})_{i,j}^{k,\ell} = R_{\sigma(i),\sigma(j)}^{\sigma(k),\sigma(\ell)}$$

for all $\sigma \in \mathbb{S}_n$ and $1 \le i, j, k, \ell \le n$.

There are two other transformations which associate QYBE solutions to QYBE solutions. These are the "transpose" and conjugation by the twist map which are denoted by $(\ )^T$ and $(\ )^\tau$ respectively. In terms of B-coordinates

$$(R^T)_{i,j}^{k,l} = R_{k,l}^{i,j}, \tag{2.7}$$
$$(R^\tau)_{i,j}^{k,l} = R_{j,i}^{l,k}. \tag{2.8}$$

We leave it as an exercise to show that these operations take QYBE solutions to QYBE solutions.

Note that $(\ )^T$ and $(\ )^\tau$ commute with each other and $(\ )^\tau$ commutes with the $\mathrm{GL}(n, k)$ action described above. See the exercises for this section. We have noted in Exercise 2.1.4 that $\mathrm{QYB}(M)$ is closed under scalar multiplication. Therefore the group of units k^* of k acts by scalar multiplication on $\mathrm{QYB}(M)$. This action commutes with the rest. Thus:

Proposition 2.3.1 *Let M be an n-dimensional vector space over the field k and suppose $\mathcal{G}$ is the free product of $\mathrm{GL}(n, k)$ and $\mathbb{Z}_2$. The action by $\mathrm{GL}(n, k)$ described above, the transformations $(\ )^u$, $(\ )^T$, and $(\ )^\tau$, and scalar multiplication give rise to an action of the direct product $\mathcal{G} \times \mathbb{Z}_2 \times k^*$ on the set $\mathrm{QYB}(M)$ of all QYBE solutions $R : M \otimes M \longrightarrow M \otimes M$.* ∎

Exercise 2.3.1 Suppose that $R : M \otimes M \longrightarrow M \otimes M$ is linear.

a) Using (2.7), show that R^T is a QYBE solution if and only if R is a QYBE solution.

b) Using (2.8), show that R^τ is a QYBE solution if and only if R is a QYBE solution.

Exercise 2.3.2 Apropos of Proposition 2.3.1, show that $(\ )^T$ does not commute with the action of $\mathrm{GL}(n, k)$ in general. [Hint: Let $n = 2$ and consider $R = f \otimes g$, where $f, g \in \mathrm{End}(M)$.]

2.4 The One-Parameter Form of the Quantum Yang–Baxter Equation

Definition 2.4.1 *Let X be a set and $Z \subseteq X \times X$. The* one-parameter form *of the QYBE is*

$$R_{(1,2)}(x)R_{(1,3)}(\varphi(x,z))R_{(2,3)}(z) = R_{(2,3)}(z)R_{(1,3)}(\varphi(x,z))R_{(1,2)}(x),$$

for all $(x, z) \in Z$, where M is a vector space over k, $R : X \longrightarrow \mathrm{End}(M \otimes M)$, and $\varphi : Z \longrightarrow X$.

For $x \in X$ we let $R_{(i,j)}(x) = R(x)_{(i,j)}$ be defined as in Section 2.1.

Definition 2.4.2 *A function $R : X \longrightarrow \mathrm{End}(M \otimes M)$ such that the conditions of the definition above are met is said to be a* one-parameter solution *to the QYBE.*

In the full generality of the definition the operation φ is not assumed to have any special properties. However, classically [Drinfel'd, 1987], [Faddeev et al., 1988], [Faddeev et al., 1989], [Jimbo, 1985] X is the set of complex numbers and φ is addition or multiplication. Other situations can arise naturally as we shall see. Usually we will assume that there is a distinguished element $1 \in X$ which acts as an identity element for φ, i.e.

$$\varphi(1, x) = x = \varphi(x, 1)$$

for all $x \in X$ such that $(1, x), (x, 1) \in Z$. We will simply call such an element an identity element for φ.

For example, consider

$$R_1(x) = \begin{pmatrix} x & 0 & 0 & 0 \\ 0 & 1 & x-1 & 0 \\ 0 & x-1 & 1 & 0 \\ 0 & 0 & 0 & x \end{pmatrix}.$$

Here we may take $X = \mathbb{C}$ and

$$\varphi(x, z) = \frac{xz - 1}{x + z - 2}$$

(as can be trivially checked using computer algebra). Thus, $Z = \{(x, z) \in X \times X \mid x + z \neq 2\}$. Note that φ is associative, but there is no identity element. For another example, let

$$R_2(x) = \begin{pmatrix} 1 & 0 & 0 & 0 \\ 0 & \frac{1}{x} & 1 - \frac{1}{x} & 0 \\ 0 & 1 - \frac{1}{x} & \frac{1}{x} & 0 \\ 0 & 0 & 0 & 1 \end{pmatrix}.$$

Here we may take $X = \mathbb{C} - \{0\}$ and

$$\varphi(x, z) = \frac{xz - 1}{x + z - 2}$$

as before. Another example is given by

$$R_3(x) = \begin{pmatrix} 1 & 0 & 0 & 0 \\ 0 & x & 1-x & 0 \\ 0 & 1-x & x & 0 \\ 0 & 0 & 0 & 1 \end{pmatrix}.$$

Here we may take $X = \mathbb{C}$ and we have

$$\varphi(x, z) = \frac{x + z - 2xz}{1 - xz}$$

so that we can take $Z = \{(x, z) \in X \times X \mid xz \neq 1\}$. Note that φ is associative and 0 is the identity element, however, only three elements (viz. 0 and $-1 \pm \sqrt{2}$) have inverses. Finally, consider

$$R_4(x) = \begin{pmatrix} 1 & 0 & 0 & 0 \\ 0 & \frac{x}{x+1} & \frac{1}{x+1} & 0 \\ 0 & \frac{1}{x+1} & \frac{x}{x+1} & 0 \\ 0 & 0 & 0 & 1 \end{pmatrix}.$$

Here we take $X = \mathbb{C} - \{-1\}$,

$$\varphi(x, z) = x + z$$

and $Z = \{(x, z) \in X \times X \mid x + z \neq -1\}$. Here, φ obviously comes from a group law. Note that these one-parameter QYBE solutions are not unrelated. In fact, its easy to see that

$$R_2(x) = \frac{1}{x} R_1(\frac{1}{x}),$$
$$R_3(x) = R_2(\frac{1}{x}),$$
$$R_4(x) = R_3\left(\frac{x}{x+1}\right).$$

See Section 8.6.1 for more on this family.

Notice if $R : X \longrightarrow \mathrm{End}(M \otimes M)$ is a solution to the one-parameter QYBE then $R(x)$ is a constant solution to the QYBE whenever $(x, x) \in Z$ and $\varphi(x, x) = x$. On the other hand, if $R : M \otimes M \longrightarrow M \otimes M$ is a constant solution to the QYBE then for any singleton set $X = \{x\}$ there is a unique one-parameter solution to the QYBE $\mathcal{R} : X \longrightarrow \mathrm{End}(M \otimes M)$ such that $\mathcal{R}(x) = R$. Therefore constant solutions can be thought of as special cases of one-parameter solutions.

As is customary, we will write $xz = \varphi(x, z)$ when no confusion is likely to arise.

2.5 The Two-Parameter Form of the Quantum Yang–Baxter Equation

Definition 2.5.1 *Let X be a set and M be a vector space over k. The* two-parameter form *of the QYBE is*

$$R_{(1,2)}(u,v)R_{(1,3)}(u,w)R_{(2,3)}(v,w) = R_{(2,3)}(v,w)R_{(1,3)}(u,w)R_{(1,2)}(u,v)$$

for all $u, v, w \in X$, where X is a set and $R : X \times X \longrightarrow \mathrm{End}(M \otimes M)$.

For $x, y \in X$ we again let $R_{(i,j)}(x,y) = R(x,y)_{(i,j)}$ be as defined in Section 2.1.

Definition 2.5.2 *A function $R : X \times X \longrightarrow \mathrm{End}(M \otimes M)$ for which the equations of the definition above are all satisfied is said to be a* two-parameter solution *to the QYBE.*

Important cases involve the complex numbers as parameter set X and an operator $R(u,v)$ that depends only on $u - v$ [Jimbo, 1985], [Kulish et al., 1981]. In this case, we can write $x = u - v$ and think of R as a one-parameter solution in x. With this change of notation the two-parameter form of the QYBE becomes

$$R_{(1,2)}(u-v)R_{(1,3)}(u-w)R_{(2,3)}(v-w) = R_{(2,3)}(v-w)R_{(1,3)}(u-w)R_{(1,2)}(u-v),$$

i.e.

$$R_{(1,2)}(x)R_{(1,3)}(x+y)R_{(2,3)}(y) = R_{(2,3)}(y)R_{(1,3)}(y+x)R_{(1,2)}(x).$$

Notice that $R(0)$ is a constant solution in this case. The interested reader can find some solutions which do not depend only on $u - v$ in [Hlavatiý, 1992] and the references given there.

Exercise 2.5.1 Suppose now that M is finite dimensional and $\{m_i\}$ is a basis for M.

a) Write
$$R(x)(m_i \otimes m_j) = R_{i,j}^{k,l}(x) m_k \otimes m_l,$$
and $y = \varphi(x,z)$. Using this notation, derive the one-parameter form of the QYBE in coordinates:
$$R_{j,k}^{s_2,s_3}(z)R_{i,s_3}^{s_1,c}(y)R_{s_1,s_2}^{a,b}(x) = R_{i,j}^{r_1,r_2}(x)R_{r_1,k}^{a,r_3}(y)R_{r_2,r_3}^{b,c}(z).$$

b) Derive the two-parameter form of the QYBE in coordinates in a similar manner.

2.6 A System of Polynomial Equations (the QYB Variety)

Let M be an n-dimensional vector space over k. As can be seen from (2.3), the set $\mathrm{QYB}(M)$ of all constant solutions $R : M \otimes M \longrightarrow M \otimes M$ to the QYBE can be thought of as a subset of $\mathrm{M}_{n^2}(k)$ described by n^6 cubic polynomial equations in n^4

unknowns. Thus it is not reasonable to expect to find such solutions R to the QYBE by brute force in general.

We ran an experiment using computer algebra in the case of Dim$M = n = 2$. Choose a basis for M and arrange the coordinates of R into a 4×4 matrix M_R as follows:

$$M_R = \begin{pmatrix} R_{1,1}^{1,1} & R_{1,1}^{1,2} & R_{1,1}^{2,1} & R_{1,1}^{2,2} \\ R_{1,2}^{1,1} & R_{1,2}^{1,2} & R_{1,2}^{2,1} & R_{1,2}^{2,2} \\ R_{2,1}^{1,1} & R_{2,1}^{1,2} & R_{2,1}^{2,1} & R_{2,1}^{2,2} \\ R_{2,2}^{1,1} & R_{2,2}^{1,2} & R_{2,2}^{2,1} & R_{2,2}^{2,2} \end{pmatrix}.$$

In our experiment we specialized to the upper-triangular case and wrote M_R as

$$M_R = \begin{pmatrix} a_1 & a_2 & a_3 & a_4 \\ 0 & b_2 & b_3 & b_4 \\ 0 & 0 & c_3 & c_4 \\ 0 & 0 & 0 & d_4 \end{pmatrix}.$$

We created the matrix M_R above as a matrix with entries from the polynomial ring in the given ten entries $a_1, \ldots, d_4$ (with rational number coefficients) and obtained the left hand and right hand sides of the QYBE from that. The differences of the left hand sides and the right hand sides were accumulated into a list and duplicates were removed. The result was a set of 20 cubic equations in 10 unknowns. This system of equations was used for input to a *factorized* Gröbner basis routine [Jenks and Sutor, 1992, p. 376]. The result was a decomposition of this variety into exactly 248 sub-varieties, many of which were actually linear systems, e.g.,

$$a_1 = 0, \quad a_3 = -\frac{1}{3}a_2,$$

$$b_2 = 0, \quad b_3 = 0, \quad b_4 = \frac{1}{2}a_2,$$

$$c_4 = \frac{1}{6}a_2, \quad d_4 = 0.$$

Some were nearly linear such as

$$a_1 = 0, \quad b_2 = 0, \quad b_3 = 0,$$

$$b_4a_2 + a_3a_2 + a_2^2 = 0, \quad c_4 + b_4 - a_3 - a_2 = 0,$$

$$c_3 = 0, \quad d_4 = 0.$$

Some sub-varieties were produced that were a bit more complicated than these. In fact, any attempt to classify solutions should take into account the $\mathcal{G} \times \mathbb{Z}_2 \times k^*$ action on the QYB variety defined in Proposition 2.3.1. This was done in [Hietarinta, 1993b] where a subset of the symmetries just mentioned was used to classify *all* two-dimensional solutions of the QYBE using computer algebra methods. This use of the symmetries allows a much finer analysis of the cases arising than is possible by the brute force method above. It seems that even with the use of the more refined methods, a complete computer calculation of degree three solutions are still out of reach at this time and other methods are needed. We will come back to these issues in Chapter 8.

2.7 The Bialgebra Associated to the Quantum Yang–Baxter Equation

We will describe the fundamental bialgebra associated to a one-parameter QYBE solution. The constant case, we have noted, can be thought of as the special case where X is a singleton set. Our discussion motivates the material presented in Chapters 3 and 4 where we go into more detail.

2.7.1 *A Module Action Associated to a Quantum Yang–Baxter Equation Solution*

Let $R : X \longrightarrow \mathrm{End}(M \otimes M)$ be a one-parameter QYBE solution where M is an n-dimensional vector space over k with basis $\{m_i\}$. There is a natural left module action on M of the free algebra A over k with on the set of formal symbols $\{t_i^j(x)\}_{1\leq i,j\leq n, x\in X}$ given by

$$t_i^j(x)\cdot m_k = R_{i,k}^{j,s}(x)m_s. \tag{2.9}$$

To see this, let C be the vector space with basis of symbols $t_i^j(x)$. We observe that $A = T(C)$ is the tensor algebra on the vector space C over k. The linear map

$$C \xrightarrow{\pi_0} \mathrm{End}(M)$$

defined by

$$t_i^j(x) \longmapsto (m_k \longmapsto R_{i,k}^{j,s}(x)m_s)$$

gives rise to an algebra map

$$T(C) \xrightarrow{\pi} \mathrm{End}(M)$$

by the universal mapping property of the tensor algebra on the vector space C. This algebra map is a representation of $A = T(C)$ which accounts for a left A-module structure on M given by (2.9). We note that

$$t_a^b(w)t_c^d(u)\cdot m_k = R_{c,k}^{d,s}(u)R_{a,s}^{b,v}(w)m_v$$

for any elements $u, w \in X$. Since R is a solution to the QYBE it is easy to see that the elements

$$R_{s_1,s_2}^{a,b}(x)t_i^{s_1}(xz)t_j^{s_2}(z) - t_{s_2}^b(z)t_{s_1}^a(xz)R_{i,j}^{s_1,s_2}(x) \tag{2.10}$$

are in Ker$\pi = I$.

Definition 2.7.1 $A(R) = T(C)/I$ *is called the* FRT-construction.

This algebra $A(R)$ occurs in [Drinfel'd, 1987, Section 11], [Faddeev et al., 1990]. Observe that

$$T(C) \xrightarrow{\pi} \mathrm{End}(M)$$

factors through $A(R)$. Thus we have a representation

$$A(R) = T(C)/I \longrightarrow \mathrm{End}(M)$$

which accounts for a left $A(R)$-module action (M, μ) on M. Also see [Majid, 1990b, Section 3].

Representations of $A(R)$ are important in mathematical physics [Faddeev, 1995], [Jimbo, 1985], [Sklyanin, 1982], [Sklyanin, 1991]. As such, it would be useful to be able to produce, for example, new representations from known ones. As we have seen in (1.14), a coproduct can be used to tensor representations. Thus it is natural to look for a coproduct for $A(R)$. Notice that $C = \oplus_{x \in X} \mathrm{C}_{n,x}(k)$ as a vector space, where $\mathrm{C}_{n,x}(k)$ is the subspace of C with basis $\{t_i^j(x)\}_{1 \leq i,j \leq n}$. Regard $\mathrm{C}_{n,x}(k)$ as the comatrix coalgebra $\mathrm{C}_n(k)$ with standard basis consisting of the $t_i^j(x)$'s. Thus

$$\Delta(t_i^j(x)) = t_s^j(x) \otimes t_i^s(x)$$

and

$$\epsilon(t_i^j(x)) = \delta_i^j.$$

Regard C as the direct sum of coalgebras and regard $T(C)$ as tensor bialgebra of C over k. An easy computation shows that I is a coideal of $T(C)$. Thus I is a bi-ideal of $T(C)$ and hence $A(R) = T(C)/I$ is, in fact, a bialgebra.

Exercise 2.7.1 Prove that the ideal $I = \mathrm{Ker}\pi$ described above is indeed a coideal of $T(C)$.

2.7.2 Comodule Coaction

In addition to the left $A(R)$-module structure

$$A(R) \otimes M \xrightarrow{\mu} M$$

on M defined above, we have a parameterized family of right $A(R)$-comodule structures

$$X \xrightarrow{\rho} \mathrm{Hom}(M, M \otimes A(R))$$

which is just given by the regular *co*representation of $\mathrm{C}_{n,x}(k)$ in M for each $x \in X$. Thus, writing $\rho(x) = \rho_x$, we have

$$\rho_x(m_i) = m_s \otimes t_i^s(x).$$

We once again extend the H-S notation, this time to the parameterized context, and write $\rho_x(m) = m^{<1>_x} \otimes m^{(2)_x}$ for $x \in X$ and $m \in M$.

It is very important to note that we can write

$$R(x)(m_i \otimes m_j) = m_i^{<1>_x} \otimes m_i^{(2)_x} \cdot m_j \tag{2.11}$$

for $1 \leq i, j \leq n$. Therefore $R(x)$ is the composite $(1 \otimes \mu)(\rho_x \otimes 1)$. In the next section we next consider this composite in a more general context.

2.8 Factoring a Quantum Yang–Baxter Solution Over a Bialgebra

Suppose that A is a bialgebra over the field k and M is a vector space over k. Let $\mu : A \otimes M \longrightarrow M$ and $\rho : M \longrightarrow M \otimes A$ be *linear* maps. We will write $\mu(a \otimes m) = a \cdot m$ and $\rho(m) = m^{<1>} \otimes m^{(2)}$ for $a \in A$ and $m \in M$.

Equation 2.11 of the previous section leads us to consider a very general composite

$$M \otimes M \xrightarrow{R_{(\mu,\rho)}} M \otimes M$$

which is

$$M \otimes M \xrightarrow{\rho \otimes 1_M} M \otimes A \otimes M \xrightarrow{1_M \otimes \mu} M \otimes M.$$

Thus

$$\begin{aligned} R_{(\mu,\rho)}(m \otimes n) &= (1_M \otimes \mu)(\rho \otimes 1_M)(m \otimes n) \\ &= (1_M \otimes \mu)(m^{<1>} \otimes m^{(2)} \otimes n) \\ &= m^{<1>} \otimes m^{(2)} \cdot n \end{aligned}$$

for all $m, n \in M$.

Now Let X be a set and consider a parameterized family of left A-module structures

$$X \xrightarrow{\mu} \mathrm{Hom}(A \otimes M, M)$$

and a parameterized family of right A-comodule structures

$$X \xrightarrow{\rho} \mathrm{Hom}(M, M \otimes A)$$

on M. Set $R(x, y) = R_{(\mu_x, \rho_y)}$. Then

$$R(x, y)(m \otimes n) = m^{<1>_y} \otimes m^{(2)_y} \cdot_x n$$

where we write $a \cdot_x n = \mu_x(a \otimes n)$ and $\rho_y(m) = m^{<1>_y} \otimes m^{(2)_y}$ for $a \in A$ and $n, m \in M$. It is natural to ask when this composite $R(x, y)$ is a two-parameter QYBE solution. In cases where either the action is constant, meaning $\mu_x = \mu_y$ for all $x, y \in X$, or the coaction is constant, meaning $\rho_x = \rho_y$ for all $x, y \in X$, we obtain an operator $R(x)$ depending on one parameter. In the case that both parameters are constant, we obtain a constant operator R defined by

$$R(m \otimes n) = (1_M \otimes \mu)(\rho \otimes 1_M)(m \otimes n) = m^{<1>} \otimes m^{(2)} \cdot n \qquad (2.12)$$

for all $m, n \in M$. It is natural to ask when this composite R is a QYBE solution. As we will see, the answer to this question leads to an interesting theory. What is involved is some form of compatibility between the module and comodule structures which we next address. In what follows we separate the cases of the constant, one-parameter, and two-parameter QYBE.

2.9 Compatibility Conditions in the Constant Case

Throughout this section A is a bialgebra over k and M is a vector space over k with a left A-module structure (M, μ) and a right A-comodule structure (M, ρ), unless otherwise specified. We will assume bases for M and A have been fixed we will denote the structure constants of μ by $\{\mu^l_{i,j}\}$ and the structure constants of ρ by $\{\rho^{i,j}_l\}$. Let

$$R = (1_M \otimes \mu)(\rho \otimes 1_M)$$

be the composite defined by (2.12).

2.9.1 The Fundamental Compatibility Condition in Coordinates

The composite R has coordinates

$$R^{k,l}_{i,j} = \rho^{k,s}_i \mu^l_{s,j}.$$

Let's write out the QYBE in these terms. The left hand side of (2.3) is

$$\rho^{a,t_3}_{s_1} \mu^b_{t_3,s_2} \rho^{s_1,t_2}_i \mu^c_{t_2,s_3} \rho^{s_2,t_1}_j \mu^{s_3}_{t_1,k}$$

while the right hand side is

$$\rho^{b,t_1}_{s_2} \mu^c_{t_1,s_3} \rho^{a,t_2}_{s_1} \mu^{s_3}_{t_2,k} \rho^{s_1,t_3}_i \mu^{s_2}_{t_3,j}.$$

Rearranging the left hand side we get

$$\mu^c_{t_2,s_3} \mu^{s_3}_{t_1,k} \mu^b_{t_3,s_2} \rho^{s_2,t_1}_j \rho^{a,t_3}_{s_1} \rho^{s_1,t_2}_i \qquad (2.13)$$

while the right hand side can be arranged as

$$\mu^c_{t_1,s_3} \mu^{s_3}_{t_2,k} \mu^{s_2}_{t_3,j} \rho^{b,t_1}_{s_2} \rho^{a,t_2}_{s_1} \rho^{s_1,t_3}_i. \qquad (2.14)$$

By associativity of the module action the left hand side of (2.3) can be written

$$\mu^c_{\theta,k} m^\theta_{t_2,t_1} \mu^b_{t_3,s_2} \rho^{s_2,t_1}_j \rho^{a,t_3}_{s_1} \rho^{s_1,t_2}_i$$

and the right hand can be written

$$\mu^c_{\theta,k} m^\theta_{t_1,t_2} \mu^{s_2}_{t_3,j} \rho^{b,t_1}_{s_2} \rho^{a,t_2}_{s_1} \rho^{s_1,t_3}_i,$$

where $\{m^l_{i,j}\}$ denotes the structure constants of multiplication in A. Now we can apply the coassociativity law for the coaction and rewrite (2.3) in the form

$$\mu^c_{\theta,k} m^\theta_{t_2,t_1} \mu^b_{t_3,s_2} \rho^{s_2,t_1}_j \Delta^{t_3,t_2}_{\theta'} \rho^{a,\theta'}_i = \mu^c_{\theta,k} m^\theta_{t_1,t_2} \mu^{s_2}_{t_3,j} \rho^{b,t_1}_{s_2} \Delta^{t_2,t_3}_{\theta'} \rho^{a,\theta'}_i,$$

where $\{\Delta^{i,j}_l\}$ denotes the structure constants of comultiplication in A. Clearly, if

$$m^\theta_{t_2,t_1} \mu^b_{t_3,s_2} \rho^{s_2,t_1}_j \Delta^{t_3,t_2}_{\theta'} = m^\theta_{t_1,t_2} \mu^{s_2}_{t_3,j} \rho^{b,t_1}_{s_2} \Delta^{t_2,t_3}_{\theta'}, \tag{2.15}$$

then R is a solution to the QYBE. We will see below that the converse does not necessarily hold.

Definition 2.9.1 *Equation (2.15) is the* fundamental compatibility condition *for the left A-module structure and the right A-comodule structure on M.*

Also see [Yetter, 1990]. In Section 4.1 we will present an algebraic condition which will ensure that the compatibility condition (2.15) holds *if and only if* the corresponding composite $R = (1_M \otimes \mu)(\rho \otimes 1_M)$ satisfies the QYBE.

Exercise 2.9.1 Suppose that A is a bialgebra over k and

$$A \otimes M \xrightarrow{\ \mu\ } M \xrightarrow{\ \rho\ } M \otimes A$$

are linear maps, where M is a vector space over k. Assume that associative axiom $\mu(m \otimes 1_M) = \mu(1_A \otimes \mu)$ and the coassociative axiom $(\rho \otimes 1_A)\rho = (1_M \otimes \Delta)\rho$ hold. Show that $R = (1_M \otimes \mu)(\rho \otimes 1_M)$ satisfies the QYBE if the compatibility condition (2.15) holds.

2.9.2 The (Co)Commutative Compatibility Condition

Note that compatibility in the form (2.15) has an interesting symmetry. Indeed, if A is commutative then $m^t_{j,i} = m^t_{i,j}$ and there results a simpler condition ensuring a solution to the QYBE. In addition, if A is cocommutative, we have $\Delta^{j,i}_t = \Delta^{i,j}_t$ and the condition

$$\mu^b_{t_3,s_2} \rho^{s_2,t_1}_j = \mu^{s_2}_{t_3,j} \rho^{b,t_1}_{s_2} \tag{2.16}$$

will ensure a solution to the QYBE.

Definition 2.9.2 *Equation (2.16) is called the* commutative cocommutative compatibility condition.

In terms of the module and comodule structures (2.16) has the very natural expression

$$(\mu \otimes 1)(1 \otimes \rho) = \rho\mu.$$

2.9.3 Compatibility Conditions in H-S Notation

It is easy to write (2.15) without reference to bases using the H-S notation. In this form the compatibility condition is

$$a_{(1)}{\cdot}m^{<1>} \otimes a_{(2)}m^{(2)} = (a_{(2)}{\cdot}m)^{<1>} \otimes (a_{(2)}{\cdot}m)^{(2)}a_{(1)} \tag{2.17}$$

for all $a \in A$ and $m \in M$. Similarly, the commutative cocommutative compatibility condition (2.16) is expressed

$$(a{\cdot}m)^{<1>} \otimes (a{\cdot}m)^{(2)} = a{\cdot}m^{<1>} \otimes m^{(2)} \tag{2.18}$$

for all $a \in A$ and $m \in M$. The proofs will be left to the reader. These expressions of compatibility and their structure constant counterparts will be used interchangeably throughout the text.

Exercise 2.9.2 Using H-S notation, prove directly that if the compatibility condition (2.17) holds then the corresponding composite $R = (1_M \otimes \mu)(\rho \otimes 1_M)$ satisfies the QYBE.

Exercise 2.9.3 Suppose that A is a commutative cocommutative bialgebra over k. Show that (2.18) implies (2.17).

Exercise 2.9.4 Suppose that A is a bialgebra over k. Let M be a right A-module and also a left A-comodule. Show that if

$$m^{(1)}{\cdot}a_{(1)} \otimes m^{<2>}{\cdot}a_{(2)} = a_{(2)}(m{\cdot}a_{(1)})^{(1)} \otimes (m{\cdot}a_{(1)})^{<2>} \tag{2.19}$$

holds for all $m \in M, a \in A$, then the linear map R given by the composite

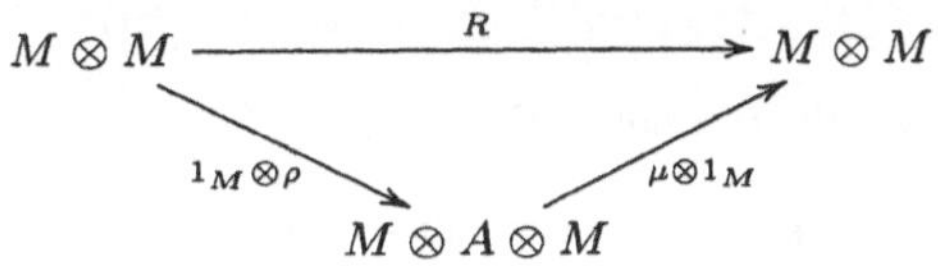

satisfies the QYBE.

2.10 Compatibility Conditions in the One-Parameter and Two-Parameter Cases

For the most part we will let the reader formulate and explore analogs of (2.15) and (2.16) in the one-parameter and two-parameter cases. See [Lambe, 1994, Sections 2.4-2.6] and the exercises below.

Let M be a vector space over k. Suppose that X is a set with binary operation and $R : X \longrightarrow \mathrm{End}(M \otimes M)$ is a function. Let $\mu : X \longrightarrow \mathrm{Hom}(A \otimes M, M)$ be a parameterized family of left A-module structures on M and $\rho : X \longrightarrow \mathrm{Hom}(M, M \otimes$

A) be a parameterized family of right A-comodule structures on M. Recall that the module action is constant if $\mu_x = \mu_y$ for all $x, y \in X$ and the comodule action is constant if $\rho_x = \rho_y$ for all $x, y \in X$. These are important cases.

Consider the case when the module structure is constant. In terms of this data, the composite $R(x) = (1_M \otimes \mu)(\rho_x \otimes 1_M)$ has coordinates

$$R_{i,j}^{k,l}(x) = \rho_i^{k,s}(x)\mu_{s,j}^{l}. \tag{2.20}$$

Exercise 2.10.1 Suppose that X has a monoid structure and the module action is constant.

a) Use the associativity law of the module action to derive the one-parameter compatibility condition

$$\begin{aligned} m_{t_2,t_1}^{\theta}\mu_{t_3,s_2}^{b}\rho_j^{s_2,t_1}(z)\rho_{s_1}^{a,t_3}(x)\rho_i^{s_1,t_2}(xz) = \\ m_{t_1,t_2}^{\theta}\mu_{t_3,j}^{s_2}\rho_{s_2}^{b,t_1}(z)\rho_{s_1}^{a,t_2}(xz)\rho_i^{s_1,t_3}(x). \end{aligned} \tag{2.21}$$

b) Show that if (2.21) holds then the corresponding composite $R(x)$ will satisfy the one-parameter QYBE.

c) The converse of (2.21) is not necessarily true. Find a counterexample.

Exercise 2.10.2 Repeat the exercise above in the case of a parameterized module structure and a constant comodule structure.

Exercise 2.10.3 Show that the one-parameter compatibility condition becomes

$$\begin{aligned} (m^{<1>_{xz}})^{<1>_x} \otimes (m^{<1>_{xz}})^{(2)_x} n^{<1>_z} \otimes n^{(2)_{xz}} n^{(2)_z} = \\ (m^{<1>_x})^{<1>_{xz}} \otimes (m^{(2)_x} n)^{<1>_z} \otimes (m^{(2)_x} n)^{(2)_z} (m^{<1>_x})^{(2)_{xz}} \end{aligned} \tag{2.22}$$

when expressed in terms of the H-S notation.

Exercise 2.10.4 Derive versions of the commutative cocommutative compatibility conditions in the one-parameter and two-parameter cases.

Exercise 2.10.5 Suppose that M has basis $\{m_j\}$ and that A has basis $\{a_i\}$. Consider a family of module structures $\mu_v : A \otimes M \longrightarrow M$ and a family of comodule structures $\rho_u : M \longrightarrow M \otimes A$. We define the structure constants $\{\mu_{i,j}^k(v)\}$ and $\{\rho_k^{i,j}(u)\}$ by

$$a_i \cdot_v m_j = \mu_{i,j}^k(v) m_k$$

and

$$\rho_u(m_k) = \rho_k^{i,j}(u) a_i \otimes m_j.$$

a) In terms of this data, show that the composite

$$R(u, v) = (1_M \otimes \mu_v)(\rho_u \otimes 1_M)$$

has coordinates

$$R(u,v)_{i,j}^{k,l} = \rho_i^{k,s}(u)\mu_{s,j}^{l}(v).$$

b) The associative law for the action is

$$\mu_{a,b}^{c}(w)\mu_{d,e}^{b}(w) = \mu_{f,e}^{c}(w)m_{a,d}^{f}$$

while the coassociative law for the coaction is

$$\rho_a^{b,c}(u)\rho_d^{a,e}(u) = \Delta_f^{c,e}\rho_d^{b,f}(u).$$

Using these, prove that if the two-parameter compatibility condition

$$m_{t_2,t_1}^{\theta}\mu_{t_3,s_2}^{b}(v)\rho_j^{s_2,t_1}(v)\Delta_{\theta'}^{t_3,t_2} = m_{t_1,t_2}^{\theta}\mu_{t_3,j}^{s_2}(v)\rho_{s_2}^{b,t_1}(v)\Delta_{\theta'}^{t_2,t_3} \tag{2.23}$$

holds, then the two-parameter QYBE is satisfied.

c) Find an example where the two-parameter QYBE is satisfied, but the two-parameter compatibility condition does not hold.

2.11 Reducing the Degree of the Quantum Yang–Baxter Variety

Suppose that A is bialgebra over k and $\mu : A \otimes M \longrightarrow M$ is a fixed left A-module structure on a vector space M over k. Then determining all linear maps $\rho : M \longrightarrow M \otimes A$ which satisfy (2.15) is actually a matter of solving a *linear* system for the unknown coefficients $\rho_j^{k,l}$.

2.11.1 From Cubic to Quadratic to Linear

If A and M are finite-dimensional, one can take advantage of one of the many computer libraries available to solve the linear system mentioned above. This is however not a complete solution as it stands. There may be solutions to the linear system which do not satisfy the coassociative axiom $(\rho \otimes 1_A)\rho = (1_M \otimes \Delta)\rho$. Note that these solutions might not give rise to QYBE solutions since coassociativity was used to show that the compatibility condition (2.15) implies $R = (1_M \otimes \mu)(\rho \otimes 1_M)$ is a solution to the QYBE. It is perhaps a bit surprising that often solutions which fail to satisfy $(\rho \otimes 1_A)\rho = (1_M \otimes \Delta)\rho$ will nonetheless give QYBE solutions. This sort of behavior was observed in the more complicated context of the QYBE with spectral parameter in [Lambe, 1994, p. 42] as we will review in Section 8.6.

2.11.2 A Curious Example

We review here a curious class class of examples [Lambe, 1996]. Let $k = \mathbb{C}$ be the field of complex numbers. Consider the group algebra $H = k[\mathbb{Z}_2]$ of the group $\mathbb{Z}_2$ with two elements. We will present an example that leads to a new procedure for finding solutions to the QYBE associated to bialgebras. The example will be kept as simple as possible for the exposition, but even this will lead us into directions that cannot be properly addresses until later chapters.

The structure of modules M over H is well-known. Indeed, note that any module M over H is completely determined by an $n \times n$ matrix A over k such that $A^2 = I$; for such A's determine all representations $H \longrightarrow \mathrm{End}(M)$.

Note also that the dual algebra H^* has generators α_i, $i = 1, 2$ such that $\alpha_i \alpha_j = \delta_{i,j}\alpha_j$. Thus, any left H-module M which also has a linear map $\rho : M \longrightarrow M \otimes H$ satisfying the coassociative condition $(\rho \otimes 1_H)\rho = (1_M \otimes \Delta)\rho$ is determined by three $n \times n$ matrices A, C^i, (using superscripts to index the C's) B such that $A^2 = I$ and $C^i C^j = \delta_{i,j} C^j$ for $i, j = 1, 2$.

We are interested in setting up the commutative cocommutative compatibility condition (2.16) for this situation. For this, it is convenient to label the matrices according to the corresponding structure constants. As usual, let $\{\rho_i^{j,k}\}$ denote the structure constants for ρ and $\{\mu_{i,j}^k\}$ denote the structure constants for the module structure. Let

$$A_s = \left[\mu_{s,row}^{col}\right], \tag{2.24}$$

$$C^s = \left[\rho_{row}^{col,s}\right]. \tag{2.25}$$

The matrix A_1 corresponds to 1, so that $A_1 = I$. The matrix A_2 corresponds to the generator of H of order 2 and so $A_2 A_2 = I$. The matrices C^i satisfy the equations given above. With this notation, the compatibility condition (2.16) becomes

$$A_2 C^i = C^i A_2.$$

To summarize, we want to solve

(associativity condition)

$$A_1 = I, \quad A_2 A_2 = I,$$

which is a quadratic system,

(coassociativity condition)

$$C^i C^i = C^i, \quad C^1 C^2 = C^2 C^1 = 0,$$

which is another quadratic system, and the

(commutative cocommutative compatibility condition)

$$A_2 C^i = C^i A_2,$$

which is a *linear* condition.

We will concentrate on one module structure on $M = \mathbb{C}^3$ given by the matrix

$$A_2 = \begin{bmatrix} -1 & 0 & 0 \\ 0 & 1 & 0 \\ 0 & 0 & 1 \end{bmatrix}. \tag{2.26}$$

We have used computer algebra again to find all solutions to the quadratic variety given by equations just listed in this case. To understand the answer, we should explain the we formed matrices

$$C^1 = \begin{bmatrix} a_1 & a_2 & a_3 \\ b_1 & b_2 & b_3 \\ c_1 & c_2 & c_3 \end{bmatrix}, \quad C^2 = \begin{bmatrix} d_1 & d_2 & d_3 \\ e_1 & e_2 & e_3 \\ f_1 & f_2 & f_3 \end{bmatrix}$$

with polynomial entries. Then, by simple matrix manipulations, we formed the appropriate combinations $C^1 A_2 - A_2 C^1$, etc. and extracted the rows to form a list of polynomials that represents the quadratic variety we want to investigate. For this, we used the Gröbner factorization algorithms to obtain a set of sub-varieties that partition the solution space. We obtained 73 such sub-varieties exactly 9 of which were quadratic. All the others were linear and not very interesting. The 9 quadratic sub-varieties are typified by the following member of that family:

$$C^1 = \begin{bmatrix} 1 & 0 & 0 \\ 0 & 0 & 0 \\ 0 & 0 & 0 \end{bmatrix}, \quad C^2 = \begin{bmatrix} 0 & 0 & 0 \\ 0 & e_2 & e_3 \\ 0 & f_2 & f_3 \end{bmatrix}$$

where

$$f_3 + e_2 - 1 = 0, \; e_3 f_2 + e_2^2 - e_2 = 0$$

so that a complete solution for this case is given by the parameterized family such that e_3 and f_2 are arbitrary,

$$f_3 = 1 - e_2, \quad \text{and} \quad e_2 = \frac{1 \pm \sqrt{-4e_3 f_2 + 1}}{2}.$$

Note that the corresponding QYBE solution is given by $R = A_s C^s = A_1 C^1 + A_2 C^2$.

Now we will observe that in fact, if we do not assume that the quadratic solutions hold in the 9 cases above, we *still* get QYBE solutions. In fact the corresponding R is given by

$$\begin{bmatrix} 1 & 0 & 0 & 0 & 0 & 0 & 0 & 0 & 0 \\ 0 & 1 & 0 & 0 & 0 & 0 & 0 & 0 & 0 \\ 0 & 0 & 1 & 0 & 0 & 0 & 0 & 0 & 0 \\ 0 & 0 & 0 & -e_2 & 0 & 0 & -e_3 & 0 & 0 \\ 0 & 0 & 0 & 0 & e_2 & 0 & 0 & e_3 & 0 \\ 0 & 0 & 0 & 0 & 0 & e_2 & 0 & 0 & e_3 \\ 0 & 0 & 0 & -f_2 & 0 & 0 & -f_3 & 0 & 0 \\ 0 & 0 & 0 & 0 & f_2 & 0 & 0 & f_3 & 0 \\ 0 & 0 & 0 & 0 & 0 & f_2 & 0 & 0 & f_3 \end{bmatrix}$$

and it can be verified directly that R satisfies the QYBE. This is a curious situation. Clearly, when the quadratic equations do not hold, the coassociative axiom $(\rho \otimes 1_H)\rho =$

$(1_M \otimes \Delta)\rho$ does not hold, but the corresponding composite $R = (1_M \otimes \mu)(\rho \otimes 1_M)$ satisfies the QYBE. When modules over a given bialgebra are known, this presents the possibility of finding QYBE solutions by simply forming the linear system given by the compatibility conditions without regard to comodule axioms for ρ. A given solution to the linear equations can easily be checked to see if the QYBE is satisfied. This computational procedure has less complexity than trying to solve the quadratic variety corresponding to coassociative solutions.

Of course, any solutions R arising from the above procedure do in fact arise as a module/comodule M over *some* bialgebra, viz., the corresponding FRT construction $A(R)$. We will see later in Corollary 4.1.1 that an algebraic "reduction" of $A(R)$ is possible and that the smaller bialgebra $\widetilde{A(R)}$ possesses an action and coaction on M that yields R. We will comment on $\widetilde{A(R)}$ where R arises from the matrix A_2 described in (2.26) at the end of Section 4.1.

3 CATEGORIES OF QUANTUM YANG-BAXTER MODULES

In this chapter we begin to explore the implications of ideas presented in Chapter 2, particularly those of Sections 2.7 and 2.8. The setting of Sections 3.10 through 3.12 can be put in a more general context [Majid, 1994], [Pareigis, 1996]. A early treatment of the commutativity of an object in a category is given in [Eckmann and Hilton, 1962, p. 241].

3.1 Various Categories

For a fixed bialgebra A over k each of the compatibility conditions given in Section 2.9 of the last chapter give rise to a category as follows. The objects are all A-modules which are also A-comodules for which the compatibility condition holds. The morphisms are simply A-module maps that are also A-comodule maps. As we shall see in later sections, these categories have certain nice algebraic properties such as the existence of a "tensor product". For object M the composite

$$M \otimes M \xrightarrow{\rho \otimes 1_M} M \otimes H \otimes M \xrightarrow{1_M \otimes \mu} M \otimes M \tag{3.1}$$

is a QYBE solution. It is clear that any endofunctor on one of these categories will give a new QYBE solution from any QYBE solution described by this composite. For this reason we will study a number of algebraic constructions in the next chapter which give rise to several families of such endofunctors. To find QYBE solutions in the first place we can make good computational use of the compatibility conditions as seen in Section 2.11.2. Computational methods will be emphasized in the following sections and chapters.

3.1.1 Left Quantum Yang–Baxter A-Modules

Definition 3.1.1 *Let A be a bialgebra over k. A* left quantum Yang–Baxter A-module *(left QYB A-module) is a triple (M, μ, ρ), where (M, μ) is a left A-module and (M, ρ) is a right A-comodule, such that the compatibility condition (2.17)*

$$a_{(1)}\cdot m^{<1>} \otimes a_{(2)}m^{(2)} = (a_{(2)}\cdot m)^{<1>} \otimes (a_{(2)}\cdot m)^{(2)}a_{(1)}$$

holds for all $a \in A, m \in M$. The category whose objects are left QYB A-modules and morphisms are linear maps which are both module and comodule maps is denoted by ${}_A\mathcal{QYB}$.

Let (M, μ, ρ) be a left QYB A-module. Recall from Section 2.9 that the composite $R = (1_M \otimes \mu)(\rho \otimes 1_M)$ is a solution to the QYBE.

Definition 3.1.2 *The solution $R = (1_M \otimes \mu)(\rho \otimes 1_M)$ to the QYBE is called the solution* associated with *or* corresponding to (M, μ, ρ).

Every solution to the QYBE in the finite-dimensional case is associated with a left QYB A-module for some algebra A over k.

Theorem 3.1.1 *Suppose that $R : M \otimes M \longrightarrow M \otimes M$ is a solution to the QYBE where M is a finite-dimensional vector space over the field k. Then M has the structure of a left QYB $A(R)$-module such that R is the associated solution to the QYBE.*

Proof: Let (M, μ) be the left $A(R)$-module structure and (M, ρ) be the right $A(R)$-comodule structure defined in Section 2.7. In light of the calculations of that section we need only establish that the compatibility condition (2.15) holds. This is left as a very instructive exercise to the reader. ∎

In Chapter 4 we shall approach Theorem 3.1.1 from a more abstract point of view and give a basis free description of the FRT construction.

Throughout the following exercises A is a bialgebra over the field k.

Exercise 3.1.1 Show that ${}_k\text{Vec}$ can be thought of as a subcategory of ${}_A\mathcal{QYB}$ where vector spaces over k are equipped with the trivial left A-module and right A-comodule structures.

Exercise 3.1.2 Suppose that L, M, and N are objects of ${}_A\mathcal{QYB}$. Show that the linear isomorphisms

$$k \otimes M \simeq M \quad \text{and} \quad M \otimes k \simeq M$$

of (1.21) are morphisms of ${}_A\mathcal{QYB}$ and show that the linear isomorphism

$$(L \otimes M) \otimes N \simeq L \otimes (M \otimes N)$$

of (1.22) is a morphism of ${}_A\mathcal{QYB}$.

Exercise 3.1.3 Suppose that M and N are objects of ${}_A\mathcal{QYB}$. Show that the linear map $\sigma_{M,N} : M \otimes N \longrightarrow N \otimes M$ defined by

$$\sigma_{M,N}(m \otimes n) = n^{<1>} \otimes n^{(2)} \cdot m$$

for all $m \in M$ and $n \in N$ is a morphism of ${}_A\mathcal{QYB}$. See [Yetter, 1990].

Exercise 3.1.4 Suppose that (M, μ, ρ) is a left QYB A-module and let $X = \mathrm{End}_{Bialg}(A)$ be the semigroup of bialgebra endomorphisms of A under function composition. For $x \in X$ define

$$M \otimes M \xrightarrow{R_x} M \otimes M$$

by

$$R_x(m \otimes n) = m^{<1>} \otimes x(m^{(2)}) \cdot n$$

for all $m, n \in M$. Show that

$$X \xrightarrow{R} \mathrm{End}(M \otimes M)$$

defined by $R(x) = R_x$ is a one-parameter solution to the QYBE, where $\varphi : X \times X \longrightarrow X$ is defined by $\varphi(x, z) = zx$ for $x, z \in X$. See [Cotta-Ramusino et al., 1993, Section 5].

This is one kind of "Baxterization" procedure. Also see [Jones, 1990] and [Liguori and Mintchev, 1992].

3.1.2 Quantum Yang–Baxter A-Modules when A is Commutative and Cocommutative

Now assume that A is a commutative and cocommutative bialgebra over the field k.

Definition 3.1.3 *Let A be a commutative cocommutative bialgebra over k. Then ${}_A\mathcal{CQYB}$ is the category whose objects are triples (M, μ, ρ), where (M, μ) is a left A-module and (M, ρ) is a right A-comodule such that the compatibility condition (2.18)*

$$(a \cdot m)^{<1>} \otimes (a \cdot m)^{(2)} = a \cdot m^{<1>} \otimes m^{(2)}$$

holds for all $a \in A, m \in M$ and whose morphisms are linear maps which are module and comodule maps.

We will leave it to the reader to define $\mathcal{CQYB}_A$ and develop the theory for ${}_A\mathcal{CQYB}$ and $\mathcal{CQYB}_A$ as the theory for ${}_A\mathcal{QYB}$ is developed in this chapter.

3.1.3 Right Quantum Yang–Baxter A-Modules

Definition 3.1.4 *Let A be a bialgebra over k.* A right quantum Yang–Baxter A-module *(right QYB A-module) is a triple (M, μ, ρ), where (M, μ) is a right A-module and (M, ρ) is a left A-comodule such that the compatibility condition (2.19)*

$$m^{(1)}a_{(1)} \otimes m^{<2>} \cdot a_{(2)} = a_{(2)}(m \cdot a_{(1)})^{(1)} \otimes (m \cdot a_{(1)})^{<2>}$$

holds for all $a \in A, m \in M$. The category whose objects are right QYB A-modules and morphisms are linear maps which are both module and comodule maps is denoted by $\mathcal{QYB}_A$.

Exercise 2.9.4 established that for a given right QYB A-module (M, μ, ρ) the composite

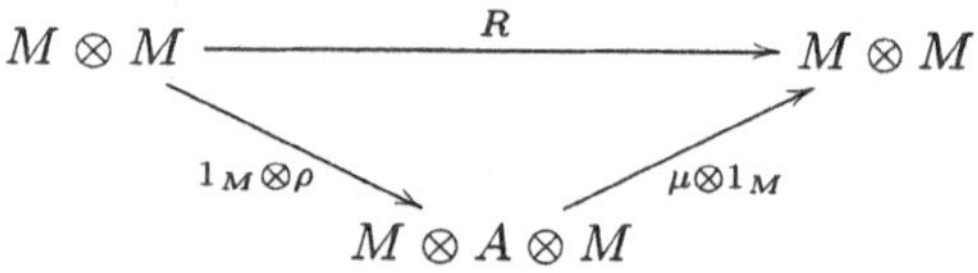

satisfies the QYBE.

Definition 3.1.5 *The solution $R = (\mu \otimes 1_M)(1_M \otimes \rho)$ to the QYBE is called the solution* associated with *or* corresponding to *the right QYB A-module (M, μ, ρ).*

In H-S notation, this composite is given by

$$R(m \otimes n) = m \cdot n^{(1)} \otimes n^{<2>} \tag{3.2}$$

for all $m, n \in M$. Choose a basis $\{m_i\}$ for M and a basis $\{a_j\}$ for A. The coordinates for R are given by

$$R^{k,l}_{i,j} = \mu^{k}_{i,s}\rho^{s,l}_{j}$$

where $m_i \cdot a_j = \mu^k_{i,j} m_k$ and $\rho(m_i) = \rho^{j,k}_i a_j \otimes m_k$.

Exercise 3.1.5 Let A be a bialgebra over k and let M be a vector space over k with a right A-module structure (M, μ) and a left A-comodule structure (M, ρ).

a) Show that (M, μ, ρ) is a right QYB A-module if and only if $(M, \mu^{op}, \rho^{cop})$ is a left QYB $A^{op\ cop}$-module (see Exercise 1.4.3).

b) Suppose that (M, μ, ρ) is a right QYB A-module. Let R be the QYBE solution associated with (M, μ, ρ) and let $\mathcal{R}$ be the QYBE solution associated with the left QYB $A^{op\ cop}$-module $(M, \mu^{op}, \rho^{cop})$. Show that $R = \mathcal{R}^{\tau}$.

3.1.4 Weak Quantum Yang–Baxter A-Modules

In light of the example of Section 2.11.2 we are motivated to define four more categories as follows.

Definition 3.1.6 *Let A be a bialgebra over k. Then a* weak left *QYB A-module is a triple (M, μ, ρ), where (M, μ) is a left A-module and $\rho : M \longrightarrow M \otimes A$ is a linear map, such that the compatibility condition (2.17)*

$$a_{(1)}\cdot m^{<1>} \otimes a_{(2)}m^{(2)} = (a_{(2)}\cdot m)^{<1>} \otimes (a_{(2)}\cdot m)^{(2)}a_{(1)},$$

holds for all $a \in A$ and $m \in M$,

$$\begin{aligned} & m^{<1><1>} \otimes m^{<1>(2)}\cdot n^{<1>} \otimes m^{(2)}n^{(2)} \\ = \; & m^{<1>} \otimes {m^{(2)}}_{(1)}\cdot n^{<1>} \otimes {m^{(2)}}_{(2)}n^{(2)}, \end{aligned} \tag{3.3}$$

and

$$\begin{aligned} & m^{<1><1>} \otimes (m^{(2)}\cdot n)^{<1>} \otimes (m^{(2)}\cdot n)^{(2)}m^{<1>(2)} \\ = \; & m^{<1>} \otimes ({m^{(2)}}_{(2)}\cdot n)^{<1>} \otimes ({m^{(2)}}_{(2)}\cdot n)^{(2)}{m^{(2)}}_{(1)}. \end{aligned} \tag{3.4}$$

hold for all $m, n \in M$. The category whose objects are weak left QYB A-modules and morphisms are linear maps which are module maps is denoted by ${}_A\widetilde{\mathcal{QYB}}$.

In a series of exercises at the end of this section, the reader will be guided in defining the category $\widetilde{\mathcal{QYB}}_A$ of weak QYB right A-modules as well as the category of weak left (right) QYB A-comodules. In the presence of (3.3) and (3.4), (2.17) is enough to ensure that the corresponding composite $R = (1_M \otimes \mu)(\rho \otimes 1_M)$ satisfies the QYBE. We have

Proposition 3.1.1 *Let (M, μ, ρ) be an object of ${}_A\widetilde{\mathcal{QYB}}$. Then $R = (1_M \otimes \mu)(\rho \otimes 1_M)$ satisfies the QYBE.*

Proof: Let $m, n, p \in M$. By virtue of (3.3) and associativity of the module action we obtain

$$\begin{aligned} & m^{<1><1>} \otimes m^{<1>(2)}\cdot n^{<1>} \otimes m^{(2)}\cdot(n^{(2)}\cdot p) \\ = \; & m^{<1>} \otimes {m^{(2)}}_{(1)}\cdot n^{<1>} \otimes {m^{(2)}}_{(2)}\cdot(n^{(2)}\cdot p). \end{aligned} \tag{3.5}$$

Similarly we obtain

$$\begin{aligned} & m^{<1><1>} \otimes (m^{(2)}\cdot n)^{<1>} \otimes (m^{(2)}\cdot n)^{(2)}\cdot(m^{<1>(2)}\cdot p) \\ = \; & m^{<1>} \otimes ({m^{(2)}}_{(2)}\cdot n)^{<1>} \otimes ({m^{(2)}}_{(2)}\cdot n)^{(2)}\cdot({m^{(2)}}_{(1)}\cdot p). \end{aligned} \tag{3.6}$$

Now $R(m \otimes n) = m^{<1>} \otimes m^{(2)} \cdot n$ for $m, n \in M$ in the H-S notation. It is easy to see that the left hand side of (3.5) is just the left hand of the QYBE applied to $m \otimes n \otimes p$. Similarly, the left hand side of (3.6) is just the right hand side of the QYBE applied to $m \otimes n \otimes p$. The right hand side of (3.5) transforms to the right hand side of (3.6) using (2.17) and associativity of the module action again. ∎

Exercise 3.1.6 We have already seen the compatibility condition (2.17) using structure constants in (2.15). Show, using structure constants, that the formulations of (3.3) and (3.4) are equivalent to

$$(\mu_{t_2,s_3}^{c}\mu_{t_1,k}^{s_3}\mu_{t_3,s_2}^{b}\rho_j^{s_2,t_1})(\rho_{s_1}^{a,t_3}\rho_i^{s_1,t_2}) = (\mu_{t_2,s_3}^{c}\mu_{t_1,k}^{s_3}\mu_{t_3,s_2}^{b}\rho_j^{s_2,t_1})(\Delta_{\theta'}^{t_3,t_2}\rho_i^{a,\theta'})$$

and

$$(\mu_{t_1,s_3}^{c}\mu_{t_2,k}^{s_3}\mu_{t_3,j}^{s_2}\rho_{s_2}^{b,t_1})(\rho_{s_1}^{a,t_2}\rho_i^{s_1,t_3}) = (\mu t_1, s_3{}^{c}\mu_{t_2,k}^{s_3}\mu_{t_3,j}^{s_2}\rho_{s_2}^{b,t_1})(\Delta_{\theta'}^{t_2,t_3}\rho_i^{a,\theta'})$$

respectively.

Exercise 3.1.7 By following the proof of Proposition 3.1.1, prove that if (2.17) and the two following conditions hold, the corresponding R will satisfy the QYBE:

$$(m_{t_2,t_1}^{\theta}\mu_{t_3,s_2}^{b}\rho_j^{s_2,t_1})(\rho_{s_1}^{a,t_3}\rho_i^{s_1,t_2}) = (m_{t_2,t_1}^{\theta}\mu_{t_3,s_2}^{b}\rho_j^{s_2,t_1})(\Delta_{\theta'}^{t_3,t_2}\rho_i^{a,\theta'})$$

and

$$(m_{t_1,t_2}^{\theta}\mu_{t_3,j}^{s_2}\rho_{s_2}^{b,t_1})(\rho_{s_1}^{a,t_2}\rho_i^{s_1,t_3}) = (m_{t_1,t_2}^{\theta}\mu_{t_3,j}^{s_2}\rho_{s_2}^{b,t_1})(\Delta_{\theta'}^{t_2,t_3}\rho_i^{a,\theta'}).$$

Find an example where (2.17) and the two conditions hold, but (3.3) and (3.4) do *not* hold.

Exercise 3.1.8 Use ideas analogous to those presented in this section to define the categories of weak left (right) QYB A-comodules. Also define the category of category $\widetilde{\mathcal{QYB}}_A$ of weak QYB right A-modules.

Exercise 3.1.9 Define a category ${}_A\mathcal{QYBE}$ of *left quantum Yang–Baxter equation modules* (QYBE-modules) with objects consisting of all left A-modules M which are also right A-comodules for which the composite (3.1) is a QYBE solution. The morphisms are simply linear maps which are both module and comodule maps. Also define the category $\mathcal{QYBE}_A$ of *right quantum Yang–Baxter equation modules* to have objects consisting of all right A-modules M which are also left A-comodules for which the composite

$$M \otimes M \xrightarrow{1_M \otimes \rho} M \otimes A \otimes M \xrightarrow{\mu \otimes 1_M} M \otimes M$$

is a QYBE solution. Again, the morphisms are simply linear maps which are both module and comodule maps.

a) As you read through the text, create a notebook that records which constructions given in ${}_A\mathcal{QYB}$ hold in ${}_A\mathcal{QYBE}$ and its variations.

b) Do the same for ${}_A\mathcal{CQYB}$ and its variations.

3.2 Congruence in ${}_A\mathcal{QYB}$

In this section we discuss the connection between the notion of congruence defined in Section 2.3 and morphisms of QYB modules. We develop this section along the lines of [Lambe and Radford, 1993, Section 3.6].

Proposition 3.2.1 *Suppose that A is a bialgebra over a field k and M, M' are vector spaces over k. Let $R \in \mathrm{End}(M \otimes M)$ and $R' \in \mathrm{End}(M' \otimes M')$ be solutions to the QYBE. Further assume that (M, μ, ρ) is a left QYB A-module and R is the associated QYBE solution. Then the following are equivalent:*

a) *R and R' are congruent.*

b) *There is a left QYB A-module structure (M', μ', ρ') on M' such that R' is the associated QYBE solution and $(M, \mu, \rho) \simeq (M', \mu', \rho')$.*

Furthermore, if $u : M \longrightarrow M'$ is an isomorphism of left QYB A-modules then R_u is the solution associated with (M', μ', ρ')

Proof: Suppose that $u : M \longrightarrow M'$ is a linear isomorphism. Then the left QYB A-module structure on M can be transferred to M' in such a way that u is an isomorphism. Thus the proof really comes down to showing that if u is an isomorphism then R_u is the QYBE solution associated with (M, μ', ρ').

To this end we use the fact that u is a module and comodule map and calculate for $m, n \in M$ that

$$\begin{aligned} R_u(u(m) \otimes u(n)) &= (u \otimes u)R(m \otimes n) \\ &= (u \otimes u)(m^{<1>} \otimes m^{(2)}.n) \\ &= u(m^{<1>}) \otimes u(m^{(2)}.n) \\ &= u(m^{<1>}) \otimes m^{(2)}.'u(n) \\ &= u(m)^{<1>} \otimes u(m)^{(2)}.'u(n). \end{aligned}$$

Since u is onto we conclude that R_u is the QYBE solution associated with (M, μ', ρ'). ∎

The following corollary is a minor variant of [Lambe and Radford, 1993, Proposition 3.6.2].

Corollary 3.2.1 *Let M and M' be finite-dimensional vector spaces over k and suppose that $R \in \mathrm{End}(M \otimes M)$ and $R' \in \mathrm{End}(M' \otimes M')$ are solutions to QYBE. Suppose that R and R' are congruent; in particular let $u : M \longrightarrow M'$ be a linear isomorphism and that $R' = R_u$. Then M and M' have the structure of left QYB $A(R)$-modules (M, μ, ρ) and (M', μ', ρ') such that:*

a) $R = (1_M \otimes \mu)(\rho \otimes 1_M)$,

b) $R' = (1_{M'} \otimes \mu')(\rho' \otimes 1_{M'})$, *and*

c) $u : M \longrightarrow M'$ *is an isomorphism of left QYB $A(R)$-modules.*

Proof: By Theorem 3.1.1 there is a left QYB $A(R)$-module structure on M so that R is the associated QYBE solution. Now M' has a unique left QYB $A(R)$-module structure which makes u an isomorphism. We apply the previous proposition at this point to finish the proof. ∎

3.3 Recollections of Various Module and Comodule Structures

We thought that it might be a convenience for the reader to review some basic module and comodule actions and their notations at this point before the reader gets too far into the real subject matter of this book. All of the comments we make in this section apply to this chapter in particular. Throughout this section A is an algebra over the field k and C is a coalgebra over k.

Let (M, μ) be a left A-module. We have noted in Exercise 1.4.3 that (M, μ^{op}) is a right A^{op}-module where

$$m \cdot^{op} a = \mu^{op}(m \otimes a) = \mu(a \otimes m) = a \cdot m$$

for all $m \in M$ and $a \in A$. The linear dual M^* of M has a right A-module structure (M^*, μ^T) which is called the (right) transpose action. We set

$$\alpha \cdot^T a = \mu^T(\alpha \otimes a) \tag{3.7}$$

for $\alpha \in M^*$ and $a \in A$. By definition

$$<\alpha \cdot^T a, m> = <\alpha, a \cdot m>$$

for all $\alpha \in M^*, a \in A$, and $m \in M$.

Suppose that M is finite-dimensional. Then the last equation has a simple expression in terms of structure constants. Let $\{m_i\}$ be a basis for M and $\{m^i\}$ be the dual basis for M^*. Let $\{a_j\}$ be a basis for A (possibly infinite). We have

$$<m^j \cdot^T a_i, m_k> = <m^j, a_i \cdot m_k> = <m^j, \mu^l_{i,k} m_l>$$

and so $(\mu^T)^j_{i,k} = \mu^j_{i,k}$.

Now suppose that (M, μ) is a right A-module. We define left A^{op}-module (M, μ^{op}) on M and a left A-module structure (M^*, μ^T) on M^* in a similar manner. In particular

$$<a \cdot^T \alpha, m> = <\alpha, m \cdot a>$$

for $a \in A, \alpha \in M^*$, and $m \in M$.

Assume that A is a finite-dimensional bialgebra over k. Then A^* is a bialgebra as well and $A \simeq A^{**}$ as bialgebras. Without qualification, the notations

$$a{\cdot}^T\alpha \quad \text{and} \quad \alpha{\cdot}^T a$$

are ambiguous. We will regard A as the "base" algebra and take the module actions to be *actions by the base algebra*. Thus $a{\cdot}^T\alpha, \alpha{\cdot}^T a \in A^*$. We set $a{\cdot}^T\alpha = a{\cdot}\alpha$ and $\alpha{\cdot}^T a = \alpha{\cdot}a$ to simplify notation. Thus

$$<a{\cdot}\alpha, b> = <\alpha, ba>$$

and

$$<\alpha{\cdot}a, b> = <\alpha, ab>$$

for all $a, b \in A$ and $\alpha \in A^*$. This notational convention applies to all bialgebra A over k.

We now turn our attention to comodules and related module structures. Suppose that (N, ρ) is a right C-comodule. We have noted in Exercise 1.4.3 that (N, ρ^{cop}) is a left C-comodule where

$$n^{(1)^{cop}} \otimes n^{<2>^{cop}} = \rho^{cop}(n) = (\tau_{N,C}\rho)(n) = n^{<2>} \otimes n^{(1)}$$

for all $n \in M$. Now the right C-comodule structure (N, ρ) on N accounts for a left C^*-module structure (N, μ_ρ) on M defined by

$$\alpha{\rightharpoonup}n = \mu_\rho(\alpha \otimes n)$$

for $\alpha \in C^*$ and $n \in N$. Recall that

$$\alpha{\rightharpoonup}n = n^{<1>}<\alpha, n^{(2)}>$$

for all $\alpha \in C^*$ and $n \in N$. This module action is called rational.

Consider the right C^*-module $(N^*, (\mu_\rho)^T)$. Recall that $N^r = (N^*)_r$ is the unique maximal rational C^*-submodule of N^*. The underlying left C-comodule structure for this action is denoted by (N^r, ρ^T). Thus we have the relationship

$$(\mu_\rho)^T = \mu_{(\rho^T)},$$

where the left hand side of the equation is interpreted as the restriction of the module action to N^r. Let $\alpha \in N^r$ and write $\rho^T(\alpha) = \alpha^{(1)} \otimes \alpha^{<2>}$. Then we have the important relation

$$\alpha^{(1)}<\alpha^{<2>}, n> = <\alpha, n^{<1>}>n^{(2)}$$

for $\alpha \in N^*$ and $n \in N$, which is (1.13).

Now suppose that (M, μ) is a *locally finite* left A-module. Then by Exercise 1.4.9 there is a right A^o-comodule structure (M, ρ_μ) on M uniquely determined by

$$a{\cdot}m = m^{<1>^o} <m^{(2)^o}, a>$$

for all $a \in A$ and $m \in M$, where we write $\rho(m) = m^{<1>^o} \otimes m^{(2)^o}$.

Finally, suppose that N and C are finite-dimensional. In terms of structure constants, if C has basis $\{c_i\}$, C^* has dual basis $\{c^i\}$, and N has basis $\{n_j\}$, we have

$$c^j \rightharpoonup n_i = (\mu_\rho)_i^{j,k} n_k = \rho_i^{k,l} \delta_l^j n_k$$

and so we see that $(\mu_\rho)_i^{j,k} = \rho_i^{k,j}$. It is easy to see by (1.13) that the transpose action comes from the coaction whose structure constants are given by $(\rho^T)_j^{i,k} = \rho_j^{i,k}$.

The reader is left with working out the details of the statements corresponding to the above for left C-comodules.

3.4 General Constructions in ${}_A\mathcal{QYB}$

Throughout this section A is a bialgebra over the field k. Most of the material in this section is found in [Lambe and Radford, 1993, Section 4].

3.4.1 Sub-objects, Quotient Objects of ${}_A\mathcal{QYB}$

Let (M, μ, ρ) be a left QYB A-module. Suppose that N is a subspace of M which is both an A-submodule and an A-subcomodule of M. Then N is a left QYB A-module with its submodule and subcomodule structures. We refer to N as a left QYB A-submodule of M.

Suppose that N is a left QYB A-submodule of M. Then M/N with its quotient module and comodule structures is a left QYB A-module. We refer to M/N as a quotient left QYB A-module. Observe that the projection $M \xrightarrow{\pi} M/N$ is a morphism.

3.4.2 Direct Sums in ${}_A\mathcal{QYB}$

Let $\{M_i\}_{i \in I}$ be an indexed family of objects of ${}_A\mathcal{QYB}$. With its direct sum module and comodule structures $M = \oplus_{i \in I} M_i$ is a left QYB A-module. We observe that M is the coproduct of the family $\{M_i\}_{i \in I}$ in the category ${}_A\mathcal{QYB}$.

3.4.3 Duals of Objects of ${}_A\mathcal{QYB}$

We will show that there is a contravariant functor $F : {}_A\mathcal{QYB} \longrightarrow \mathcal{QYB}_A$ which establishes a duality between the full subcategories of finite-dimensional objects of ${}_A\mathcal{QYB}$ and $\mathcal{QYB}_A$ respectively.

Suppose that (M, μ, ρ) is a left QYB A-module. Recall from Section 3.3 that M^* has a right A-module structure (M^*, μ^T) and M^r has a left A-comodule structure

(M^r, ρ^T). Recall that the comodule structures (M, ρ) and (M^r, ρ^T) are related by

$$\alpha^{(1)}<\alpha^{<2>}, m> = <\alpha, m^{<1>}>m^{(2)}$$

for all $\alpha \in M^r$ and $m \in M$.

Now M^r has a unique subspace $M^{[r]}$ which is maximal with respect to being a right A-submodule of M^* and a left A-subcomodule of M^r. By slight abuse of notation we let $(M^{[r]}, \mu^T)$ and $(M^{[r]}, \rho^T)$ denote $M^{[r]}$ with these submodule and subcomodule structures respectively.

Definition 3.4.1 *The triple* $(M^{[r]}, \mu^T, \rho^T)$ *is called the* dual of the left *QYB A-module* (M, μ, ρ).

The notion of dual of a right QYB A-module is defined in a similar manner and is denoted by $M^{[r]}$ also.

Proposition 3.4.1 *Suppose that A is a bialgebra over the field k and that* (M, μ, ρ) *is a left QYB A-module. Then:*

a) $(M^{[r]}, \mu^T, \rho^T)$ *is a right QYB A-module.*

b) *Let R be the QYBE solution associated with* (M, μ, ρ) *and let* $\mathcal{R}$ *be the QYBE solution associated with* $(M^{[r]}, \mu^T, \rho^T)$. *Then* $\mathcal{R} = (R^\tau)^*|_{M^{[r]}\otimes M^{[r]}}$.

Proof: To show part a) we first regard M as a subspace of M^{**} by $<m, \alpha> = <\alpha, m>$ for all $\alpha \in M^*$ and $m \in M$. For $m \in M$, $a \in A$, and $\alpha \in M^r$ we compute

$$\begin{aligned}
&(1_A \otimes m)(a_{(2)}(\alpha{\cdot}^T a_{(1)})^{(1)} \otimes (\alpha{\cdot}^T a_{(1)})^{<2>}) \\
&= a_{(2)}(\alpha{\cdot}^T a_{(1)})^{(1)}<(\alpha{\cdot}^T a_{(1)})^{<2>}, m> \\
&= a_{(2)}(<\alpha{\cdot}^T a_{(1)}, m^{<1>}>m^{(2)}) \\
&= <\alpha, a_{(1)}{\cdot}m^{<1>}>a_{(2)}m^{(2)} \\
&= (\alpha \otimes 1_A)(a_{(1)}{\cdot}m^{<1>} \otimes a_{(2)}m^{(2)})
\end{aligned}$$

and likewise

$$\begin{aligned}
&(1_A \otimes m)(\alpha^{(1)}a_{(1)} \otimes \alpha^{<2>}{\cdot}^T a_{(2)}) \\
&= \alpha^{(1)}a_{(1)}<\alpha^{<2>}{\cdot}^T a_{(2)}, m> \\
&= \alpha^{(1)}a_{(1)}<\alpha^{<2>}, a_{(2)}{\cdot}m> \\
&= (<\alpha, (a_{(2)}{\cdot}m)^{<1>}>(a_{(2)}{\cdot}m)^{(2)})a_{(1)} \\
&= (\alpha \otimes 1_A)((a_{(2)}{\cdot}m)^{<1>} \otimes (a_{(2)}{\cdot}m)^{(2)}a_{(1)}).
\end{aligned}$$

Since M is a dense subspace of M^{**} it follows that

$$a_{(2)}(\alpha \cdot^T a_{(1)})^{(1)} \otimes (\alpha \cdot^T a_{(1)})^{<2>} = \alpha^{(1)} a_{(1)} \otimes \alpha^{<2>} \cdot^T a_{(2)}$$

for all $a \in A$ and $\alpha \in M^*$. Thus part a) is established.

To show part b) we compute for $\alpha, \beta \in M^{[r]}$ and $m, n \in M$ that

$$\begin{aligned} <\mathcal{R}(\alpha \otimes \beta), m \otimes n> &= <\alpha \cdot^T \beta^{(1)} \otimes \beta^{<2>}, m \otimes n> \\ &= <\alpha \cdot^T \beta^{(1)}, m><\beta^{<2>}, n> \\ &= <\alpha, \beta^{(1)} \cdot m><\beta^{<2>}, n> \\ &= <\alpha, n^{(2)} \cdot m><\beta, n^{<1>}> \\ &= <\alpha \otimes \beta, n^{(2)} \cdot m \otimes n^{<1>}> \\ &= <\alpha \otimes \beta, R^{\tau}(m \otimes n)>. \end{aligned}$$

This proves part b), and the proposition follows. ▮

Let $f : M \longrightarrow N$ be a morphism of left QYB Q-modules and let $f^* : N^* \longrightarrow M^*$ be the transpose of f regarded as a linear map of the underlying vector spaces. Set $f^{[r]} = f|_{N^{[r]}}$. Using part d) of Proposition 1.4.1 it is easy to see that $f^{[r]}(N^{[r]}) \subseteq M^{[r]}$ and that $f^{[r]} : N^{[r]} \longrightarrow M^{[r]}$ is a morphisms of right QYB A-modules. It follows that $F : {}_A\mathcal{QBY} \longrightarrow \mathcal{QYB}_A$ given by $(M, \mu, \rho) \longmapsto (M, \mu^T, \rho^T)$ and $f \longmapsto f^{[r]}$ is a contravariant functor. Since $M^* = M^{[r]}$ when M is finite-dimensional the functor F establishes a duality between the full subcategories of finite-dimensional objects of these categories.

It is very instructive to consider part b) of the preceding proposition in in terms of coordinates and structure constants in the finite-dimensional case. Fix bases for A and M. Let (M, μ) be a right A-module and (M, ρ) be a left A-comodule structure on M. Then

$$(\mu^T)^j_{i,k} = \mu^j_{k,i}$$

and

$$(\rho^T)^{i,l}_k = \rho^{l,i}_k.$$

Computing the composite $\mathcal{R}$ given by

$$M^* \otimes M^* \xrightarrow{\rho^T \otimes 1_M} M^* \otimes A \otimes M^* \xrightarrow{1_M \otimes \mu^T} M^* \otimes M^*$$

we see that

$$m^i \otimes m^j \longmapsto \rho^{q,i}_k \mu^j_{s,q} m^k \otimes m^s$$

so that

$$\mathcal{R}^{i,j}_{k,l} = \rho^{q,i}_k \mu^j_{l,q}.$$

Computing the composite R given by

$$M \otimes M \xrightarrow{1 \otimes \rho} M \otimes A \otimes M \xrightarrow{\mu \otimes 1} M \otimes M$$

we see that

$$m^i \otimes m^j \longmapsto \rho_j^{k,l} \mu_{i,k}^q m_q \otimes m_l$$

so that

$$R_{i,j}^{k,l} = \rho_j^{q,l} \mu_{i,q}^k$$

and hence in the notation of Section 2.3

$$\mathcal{R} = R^{\tau T}.$$

Exercise 3.4.1 Formulate the definition of dual of a right QYB A-module and derive functorial properties of the dual.

3.4.4 Structure Induced from Objects of ${}_A\mathcal{QYB}$

The "pullback" action is defined for QYB modules in certain instances. The easy proof of the following, a minor variation on [Lambe and Radford, 1993, Proposition 4.5.1], is left as an exercise.

Proposition 3.4.2 *Suppose that A is a bialgebra over the field k.*

a) *Let $M = (M, \mu, \rho)$ be a left QYB A-module and suppose that $f : A \longrightarrow A$ is a bialgebra automorphism. Then $M_f = (M, \mu_f, \rho_{f^{-1}})$ is a left QYB A-module, where $a \cdot_f m = f(a) \cdot m$ and $\rho_{f^{-1}}(m) = m^{<1>} \otimes f^{-1}(m^{(2)})$ for $a \in A$ and $m \in M$. Furthermore the QYBE solution corresponding to M is the same as the QYBE solution corresponding to M_f.*

b) *Let $M = (M, \mu, \rho)$ be a right QYB A-module and suppose that $f : A \longrightarrow A^{op\ cop}$ is a bialgebra isomorphism. Then $M_f = (M, \mu_f, \rho_{f^{-1}})$ is a left QYB A-module, where $a \cdot_f m = m \cdot f(a)$ and $\rho_{f^{-1}}(m) = m^{<2>} \otimes f^{-1}(m^{(1)})$ for $a \in A$ and $m \in M$. Furthermore the QYBE solution corresponding to M is the same as the QYBE solution corresponding to M_f.* ∎

3.5 Constructions in ${}_H\mathcal{QYB}$ when H^{op} has an Antipode

In this section H is a bialgebra over the field k such that the bialgebra H^{op} has an antipode ς. In this case objects of ${}_H\mathcal{QYB}$ can be viewed quite naturally in terms of left H and H^*-module structures. We show that products exist in ${}_H\mathcal{QYB}$, and that $\mathrm{Hom}(M, N)$ can be regarded as an object of ${}_H\mathcal{QYB}$ in several different ways, where M, N are objects of ${}_H\mathcal{QYB}$ and M is finite-dimensional.

3.5.1 Equivalent Formulations of the Compatibility Condition (2.17) when H^{op} has an Antipode

We will express the compatibility condition (2.17) in terms of left H and H^* module actions in this section. These formulations of (2.17) arise from rules for exchanging the order of module multiplications. Our lemma is [Lambe and Radford, 1993, Lemma 5.1.1].

Lemma 3.5.1 *Suppose that H is a bialgebra over the field k such that the bialgebra H^{op} is a Hopf algebra with antipode ς. Let M be a vector space over k with left H-module structure (M, μ) and right H-comodule structure (M, ρ). Then the following are equivalent:*

a) $a_{(1)}{\cdot}m^{<1>}\otimes a_{(2)}m^{(2)} = (a_{(2)}{\cdot}m)^{<1>}\otimes(a_{(2)}{\cdot}m)^{(2)}a_{(1)}$ *for all $a \in H$ and $m \in M$.*

b) $\rho(a{\cdot}m) = a_{(2)}{\cdot}m^{<1>} \otimes a_{(3)}m^{(2)}\varsigma(a_{(1)})$ *for all $a \in H$ and $m \in M$.*

c) $\alpha{\rightharpoonup}(a{\cdot}m) = a_{(2)}{\cdot}((\varsigma(a_{(1)}){\cdot}\alpha{\cdot}a_{(3)}){\rightharpoonup}m)$ *for all $a \in H$, $\alpha \in H^*$, and $m \in M$.*

d) $a{\cdot}(\alpha{\rightharpoonup}m) = (a_{(1)}{\cdot}\alpha{\cdot}\varsigma(a_{(3)})){\rightharpoonup}(a_{(2)}{\cdot}m)$ *for all $a \in H$, $\alpha \in H^*$, and $m \in M$.*

Proof: First suppose part a) holds. Then

$$\begin{aligned}
\rho(a{\cdot}m) &= (a{\cdot}m)^{<1>} \otimes (a{\cdot}m)^{(2)} \\
&= (a_{(2)}{\cdot}m)^{<1>} \otimes (a_{(2)}{\cdot}m)^{(2)}(\epsilon(a_{(1)})1) \\
&= (a_{(3)}{\cdot}m)^{<1>} \otimes (a_{(3)}{\cdot}m)^{(2)}a_{(2)}\varsigma(a_{(1)}) \\
&= a_{(2)}{\cdot}m^{<1>} \otimes a_{(3)}m^{(2)}\varsigma(a_{(1)})
\end{aligned}$$

for all $a \in H$ and $m \in M$ which shows that part b) holds. Therefore part a) implies part b).

To show that part c) follows from part b), we assume part b) holds and note that

$$\begin{aligned}
\alpha{\rightharpoonup}(a{\cdot}m) &= a_{(2)}{\cdot}m^{<1>}<\alpha, a_{(3)}m^{(2)}\varsigma(a_{(1)})> \\
&= a_{(2)}{\cdot}m^{<1>}<\varsigma(a_{(1)}){\cdot}\alpha{\cdot}a_{(3)}, m^{(2)}> \\
&= a_{(2)}{\cdot}((\varsigma(a_{(1)}){\cdot}\alpha{\cdot}a_{(3)}){\rightharpoonup}m)
\end{aligned}$$

for all $\alpha \in H^*, a \in H$, and $m \in M$.

Suppose that part c) holds. Then for $a \in H, \alpha \in H^*$, and $m \in M$ we have

$$\begin{aligned}
&(a_{(1)}{\cdot}\alpha{\cdot}\varsigma(a_{(3)})){\rightharpoonup}(a_{(2)}{\cdot}m) \\
&= a_{(2)(2)}{\cdot}(((\varsigma(a_{(2)(1)}){\cdot}(a_{(1)}{\cdot}\alpha{\cdot}\varsigma(a_{(3)})){\cdot}a_{(2)(3)}){\rightharpoonup}m) \\
&= a_{(3)}{\cdot}(((\varsigma(a_{(2)})a_{(1)}){\cdot}\alpha{\cdot}(\varsigma(a_{(5)})a_{(4)})){\rightharpoonup}m) \\
&= a_{(2)}{\cdot}(((\epsilon(a_{(1)})1){\cdot}\alpha{\cdot}(\epsilon(a_{(3)})1)){\rightharpoonup}m) \\
&= a{\cdot}(\alpha{\rightharpoonup}m)
\end{aligned}$$

which gives the equation of part d).

To complete the proof is a matter of showing that part d) implies part a). Suppose that part d) holds. Then

$$\begin{aligned}
&(1_H \otimes \alpha)(a_{(1)}\cdot m^{<1>} \otimes a_{(2)}m^{(2)}) \\
&= a_{(1)}\cdot m^{<1>} <\alpha, a_{(2)}m^{(2)}> \\
&= a_{(1)}\cdot((\alpha\cdot a_{(2)}) \rightharpoonup m) \\
&= (a_{(1)(1)}\cdot(\alpha\cdot a_{(2)})\cdot\varsigma(a_{(1)(3)})) \rightharpoonup (a_{(1)(2)}\cdot m) \\
&= (a_{(1)}\cdot\alpha\cdot(a_{(4)}\varsigma(a_{(3)}))) \rightharpoonup (a_{(2)}\cdot m) \\
&= (a_{(1)}\cdot\alpha\cdot(\epsilon(a_{(3)})1)) \rightharpoonup (a_{(2)}\cdot m) \\
&= (a_{(2)}\cdot m)^{(1)} <\alpha, (a_{(2)}\cdot m)^{(2)} a_{(1)}> \\
&= (1_H \otimes \alpha)((a_{(2)}\cdot m)^{<1>} \otimes (a_{(2)}\cdot m)^{(2)} a_{(1)})
\end{aligned}$$

for all $a \in H, m \in M$, and $\alpha \in H^*$. Thus part d) part implies a) and the proof of the lemma is finished. ∎

Let M be a vector space over k which has a left H-module structure (M, μ) and a left H^*-module structure (M, μ'). Writing $\mu'(a \otimes m) = a \cdot' m$, note that parts c) and d) of Lemma 3.5.1 are equivalent in this generality, i.e.

$$\alpha\cdot'(a\cdot m) = a_{(2)}\cdot((\varsigma(a_{(1)})\cdot\alpha\cdot a_{(3)})\cdot' m) \tag{3.8}$$

holds for all $\alpha \in H^*, a \in H$, and $m \in M$ if and only if

$$a\cdot(\alpha\cdot' m) = (a_{(1)}\cdot\alpha\cdot\varsigma(a_{(3)}))\cdot'(a_{(2)}\cdot m) \tag{3.9}$$

holds for all $a \in H, \alpha \in H^*$, and $m \in M$.

3.5.2 The Rational Part of a Left H, H^*-Module when (3.8) or (3.9) is Satisfied

A left QYB H-module can be described as a vector space M which has a left H module structure (M, μ) and a left rational H^*-module structure (M, μ') such that (3.8), or equivalently (3.9), is satisfied.

Recall from Sections 1.4 and 3.3 that a left C^*-module M, where C is a coalgebra over k, has a unique maximal submodule M_r with an underlying right C-comodule structure. When M is rational $M = M_r$. Our initial observation suggests a more general result which is [Lambe and Radford, 1993, Proposition 5.2.1].

Proposition 3.5.1 *Suppose that H be a bialgebra over the field k such that H^{op} is a Hopf algebra with antipode ς. Suppose that M is a vector space over k with a left H-module structure (M, μ) and a left H^*-module structure (M, μ') such that (3.8) or (3.9) holds for all $\alpha \in H^*, a \in H$, and $m \in M$. Let (M_r, ρ') be the underlying* *right H-comodule structure derived from (M, μ'). Then (M_r, μ, ρ') is a left QYB H-module.*

Proof: Let $m \in M_r$ and $a \in H$. By (3.8) it suffices to show that $a{\cdot}m \in M_r$. Write $\rho'(m) = \sum_{i=1}^{n} m_i \otimes h_i \in M \otimes H$. Then

$$\begin{aligned}
\alpha \cdot' (a{\cdot}m) &= a_{(2)}{\cdot}((\varsigma(a_{(1)}){\cdot}\alpha{\cdot}a_{(3)}) \cdot' m) \\
&= \sum_{i=1}^{n} a_{(2)}{\cdot}(m_i{<}\varsigma(a_{(1)}){\cdot}\alpha{\cdot}a_{(3)}, h_i{>}) \\
&= \sum_{i=1}^{n} a_{(2)}{\cdot}m_i{<}\alpha, a_{(3)}h_i\varsigma(a_{(1)}){>} \\
&= \sum (1_M \otimes \alpha)(a_{(2)}{\cdot}m_i \otimes a_{(3)}h_i\varsigma(a_{(1)}))
\end{aligned}$$

for all $\alpha \in H^*$. Therefore $a{\cdot}m \in M_r$; in fact $\rho'(a{\cdot}m) = a_{(2)}{\cdot}m_i \otimes a_{(3)}h_i\varsigma(a_{(1)})$. ∎

The rational part of the linear dual of a right QYB H-module is a left QYB H-module when H^{op} is a Hopf algebra. The following corollary is slight improvement to [Lambe and Radford, 1993, Corollary 5.2.2].

Corollary 3.5.1 *Suppose that H is a bialgebra over the field k such that H^{op} is a Hopf algebra with antipode ς. Let (M, μ, ρ) be a right QYB H-module. Regard M^* as a left H-module (M^*, μ^T) and as a left H^*-module $(M^*, (\mu_\rho)^T)$ under the transpose actions. Then (M^r, μ^T, ρ^T) is a left QYB H-module. In particular $M^r = M^{[r]}$.*

Proof: For $m \in M$ and $a \in H$

$$m^{(1)}a_{(1)} \otimes m^{<2>}{\cdot}a_{(2)} = a_{(2)}(m{\cdot}a_{(1)})^{(1)} \otimes (m{\cdot}a_{(1)})^{<2>}$$

by assumption; hence

$$(m{\cdot}a){\leftharpoonup}\alpha = (m{\leftharpoonup}(a_{(1)}{\cdot}\alpha{\cdot}\varsigma(a_{(3)}))){\cdot}a_{(2)}$$

for all $m \in M, a \in H$, and $\alpha \in H^*$. Let $\mu' = (\mu_\rho)^T$. This last calculation implies that

$$a{\cdot}^T(\alpha{\cdot}'\beta) = (a_{(1)}{\cdot}\alpha{\cdot}\varsigma(a_{(3)})){\cdot}'(a_{(2)}{\cdot}^T\beta)$$

for all $a \in H, \alpha \in H^*$, and $\beta \in M^*$. The corollary now follows by Proposition 3.5.1. ∎

3.5.3 Direct Products in $_H\mathcal{QYB}$ when H^{op} has an Antipode

As a corollary to the previous proposition we have [Lambe and Radford, 1993, Corollary 5.3.1]:

Corollary 3.5.2 *Suppose that H is a bialgebra over the field k such that H^{op} is a Hopf algebra with antipode ς. Then products exist in ${}_H\mathcal{QYB}$.*

Proof: Suppose $\{M_i\}_{i\in I}$ is an indexed family of left QYB H-modules. Then each M_i is a left H-module and a (rational) left H^*-module such that (3.9) holds. Consider $M = \prod_{i\in I} M_i$ with the H-module and H^*-module structures. Then clearly (3.9) is satisfied for M with these structures since it is satisfied componentwise. By Proposition 3.5.1 the product of the family $\{M_i\}_{i\in I}$ in the category ${}_H\mathcal{QYB}$ is $M_r = (\prod_{i\in I} M_i)_r$. ∎

3.5.4 Sub-Objects of Objects of ${}_\mathrm{H}\mathcal{QYB}$ when H^{op} has an Antipode

In this section we deduce a basic result on subobjects of objects of ${}_H\mathcal{QYB}$ which is similar to a very basic result on sub-Hopf modules of a Hopf module. Hopf modules are very important in the theory of Hopf algebras.

Definition 3.5.1 *Let A be a bialgebra over the field k. A* left A-Hopf module *is a triple (M, μ, ρ), where (M, μ) is a left A-module and (M, ρ) is a left A-comodule such that*

$$\rho(a{\cdot}m) = a_{(1)}m^{(1)} \otimes a_{(2)}{\cdot}m^{<2>}$$

for all $a \in A$ and $m \in M$.

Let H be a Hopf algebra over k and let M be a left H-Hopf module (for a discussion of Hopf-modules see Section 6.2). The subcomodule generated by a submodule of M is a sub-Hopf module, and the submodule generated by a subcomodule of M is a sub-Hopf module. See [Sweedler, 1969, pp. 106-107] for details. There is an similar result for left QYB H-modules. The proof follows the lines of the proof of the corresponding statement for Hopf modules. The result is [Lambe and Radford, 1993, Proposition 5.4.1].

Proposition 3.5.2 *Suppose that H is a bialgebra over the field k such that H^{op} is a Hopf algebra with antipode ς. Let (M, μ, ρ) be a left QYB H-module. Then:*

a) *If N is a subcomodule of (M, ρ) then $H{\cdot}N$ is a QYB submodule of M.*

b) *If N is a submodule of (M, μ) then $H^*{\rightharpoonup}N$ is a QYB submodule of M.*

Proof: The subcomodules of M are precisely the H^*-submodules of (M, μ_ρ) by part c) of Proposition 1.4.1. Thus part a) follows by part c) of Lemma 3.5.1, and part b) follows by part d) of the same result. ∎

3.6 The Relationship Between QYBE Solutions R and R^τ

Let M be a vector space over the field k and suppose that $R : M \otimes M \longrightarrow M \otimes M$ is a solution to the QYBE. Recall that $R^\tau = \tau_{M,M} R \tau_{M,M}$ is a solution to the QYBE as well by Proposition 2.2.1.

Suppose further that M has a left QYB A-module structure (M, μ, ρ) such that R is the associated QYBE solution, where A is a bialgebra over k. Regard A as a coalgebra. We recall from Sections 1.4.2 and 3.3 that the right A-comodule (M, ρ) gives rise to a left (rational) action on M by the dual algebra A^* described by

$$\alpha \rightharpoonup m = m^{<1>} < \alpha, m^{(2)} > \tag{3.10}$$

for all $\alpha \in A^*$ and $m \in M$. Now regard A as an algebra. Assume further that (M, μ) is a *locally finite* left A-module. Then by part b) of Exercise 1.4.9 there is a unique right A^o-comodule structure (M, ρ_μ) on M determined by

$$a \cdot m = m^{<1>^o} < m^{(2)^o}, a > \tag{3.11}$$

for all $a \in A$ and $m \in M$, where we write $\rho_\mu(m) = m^{<1>^o} \otimes m^{(2)^o}$. Since A is a bialgebra it follows by Proposition 1.5.2 that A^o is a bialgebra which, as an algebra, is a subalgebra of the dual algebra A^*. We let (M, μ_ρ) denote the restriction of the rational left A^*-module action on M described by (3.10) to A^o.

The next theorem applies in particular to R and R^τ in the finite-dimensional case. It is a generalization of [Lambe and Radford, 1993, Theorem 7.3].

Theorem 3.6.1 *Suppose that A is a bialgebra over the field k and let (M, μ, ρ) be a left QYB A-module with associated QYBE solution R. Suppose further that (M, μ) is a locally finite left A-module. Then:*

a) *(M, μ_ρ, ρ_μ) is a left QYB A^o-module.*

b) *R^τ is the QYBE solution associated with (M, μ_ρ, ρ_μ).*

Proof: To show part a) we will think of A as a subspace of A^{**} by $<a, \alpha> = <\alpha, a>$ for all $a \in A$ and $\alpha \in A^*$. Recall that the coproduct of A^o is characterized by the equation $<\alpha, ab> = <\alpha_{(1)}, a><\alpha_{(2)}, b>$ for all $\alpha \in A^*$ and $a, b \in A$. For $\alpha \in A^*, a \in A$, and $m \in M$ we compute, using (3.11) and (3.10), that

$$\begin{aligned}
&(1_M \otimes a)((\alpha_{(1)} \rightharpoonup m^{<1>^o}) \otimes \alpha_{(2)} m^{(2)^o}) \\
&= (\alpha_{(1)} \rightharpoonup m^{<1>^o}) < \alpha_{(2)} m^{(2)^o}, a > \\
&= (\alpha_{(1)} \rightharpoonup m^{<1>^o}) < \alpha_{(2)}, a_{(1)} > < m^{(2)^o}, a_{(2)} > \\
&= (\alpha_{(1)} \rightharpoonup (a_{(2)} \cdot m)) < \alpha_{(2)}, a_{(1)} > \\
&= (a_{(2)} \cdot m)^{<1>} < \alpha_{(1)}, (a_{(2)} \cdot m)^{(2)} > < \alpha_{(2)}, a_{(1)} >
\end{aligned}$$

$$= (a_{(2)}\cdot m)^{<1>}<\alpha, (a_{(2)}\cdot m)^{(2)}a_{(1)}>$$
$$= (1_M \otimes \alpha)((a_{(2)}\cdot m)^{<1>} \otimes (a_{(2)}\cdot m)^{(2)}a_{(1)}).$$

Likewise

$$\begin{aligned}
&(1_M \otimes a)((\alpha_{(2)} \rightharpoonup m)^{<1>^\circ} \otimes (\alpha_{(2)} \rightharpoonup m)^{(2)^\circ}\alpha_{(1)}) \\
&= (\alpha_{(2)} \rightharpoonup m)^{<1>^\circ}<(\alpha_{(2)} \rightharpoonup m)^{(2)^\circ}\alpha_{(1)}, a> \\
&= (\alpha_{(2)} \rightharpoonup m)^{<1>^\circ}<(\alpha_{(2)} \rightharpoonup m)^{(2)^\circ}, a_{(1)}><\alpha_{(1)}, a_{(2)}> \\
&= (a_{(1)}\cdot(\alpha_{(2)} \rightharpoonup m))<\alpha_{(1)}, a_{(2)}> \\
&= a_{(1)}\cdot m^{<1>}<\alpha_{(2)}, m^{(2)}><\alpha_{(1)}, a_{(2)}> \\
&= a_{(1)}\cdot m^{<1>}<\alpha, a_{(2)}m^{(2)^\circ}> \\
&= (1_M \otimes \alpha)(a_{(1)}\cdot m^{<1>} \otimes a_{(2)}m^{(2)}).
\end{aligned}$$

Since A is a dense subalgebra of A^{**} we conclude that

$$(\alpha_{(2)} \rightharpoonup m)^{<1>^\circ} \otimes (\alpha_{(2)} \rightharpoonup m)^{(2)^\circ}\alpha_{(1)} = (\alpha_{(1)} \rightharpoonup m^{<1>^\circ}) \otimes \alpha_{(2)}m^{(2)^\circ},$$

and thus part a) is established.

To show part b) we let $\mathcal{R}$ be the solution to the QYBE associated with (M, μ_ρ, ρ_μ). For $m, n \in M$ we calculate

$$\begin{aligned}
\mathcal{R}(m \otimes n) &= m^{<1>^\circ} \otimes (m^{(2)^\circ} \rightharpoonup n) \\
&= m^{<1>^\circ} \otimes n^{<1>}<m^{(2)^\circ}, n^{(2)}> \\
&= n^{(2)}\cdot m \otimes n^{<1>} \\
&= R^\tau(m \otimes n).
\end{aligned}$$

Therefore $\mathcal{R} = R^\tau$ and part b) is established. This completes the proof of the theorem. ∎

Passing from (M, μ, ρ) to (M, μ_ρ, ρ_μ) converts questions about the module structure of (M, μ) to questions about the comodule structure of (M, ρ_μ) and converts questions about the comodule structure (M, ρ) to questions about the module structure of (M, μ_ρ). This technique can be useful in classifying QYBE solutions. See Exercise 8.3.5.

3.7 QYB Structures on H when H^op is a Hopf Algebra

Throughout this section H is a bialgebra over the field k such that the bialgebra H^{op} has an antipode ς. Finite-dimensional Hopf algebras over k have this property by Theorem 1.6.2.

In this section we consider two ways in which H itself can be given the structure of a left QYB H-module. In the first case the module action is the multiplication of H. In the second the comodule action is the comultiplication of H. We show such structures exist and describe them all.

Suppose that H is an infinite-dimensional Hopf algebra over k. Let m be the multiplication of H. There are left QYB H-module structures on H of the form (H, m, ρ) which are of theoretical interest. The only finite-dimensional left ideal of H is (0) by [Sweedler, 1969, Exercise 4, p. 108]. Thus the non-zero QYB submodules of H are infinite-dimensional.

Interesting in its own right is that more general adjoint actions and coadjoint actions arise in connection with QYB structures on H.

3.7.1 Generalized Coadjoint Action

Let m be the multiplication of H. In this section we find all left QYB H-module structures of the type (H, m, ρ) on H. These structures ρ are generalized coadjoint actions. We follow [Lambe and Radford, 1993, Section 8.1]. Our first result is [Lambe and Radford, 1993, Lemma 8.1.1].

Lemma 3.7.1 *Suppose that H is a bialgebra over a field k such that H^{op} is a Hopf algebra with antipode ς. Let m denote the multiplication of H. Assume that (H, m, ρ) is a left QYB H-module and write $\rho(1) = \sum_{i=1}^{n} x^i \otimes y^i$. Then:*

$$\rho(h) = \sum_{i=1}^{n} h_{(2)}x^i \otimes h_{(3)}y^i\varsigma(h_{(1)})$$

for all $h \in H$.

Proof: The compatibility condition (2.17) in this case is

$$h_{(1)}a^{<1>} \otimes h_{(2)}a^{(2)} = (h_{(2)}a)^{<1>} \otimes (h_{(2)}a)^{(2)}h_{(1)}$$

for $h, a \in H$, where $\rho(a) = a^{<1>} \otimes a^{(2)}$. We thus calculate

$$\begin{aligned} h^{<1>} \otimes h^{(2)} &= {h_{(2)}}^{<1>} \otimes {h_{(2)}}^{(2)}\epsilon(h_{(1)}) \\ &= {h_{(3)}}^{<1>} \otimes {h_{(3)}}^{(2)}h_{(2)}\varsigma(h_{(1)}) \\ &= (h_{(3)}1)^{<1>} \otimes (h_{(3)}1)^{(2)}h_{(2)}\varsigma(h_{(1)}) \\ &= h_{(2)}1^{<1>} \otimes h_{(3)}1^{(2)}\varsigma(h_{(1)}) \end{aligned}$$

for all $h \in H$. ∎

We next find all left QYB H-module structures of the form (H, m, ρ). These structures are in one-one correspondence with certain sums $\sum_i x^i \otimes y^i \in H \otimes H$.

What they are is the content of [Lambe and Radford, 1993, Proposition 8.1.2], the proposition which follows.

Proposition 3.7.1 *Suppose that H is a bialgebra over a field k such that H^{op} is a Hopf algebra with antipode ς. Let $\sum_{i=1}^{n} x^i \otimes y^i \in H \otimes H$ and define $\rho : H \longrightarrow H \otimes H$ by $\rho(h) = \sum_{i=1}^{n} h_{(2)}x^i \otimes h_{(3)}y^i\varsigma(h_{(1)})$ for $h \in H$. Then:*

a) *(H, ρ) is a right H-comodule if and only if*

   i) $\sum_i x^i\epsilon(y^i) = 1$ *and*

   ii) $\sum_i x^i \otimes \Delta(y^i) = \sum_i x^i_{(2)}x^j \otimes x^i_{(3)}y^j\varsigma(x^i_{(1)}) \otimes y^i$.

b) *If (H, ρ) is a right H-comodule, then (H, m, ρ) is a left QYB H-module, where m is the multiplication of H.*

Proof: We first show part b). For $h, a \in H$ we note that

$$h_{(1)}a^{<1>} \otimes h_{(2)}a^{(2)} = \sum_i h_{(1)}a_{(2)}x^i \otimes h_{(2)}a_{(3)}y^i\varsigma(a_{(1)})$$

and

$$\begin{aligned}
&(h_{(2)}a)^{<1>} \otimes (h_{(2)}a)^{(2)}h_{(1)} \\
&= \sum_i (h_{(2)}a)_{(2)}x^i \otimes (h_{(2)}a)_{(3)}y^i\varsigma((h_{(2)}a)_{(1)})h_{(1)} \\
&= \sum_i h_{(2)(2)}a_{(2)}x^i \otimes h_{(2)(3)}a_{(3)}y^i\varsigma(h_{(2)(1)}a_{(1)})h_{(1)} \\
&= \sum_i h_{(3)}a_{(2)}x^i \otimes h_{(4)}a_{(3)}y^i\varsigma(h_{(2)}a_{(1)})h_{(1)} \\
&= \sum_i h_{(3)}a_{(2)}x^i \otimes h_{(4)}a_{(3)}y^i\varsigma(a_{(1)})\varsigma(h_{(2)})h_{(1)} \\
&= \sum_i h_{(2)}a_{(2)}x^i \otimes h_{(3)}a_{(3)}y^i\varsigma(a_{(1)})(\epsilon(h_{(1)})1) \\
&= \sum_i h_{(1)}a_{(2)}x^i \otimes h_{(2)}a_{(3)}y^i\varsigma(a_{(1)}).
\end{aligned}$$

Thus the compatibility condition (2.17) holds. Note that we have not used properties of the sum $\sum_i x^i \otimes y^i$ in the calculation. We have shown that (H, m, ρ) is a left QYB H-module under the assumption that (H, ρ) is a right H-comodule.

To show part a), first suppose that (H, ρ) is a right H-comodule. By definition $\rho(1) = \sum_i x^i \otimes y^i$. Since $(1_H \otimes \epsilon)\rho = 1_H$ it follows that $\sum_i x^i\epsilon(y^i) = 1$. Since $(1_H \otimes \Delta)\rho = (\rho \otimes 1_H)\rho$, we compute that $\sum_i x^i \otimes \Delta(y^i) = \sum_i \rho(x^i) \otimes y^i = \sum_i x^i_{(2)}x^j \otimes x^i_{(3)}y^j\varsigma(x^i_{(1)}) \otimes y^i$. Thus conditions i) and ii) hold.

Conversely, suppose conditions i) and ii) hold. Suppose that $h \in H$. By i) we have

$$\begin{aligned} h^{<1>}\epsilon(h^{(2)}) &= \sum_i h_{(2)}x^i\epsilon(h_{(3)}y^i\varsigma(h_{(1)})) \\ &= \sum_i \epsilon(h_{(1)})h_{(2)}\epsilon(h_{(3)})x^i\epsilon(y^i) \\ &= h1 \\ &= h \end{aligned}$$

for all $h \in H$. That $(1_H \otimes \Delta)\rho(h) = (\rho \otimes 1_H)\rho(h)$ follows from a calculation based on ii). ∎

Let $g \in G(H)$. Then $\sum_i x^i \otimes y^i = 1 \otimes g$ satisfies conditions i) and ii) of Proposition 3.7.1. Different grouplike elements determine different QYB module structures on H. When $g = 1$, the formula of Proposition 3.7.1 is $\rho(h) = h_{(2)} \otimes h_{(3)}\varsigma(h_{(1)})$, which describes a fundamental right coadjoint action of H on itself. Suppose $g^2 = 1$ and $g \neq 1$. Suppose $a, b \in G(H)$ commute with g and that the characteristic of k is not 2. Under these circumstances

$$\sum_i x^i \otimes y^i = \frac{1}{2}(1+g) \otimes a + \frac{1}{2}(1-g) \otimes b$$

satisfies conditions i) and ii) of Proposition 3.7.1. Consequently there are interesting left QYB H-module structures of the type (H, m, ρ) on H, even in the case H is commutative and cocommutative.

There are elementary solutions $R : M \otimes M \longrightarrow M \otimes M$ to the quantum Yang–Baxter equation in the infinite-dimensional case where the only subspace V of M such that $R(V \otimes V) \subseteq V \otimes V$ is (0). For example, let G be an infinite abelian group and let $H = k[G]$ be the group algebra of G over k. Let $g \in G$. We have observed that (H, m, ρ) is a left QYB H-module, where $\rho(h) = h_{(2)}1 \otimes h_{(3)}g\varsigma(h_{(1)})$ for all $h \in H$. Since H is commutative and cocommutative it follows that $\rho(h) = h \otimes g$ for $h \in H$. Let R be the associated QYBE solution. Then R is invertible and is given by $R(a \otimes h) = a \otimes gh$ for $a, h \in H$.

Now suppose that V is a non-zero finite-dimensional subspace of H which satisfies $R(V \otimes V) \subseteq V \otimes V$. Let $S \subseteq G$ be the smallest subset such that $V \subseteq k[S]$, the linear span of S. Now S is finite since V is finite-dimensional. Since $R(V \otimes V) = V \otimes gV$, it follows that $gV = V$. Since $gS \in G(H)$ and $G(H)$ is linearly independent by Lemma 1.1.1, it follows that $gS = S$. Since $V \neq (0)$ it follows that $S \neq \emptyset$. Choose an $a \in S$. Since $a, ga, g^2a, \ldots \in S$ we conclude that g has finite order. Consequently, when g has infinite order, $R(V \otimes V) \subseteq V \otimes V$ implies that $V = (0)$ or V is infinite-dimensional.

3.7.2 Generalized Adjoint Action

We consider now the problem of finding left H-module structures (H, μ) on H such that (H, μ, Δ) is a left QYB H-module, where Δ is the comultiplication of H. These module actions turn out to be generalizations of a fundamental adjoint action of H on itself. We first show [Lambe and Radford, 1993, Lemma 8.2.1]:

Lemma 3.7.2 *Suppose that H is a bialgebra over the field k such that H^{op} is a Hopf algebra with antipode ς. Let Δ be the comultiplication of H. Assume that (H, μ, Δ) is a left QYB H-module. Then*

$$h{\cdot}a = \epsilon(h_{(2)}{\cdot}a_{(1)})h_{(3)}a_{(2)}\varsigma(h_{(1)})$$

for all $h, a \in H$.

Proof: The compatibility condition (2.17) is

$$h_{(1)}{\cdot}a_{(1)} \otimes h_{(2)}a_{(2)} = (h_{(2)}{\cdot}a)_{(1)} \otimes (h_{(2)}{\cdot}a)_{(2)}h_{(1)}$$

for all $a, h \in H$. The calculation

$$\begin{aligned} &\epsilon(h_{(2)}{\cdot}a_{(1)})h_{(3)}a_{(2)}\varsigma(h_{(1)}) \\ &\quad= \epsilon((h_{(3)}{\cdot}a)_{(1)})(h_{(3)}{\cdot}a)_{(2)}h_{(2)}\varsigma(h_{(1)}) \\ &\quad= (h_{(3)}{\cdot}a)h_{(2)}\varsigma(h_{(1)}) \\ &\quad= (h_{(2)}{\cdot}a)(\epsilon(h_{(1)})1) \\ &\quad= h{\cdot}a \end{aligned}$$

finishes the proof. ∎

The equation in Lemma 3.7.2 can be expressed in terms of a functional $p : H \otimes H \longrightarrow k$ defined by $p(h \otimes a) = \epsilon(h{\cdot}a)$ for $h, a \in H$. The description of those functionals p which determine QYB H-module structures of the type (H, μ, Δ) is found in [Lambe and Radford, 1993, Proposition 8.2.2] which we now prove.

Proposition 3.7.2 *Suppose that H is a bialgebra over a field k such that H^{op} has an antipode ς. Let $p : H \otimes H \longrightarrow k$ be a functional and set $h{\cdot}a = p(h_{(2)} \otimes a_{(1)})h_{(3)}a_{(2)}\varsigma(h_{(1)})$ for $h, a \in H$. Then:*

a) *(H, μ) is a left H-module if and only if*

   i) *$p(1 \otimes a) = \epsilon(a)$ and*

   ii) *$p(h'h \otimes a) = p(h_{(2)} \otimes a_{(1)})p(h' \otimes h_{(3)}a_{(2)}\varsigma(h_{(1)}))$ for all $a, h', h \in H$.*

b) *If (H, μ) is a left H-module, then (H, μ, Δ) is a left QYB H-module, where Δ is the comultiplication of H.*

Proof: First we show part b). Suppose that (H, μ) is a left H-module and let $h, a \in H$. Then

$$\begin{aligned} & h_{(1)} \cdot a_{(1)} \otimes h_{(2)} a_{(2)} \\ = \;& p(h_{(1)(2)} \otimes a_{(1)(1)}) h_{(1)(3)} a_{(1)(2)} \varsigma(h_{(1)(1)}) \otimes h_{(2)} a_{(2)} \\ = \;& p(h_{(2)} \otimes a_{(1)}) h_{(3)} a_{(2)} \varsigma(h_{(1)}) \otimes h_{(4)} a_{(3)}. \end{aligned}$$

Since

$$\Delta(h \cdot a) = p(h_{(3)} \otimes a_{(1)}) h_{(4)} a_{(2)} \varsigma(h_{(2)}) \otimes h_{(5)} a_{(3)} \varsigma(h_{(1)})$$

we calculate

$$\begin{aligned} & (h_{(2)} \cdot a)_{(1)} \otimes (h_{(2)} \cdot a)_{(2)} h_{(1)} \\ = \;& p(h_{(2)(3)} \otimes a_{(1)}) h_{(2)(4)} a_{(2)} \varsigma(h_{(2)(2)}) \otimes h_{(2)(5)} a_{(3)} \varsigma(h_{(2)(1)}) h_{(1)} \\ = \;& p(h_{(4)} \otimes a_{(1)}) h_{(5)} a_{(2)} \varsigma(h_{(3)}) \otimes h_{(6)} a_{(3)} \varsigma(h_{(2)}) h_{(1)} \\ = \;& p(h_{(3)} \otimes a_{(1)}) h_{(4)} a_{(2)} \varsigma(h_{(2)}) \otimes h_{(5)} a_{(3)} (\epsilon(h_{(1)}) 1) \\ = \;& p(h_{(2)} \otimes a_{(1)}) h_{(3)} a_{(2)} \varsigma(h_{(1)}) \otimes h_{(4)} a_{(3)} \end{aligned}$$

from which part b) follows. Observe that we have not used properties of the functional p in the calculation.

To show part a), we assume that (H, μ) is a left H-module and let $h, h', a \in H$. Since $\epsilon(h \cdot a) = p(h \otimes a)$, condition i) follows from $\epsilon(a) = \epsilon(1 \cdot a)$. Condition ii) follows from the relation $\epsilon(h'h \cdot a) = \epsilon(h \cdot (h \cdot a))$. The reader is left with showing that conditions i) and ii) imply (H, μ) is a left H-module structure on H. ∎

Suppose $\eta : H \longrightarrow k$ be an algebra homomorphism, which is to say $\eta \in G(H^o)$. Then $p : H \otimes H \longrightarrow k$ defined by $p(h \otimes a) = \eta(h)\epsilon(h)$ satisfies conditions i) and ii) of Proposition 3.7.2. If $p = \epsilon$ then $h \cdot a = h_{(2)} a \varsigma(h_{(1)})$ for all $h, a \in H$. This describes a basic left adjoint action of H on itself.

3.8 Tensor Product in ${}_A\mathcal{QYB}$

Suppose that A is a bialgebra over the field k and let M and N be left QYB A-modules. We show that the tensor product of the underlying A-module structures can be given the structure of a left QYB A-module in two natural ways.

Let M and N be left A-modules. Then $M \otimes N$ is a left A-module with the tensor product A-module structure which is (1.14)

$$a \cdot (m \otimes n) = a_{(1)} \cdot m \otimes a_{(2)} \cdot n$$

for all $a \in A$ and $m, n \in M$. For A^{cop} this tensor product module structure is $a \cdot (m \otimes n) = a_{(2)} \cdot m \otimes a_{(1)} \cdot n$ when expressed in terms of A.

Now suppose that M and N are right A-comodules. Then the tensor product $M\otimes N$ is a right A-comodule with the tensor product comodule A-structure given by (1.15)

$$\rho(m\otimes n) = (m^{<1>}\otimes n^{<1>})\otimes m^{(2)}n^{(2)}$$

for all $m\in M$ and $n\in N$. For A^{op} this tensor product comodule structure is $\rho(m\otimes n) = (m^{<1>}\otimes n^{<1>})\otimes n^{(2)}m^{(2)}$ when expressed in terms of A.

This section is derived from [Lambe and Radford, 1993, Sections 4.3, 5.4]. The next proposition is [Lambe and Radford, 1993, Proposition 4.3.1].

Proposition 3.8.1 *Suppose that A is a bialgebra over a field k, and let M and N be left QYB A-modules. Then $M\otimes N$ is a left QYB A-module, where*

a) $a{\cdot}(m\otimes n) = a_{(1)}{\cdot}m\otimes a_{(2)}{\cdot}n$ *and* $\rho(m\otimes n) = (m^{<1>}\otimes n^{<1>})\otimes n^{(2)}m^{(2)}$

or

b) $a{\cdot}(m\otimes n) = a_{(2)}{\cdot}m\otimes a_{(1)}{\cdot}n$ *and* $\rho(m\otimes n) = (m^{<1>}\otimes n^{<1>})\otimes m^{(2)}n^{(2)}$

for all $a\in A, m\in M$ and $n\in N$.

Proof: We show part a). The proof of part b) is similar and is left as an exercise to the reader. The need for the tensor product A^{cop}-comodule on $M\otimes N$ in connection with the tensor product A-module structure will reveal itself in the proof.

Let $a\in A, m\in M$ and $n\in N$. Then

$$\begin{aligned}
&a_{(1)}{\cdot}(m\otimes n)^{<1>}\otimes a_{(2)}(m\otimes n)^{(2)}\\
&= a_{(1)}{\cdot}(m^{<1>}\otimes n^{<1>})\otimes a_{(2)}n^{(2)}m^{(2)}\\
&= a_{(1)}{\cdot}m^{<1>}\otimes a_{(2)}{\cdot}n^{<1>}\otimes a_{(3)}n^{(2)}m^{(2)}\\
&= a_{(1)}{\cdot}m^{<1>}\otimes (a_{(3)}{\cdot}n)^{<1>}\otimes (a_{(3)}{\cdot}n)^{(2)}a_{(2)}m^{(2)}\\
&= (a_{(2)}{\cdot}m)^{<1>}\otimes (a_{(3)}{\cdot}n)^{<1>}\otimes (a_{(3)}{\cdot}n)^{(2)}(a_{(2)}{\cdot}m)^{(2)}a_{(1)}\\
&= (a_{(2)}{\cdot}(m\otimes n))^{<1>}\otimes (a_{(2)}{\cdot}(m\otimes n))^{(2)}a_{(1)},
\end{aligned}$$

which completes the proof of part a). ∎

Note that the solutions R to the QYBE associated to $M\otimes N$ resulting from parts a) and b) of Proposition 3.8.1 are formally different. In the first case

$$\begin{aligned}
&R((m\otimes n)\otimes(m'\otimes n'))\\
&= m^{<1>}\otimes n^{<1>}\otimes((n^{(2)}{}_{(1)}m^{(2)}{}_{(1)}){\cdot}m')\otimes((n^{(2)}{}_{(2)}m^{(2)}{}_{(2)}){\cdot}n')
\end{aligned}$$

and in the second case

$$\begin{aligned}
&R((m\otimes n)\otimes(m'\otimes n'))\\
&= m^{<1>}\otimes n^{<1>}\otimes((m^{(2)}{}_{(2)}n^{(2)}{}_{(2)}){\cdot}m')\otimes((m^{(2)}{}_{(1)}n^{(2)}{}_{(1)}){\cdot}n')
\end{aligned}$$

for all $m, n, m', n' \in M$.

Observe that the two ways of turning the tensor product of QYB modules into a QYB module described in Proposition 3.8.1 determine 2^{n-1} ways of turning an n-fold tensor product into a QYB module.

Exercise 3.8.1 Observe that the proof of Proposition 3.8.1 that tensor product preserves the compatibility condition (2.17) uses neither the associativity of the module structure nor the coassociativity of the comodule structure.

Exercise 3.8.2 Suppose that M and $\bar{M}$ are left A-modules and right A-comodules over a bialgebra A over k. Let M ($\bar{M}$) have basis $\{m_i\}$ ($\{\bar{m}_i\}$). We have noted that the composite

$$R = (1_M \otimes \mu)(\rho \otimes 1_M)$$

has coordinates

$$R^{k,l}_{i,j} = \rho^{k,s}_i \mu^l_{s,j}.$$

Show that the structure constants for the module structure

$$a\cdot(m \otimes \bar{m}) = a_{(1)}\cdot m \otimes a_{(2)}\cdot \bar{m}$$

are given by

$$\hat{\mu}^{(l,l')}_{s,(j,j')} = \Delta^{r,t}_s \mu^l_{r,j} \bar{\mu}^{l'}_{t,j'}$$

where the structure constants for the coalgebra structure of A are given by the $\Delta^{r,t}_s$. Under the same hypothesis, if the comodule structure for the tensor product is given by

$$\hat{\rho}^{(k,k'),s}_{(i,i')} = \rho^{k,l}_i \bar{\rho}^{k',q}_{i'} m^s_{q,l}$$

where the structure constants for the algebra structure of A are given by the $m^s_{q,l}$. Finally, putting these results together, show that if the compatibility conditions holds so that R is a QYBE solution, then the tensor product QYBE solution $\hat{R} = R \otimes \bar{R}$ has coordinates

$$\hat{R}^{(k,k'),(l,l')}_{(i,i'),(j,j')} = \rho^{k,t_1}_i \bar{\rho}^{k',q}_{i'} m^s_{q,t_1} \Delta^{r,t_2}_s \mu^l_{r,j} \bar{\mu}^{l'}_{t_2,j'}.$$

Exercise 3.8.3 Let R be a constant QYBE solution on M and let $\{m_i\}$ be a basis of M. Consider again the representation defined by (2.9)

$$\begin{array}{ccc} A(R) & \xrightarrow{\mu} & \mathrm{End}(M) \\ t^j_i & \longmapsto & (m_k \longmapsto R^{j,s}_{i,k} m_s). \end{array}$$

Since $A(R)$ is a bialgebra, we have a tensor product representation determined on generators by

$$t^j_i\cdot(m_k \otimes m_l) = t^j_r\cdot m_k \otimes t^r_i\cdot m_l.$$

Using Exercise 3.8.2, prove that the tensor product of R with itself is given by the action of $t^{j'}_{i'} t^j_i$ on $M \otimes M$ given using the action above; prove that the coordinates of the tensor product solution $R \otimes R$ are given by $t^{j'}_{i'} t^j_i\cdot(m_k \otimes m_l)$.

3.8.1 The Tensor Algebra

Let H be a bialgebra over the field k. The tensor algebra $T(M)$ of a left QYB H-module M has two natural left QYB H-module structures. When H is commutative and cocommutative the symmetric algebra $S(M)$ of M has the quotient left QYB H-module structure. To make $T(M)$ a left QYB H-module module we give it the structure of a module algebra and a module coalgebra according to [Lambe and Radford, 1993, Proposition 4.3.2], which is the first result of this section.

Proposition 3.8.2 *Suppose that H is a bialgebra and that A is an algebra over the field k. Assume further that A has a left H-module algebra structure and a right H^{op}-comodule algebra structure. Suppose that A is generated as an algebra by a subspace M which is a submodule and a subcomodule of A. If M is a left QYB H-module, then A is a left QYB H-module.*

Proof: Since A is a left H-module algebra and a right H^{op}-comodule algebra, the set of all $a \in A$ such that $h_{(1)}{\cdot}a^{<1>} \otimes h_{(2)}a^{(2)} = (h_{(2)}{\cdot}a)^{<1>} \otimes (h_{(2)}{\cdot}a)^{(2)}h_{(1)}$ for all $h \in H$ is a subalgebra of A. This completes our proof. ∎

Let (M, μ) be a left H-module. Then (M, μ) extends uniquely to a left H-module algebra structure $(T(M), \mu_T)$ on the tensor algebra $T(M)$ of the vector space M. Now suppose that (M, ρ) is a right H-comodule. Then (M, ρ) extends uniquely to a right H^{op}-comodule algebra structure $(T(M), \rho_T)$. As a corollary to Proposition 3.8.2 we have [Lambe and Radford, 1993, Corollary 4.3.3]:

Corollary 3.8.1 *Suppose that H is a bialgebra over a field k and let (M, μ, ρ) be a left QYB H-module. Then $(T(M), \mu_T, \rho_T)$ is a left QYB H-module.*

∎

The subspace $\mathcal{I}$ of differences $m \otimes n - n \otimes m$ for all $m, n \in M$ is a subcomodule of $T(M)$ when H is commutative. When H is cocommutative $\mathcal{I}$ is a submodule. Any subspace of $T(M)$ which is a submodule and a subcomodule generates an ideal which is a submodule and a subcomodule since $T(M)$ is an H-module algebra and H^{op}-comodule algebra. Consequently the symmetric algebra $S(M) = T(M)/I$ of M has the quotient left QYB H-module structure when H is commutative and cocommutative, where I is the ideal of $T(M)$ generated by $\mathcal{I}$.

Exercise 3.8.4 Formulate and prove the version of Proposition 3.8.2 when A is a left H^{cop}-module algebra and A is a right H-comodule algebra.

3.8.2 $\mathrm{Hom}(\mathrm{M}, \mathrm{N})$ and Quantum Yang–Baxter Submodules when H^{op} has an Antipode

Suppose that H is a bialgebra over the field k such that H^{op} has antipode ς. Suppose that M, N have the structure of left QYB H-modules, where M is finite-dimensional.

By the dual version of Corollary 3.5.1 we conclude that $M^* = M^r$ has the structure of a right QYB H-module. By part b) of Proposition 3.4.2 there is a left QYB H-module structure on M^* with $f = \varsigma$. Thus $\mathrm{Hom}(M, N) = M^* \otimes N$ has the structure of a left QYB H-module by part a) of Proposition 3.8.1. Note that if, in fact, H is a Hopf algebra with bijective antipode then Propositions 3.4.2 and 3.8.1 describe four formally different left QYB H-module structures for $\mathrm{Hom}(M, N)$.

3.9 Tensor Product of Parameterized Quantum Yang–Baxter Solutions

Let A be a bialgebra over k and let $R \in \mathrm{End}(M \otimes M)$ be a constant QYBE solution which is associated to a left QYB A-module M. As usual, we denote the module structure by $\mu : A \otimes M \longrightarrow M$ and comodule structure by $\rho : M \longrightarrow M \otimes A$. The following line observations led to the definition of a tensor product for one parameter QYBE solutions that was given in [Lambe, 1994].

Exercise 3.9.1 Assume the hypotheses of Exercise 3.8.2. Thus, the coordinates of the tensor product solution $\hat{R} = R \otimes \bar{R}$ are given by

$$\Delta_s^{t_1,t_2} \mu_{t_1,i_1}^{k_1} \bar{\mu}_{t_2,i_2}^{k_2} \rho_{j_1}^{l_1,r_1} \bar{\rho}_{j_2}^{l_2,r_2} m_{r_2,r_1}^{s}. \tag{3.12}$$

a) Prove that in coordinates, the bialgebra axiom Δ preserves products is equivalent to

$$\Delta_s^{t_1,t_2} m_{r_2,r_1} = \Delta_{r_2}^{a,b} \Delta_{r_1}^{c,d} m_{a,c}^{t_1} m_{b,d}^{t_2}. \tag{3.13}$$

b) Using the associativity of the module action and the coassociativity of the comodule action, prove that (3.12) is equivalent to

$$R_{i_1,\bar{\theta}}^{\theta',l_1} \mu_{a,\theta'}^{k_1} \bar{\rho}_{\theta}^{l_2,a} \bar{\mu}_{d,i_2}^{\theta''} \rho_{j_1}^{\bar{\theta},d} \bar{R}_{\theta'',j_2}^{k_2,\theta}. \tag{3.14}$$

c) Observe that if $\bar{R} = R$ then the tensor square of R is given by

$$R_{i_1,\bar{\theta}}^{\theta',l_1} R_{\theta',\theta}^{k_1,l_2} R_{i_2,j_1}^{\theta'',\bar{\theta}} R_{\theta'',j_2}^{k_2,\theta}. \tag{3.15}$$

If $R : X \longrightarrow \mathrm{End}(M \otimes M)$ is a one-parameter QYBE solution, then using (3.15) to define a function $\hat{R} : X \longrightarrow \mathrm{End}((\otimes^2 M) \otimes (\otimes^2 M))$, i.e.

$$\hat{R}_{(i_1,i_2),(j_1,j_2)}^{(l_1,l_2),(k_1,k_2)}(x) = R_{i_1,\bar{\theta}}^{\theta',l_1}(x) R_{\theta',\theta}^{k_1,l_2}(x) R_{i_2,j_1}^{\theta'',\bar{\theta}}(x) R_{\theta'',j_2}^{k_2,\theta}(x) \tag{3.16}$$

one obtains a new one parameter QYBE solution. This will be reviewed in Section 7.5.

Exercise 3.9.2 Prove directly that (3.16) satisfies the one-parameter QYBE if R does.

Exercise 3.9.3 Investigate what the conditions are for the tensor product of two parameterized linear operators to satisfy the two-parameter QYBE. Do the same for the tensor product of a constant operator and a parameterized operator in the one-parameter case.

3.10 Algebras of ${}_H\mathcal{QYB}$

Throughout this section H is a bialgebra over the field k.

Definition 3.10.1 *An* algebra of ${}_H\mathcal{QYB}$ *is a* 5*-tuple* (A, μ, ρ, m, η)*, where* (A, μ, ρ) *is an object of* ${}_H\mathcal{QYB}$*, and* (A, μ, m, η)*,* (A, ρ, m, η) *are algebras of* ${}_H$Mod, Comod$^{H^{op}}$ *respectively.*

Thus an algebra A of ${}_H\mathcal{QYB}$ can be thought of as an algebra A over k together with the structure of an object of ${}_H\mathcal{QYB}$ such that

$$h{\cdot}(ab) = (h_{(1)}{\cdot}a)(h_{(2)}{\cdot}b), \quad h{\cdot}1 = \epsilon(h)1,$$

$$\rho(ab) = a^{<1>}b^{<1>} \otimes b^{(2)}a^{(2)}, \quad \rho(1_A) = 1_A \otimes 1_H$$

hold for all $h \in H$ and $a, b \in A$.

Definition 3.10.2 *An algebra* A *of* ${}_H\mathcal{QYB}$ *is* commutative *if* $m = m\sigma_{A,A}$.

The map $\sigma_{M,N}$ is described in Exercise 3.1.3. This is what it means for algebras over k to be commutative when $\tau_{A,A}$ replaces $\sigma_{A,A}$. Thus an algebra of ${}_H\mathcal{QYB}$ is commutative if and only if

$$ab = b^{<1>}(b^{(2)}{\cdot}a)$$

for all $a, b \in A$.

Suppose that A and B are algebras of ${}_H\mathcal{QYB}$. We define a product $m_{A\otimes B}$ on the object $A \otimes B$ of ${}_H\mathcal{QYB}$ as we would for algebras over k, replacing the twist map $\tau_{A,B}$ with $\sigma_{A,B}$. Thus $m_{A\otimes B} = (m_A \otimes m_B)(1_A \otimes \sigma_{B,A} \otimes 1_B)$, or

$$(a \otimes b)(c \otimes d) = a(c^{<1>}) \otimes (c^{(2)}{\cdot}b)d$$

for all $a, b, c, d \in A$. With unit

$$1_{A\otimes B} = 1_A \otimes 1_B$$

and this rule for multiplication $A \otimes B$ is an algebra of ${}_H\mathcal{QYB}$. In particular $A \otimes B$ with this structure is an algebra over k. Observe that

$$A \xrightarrow{\iota_A} A \otimes B \quad \text{and} \quad B \xrightarrow{\iota_B} A \otimes B$$

defined by $\iota_A(a) = a \otimes 1$ and $\iota_B(b) = 1 \otimes b$ for all $a \in A$ and $b \in B$ are algebra morphisms.

Exercise 3.10.1 Let M be an object of ${}_H\mathcal{QYB}$. Determine whether or not the free algebra of ${}_H\mathcal{QYB}$ on M exists.

Exercise 3.10.2 Let A be an algebra of ${}_H\mathcal{QYB}$. Show that A^{op} is an algebra of ${}_H\mathcal{QYB}$, where the unit of A^{op} is that of A, and the product of A^{op} is given by $m^{op} = m\sigma_{A,A}$.

3.11 Coalgebras, Bialgebras, and Hopf Algebras of ${}_H\mathcal{QYB}$

Throughout this section H is a bialgebra over the field k.

Definition 3.11.1 *A coalgebra of ${}_H\mathcal{QYB}$ is a 5-tuple $(C, \mu, \rho, \Delta, \epsilon)$, where (C, μ, ρ) is an object of ${}_H\mathcal{QYB}$ and $(C, \mu, \Delta, \epsilon)$, $(C, \rho, \Delta, \epsilon)$ are coalgebras of ${}_H$Mod, Comod$^{H^{op}}$ respectively.*

Thus a coalgebra of ${}_H\mathcal{QYB}$ can be thought of as a coalgebra C over k such that

$$\Delta(h{\cdot}c) = h_{(1)}{\cdot}c_{(1)} \otimes h_{(2)}{\cdot}c_{(2)}, \quad \epsilon(h{\cdot}c) = \epsilon(h)\epsilon(c),$$

$$\Delta(c^{<1>}) \otimes c^{(2)} = ({c_{(1)}}^{<1>} \otimes {c_{(2)}}^{<1>}) \otimes {c_{(2)}}^{(2)}{c_{(1)}}^{(2)}, \quad \epsilon(c^{<1>})c^{(2)} = \epsilon(c)1_H$$

hold for all $h \in H$ and $c \in C$.

Definition 3.11.2 *A coalgebra C of ${}_H\mathcal{QYB}$ is* cocommutative *if $\Delta = \sigma_{C,C}\Delta$, or equivalently*

$$c_{(1)} \otimes c_{(2)} = {c_{(2)}}^{<1>} \otimes {c_{(2)}}^{(2)}{\cdot}c_{(1)}$$

for all $c \in C$.

Suppose that C and D are coalgebras of ${}_H\mathcal{QYB}$. Then we define a coproduct $\Delta = \Delta_{C\otimes D} = (1_C \otimes \sigma_{C,D} \otimes 1_D)(\Delta_C \otimes \Delta_D)$ and counit map $\epsilon = \epsilon_{C\otimes D} = \epsilon_C \otimes \epsilon_D$ on the object $C \otimes D$ of ${}_H\mathcal{QYB}$. Thus

$$\Delta(c \otimes d) = (c_{(1)} \otimes {d_{(1)}}^{<1>}) \otimes (({d_{(1)}}^{(2)}{\cdot}c_{(2)}) \otimes d_{(2)})$$

and

$$\epsilon(c \otimes d) = \epsilon(c)\epsilon(d)$$

for all $c \in C$ and $d \in D$. With these structures the object $C \otimes D$ is a coalgebra of ${}_H\mathcal{QYB}$. In particular $C \otimes D$ is a coalgebra over k. Notice that the maps

$$C \otimes D \xrightarrow{\pi_C} C \quad \text{and} \quad C \otimes D \xrightarrow{\pi_D} D$$

defined by $\pi_C(c \otimes d) = c\epsilon(d)$ and $\pi_D(c \otimes d) = \epsilon(c)d$ for all $c \in C$ and $d \in D$ are coalgebra morphisms.

Suppose that A is both an algebra and a coalgebra of ${}_H\mathcal{QYB}$. Then the coalgebra structure morphisms Δ and ϵ are algebra morphisms if and only the algebra structure morphisms m and η are coalgebra morphisms.

Definition 3.11.3 *A bialgebra of ${}_H\mathcal{QYB}$ is a 7-tuple $(A, \mu, \rho, m, \eta, \Delta, \epsilon)$, where (A, μ, ρ, m, η) is an algebra of ${}_H\mathcal{QYB}$ and $(A, \mu, \rho, \Delta, \epsilon)$ is a coalgebra of ${}_H\mathcal{QYB}$, such that Δ and ϵ are algebra morphisms, or equivalently m and η are coalgebra morphisms.*

Note that ϵ is an algebra morphism if and only if

$$\epsilon(ab) = \epsilon(a)\epsilon(b) \quad \text{and} \quad \epsilon(1) = 1,$$

and Δ is an algebra morphism if and only if

$$\Delta(ab) = a_{(1)}b_{(1)}{}^{<1>} \otimes (b_{(1)}{}^{(2)} \cdot a_{(2)})b_{(2)} \quad \text{and} \quad \Delta(1) = 1 \otimes 1$$

for all $a, b \in A$. It is convenient to denote objects of ${}_H$QYB by their underlying vector spaces.

Definition 3.11.4 *A bialgebra A of ${}_H\mathcal{QYB}$ is a* Hopf algebra *of ${}_H\mathcal{QYB}$ if there is a morphism $s : A \longrightarrow A$ such that*

$$a_{(1)}s(a_{(2)}) = \epsilon(a)1 = a_{(1)}s(a_{(2)})$$

for all $a \in A$.

Note that if A is a bialgebra (respectively Hopf algebra) of ${}_H\mathcal{QYB}$ and A has the trivial left H-module and right H-comodule structures then A is a bialgebra (respectively Hopf algebra) over k.

Exercise 3.11.1 Let C be a coalgebra of ${}_H\mathcal{QYB}$. Determine whether or not the free bialgebra of ${}_H\mathcal{QYB}$ on C exists.

Exercise 3.11.2 Let C be a coalgebra of ${}_H\mathcal{QYB}$. Show that C^{cop} is a coalgebra of ${}_H\mathcal{QYB}$, where the counit of C^{cop} is that of C and the coproduct of C^{cop} is given by $\Delta^{cop} = \sigma_{C,C}\Delta$.

3.12 Smash Biproducts Associated to the Category ${}_H{}^H\mathcal{QYB}$

Suppose that H is a bialgebra over k. Replacing H by H^{op}, H^{cop} and $H^{op\ cop}$ give important variations of the compatibility condition (2.17). We will consider the variation arising from H^{cop} in this section and point out its implications for smash and cosmash products.

Suppose that M is a vector space over k. Then there is a one-one correspondence between the left H-comodules and right H^{cop}-comodules given by $(M, \rho) \longmapsto (M, \rho^{cop})$, where $\rho^{cop} = \tau_{H,M}\rho$. Let ${}_H{}^H\mathcal{QYB}$ be the category whose objects are triples (M, μ, ρ), where (M, μ, ρ^{cop}) is an object of ${}_H\mathcal{QYB}$, and whose morphisms are maps which are simultaneously left H-module and left H-comodule maps. Let M be an object of ${}_H{}^H\mathcal{QYB}$. The compatibility condition (2.17) translates to

$$h_{(1)}m^{(1)} \otimes h_{(2)}\cdot m^{<2>} = (h_{(1)}\cdot m)^{(1)}h_{(2)} \otimes (h_{(1)}\cdot m)^{<2>}$$

for all $h \in H$ and $m \in M$. We modify the maps $\sigma_{M,N}$ for the category ${}_H{}^H\mathcal{QYB}$ to obtain suitable "twist" morphisms for ${}_H{}^H\mathcal{QYB}$. For objects M and N of ${}_H{}^H\mathcal{QYB}$ let $\varsigma_{M,N} = \tau_{M,N}\sigma_{N,M}\tau_{M,N}$. Then $\varsigma_{M,N}$ is a morphism of ${}_H{}^H\mathcal{QYB}$ which is given by

$$\varsigma_{M,N}(m \otimes n) = m^{(1)}\cdot n \otimes m^{<2>}$$

for $m \in M$ and $n \in N$. See [Yetter, 1990].

Observe that k is an object of ${}_H{}^H\mathcal{QYB}$ with the trivial structures. If M and N are objects of ${}_H{}^H\mathcal{QYB}$ then the tensor product $M \otimes N$ over k is an object of ${}_H{}^H\mathcal{QYB}$ where

$$h\cdot(m \otimes n) = h_{(1)}\cdot m \otimes h_{(2)}\cdot n$$

and

$$\rho(m \otimes n) = m^{(1)}n^{(1)} \otimes (m^{<2>} \otimes n^{<2>})$$

for $m \in M$ and $n \in N$. Notice that the tensor product of linear maps $f \otimes g : M \otimes N \longrightarrow M' \otimes N'$ is a morphism of ${}_H{}^H\mathcal{QYB}$ whenever $f : M \longrightarrow M'$ and $g : N \longrightarrow N'$ are morphisms of ${}_H{}^H\mathcal{QYB}$. The linear maps of (1.21) and (1.22) are morphisms of ${}_H\mathcal{QYB}$.

We define algebras, coalgebras, bialgebras, and Hopf algebras of ${}_H{}^H\mathcal{QYB}$ in the same way as we do for ${}_H\mathcal{QYB}$.

Definition 3.12.1 *Suppose that A is an algebra of ${}_H{}^H\mathcal{QYB}$. Then A is a* commutative *algebra of ${}_H{}^H\mathcal{QYB}$ if $m = m\varsigma_{A,A}$, or equivalently*

$$ab = (b^{(1)}\cdot a)b^{<2>}$$

for all $a, b \in A$.

Recall that the smash product $A \otimes H$ over k of Exercise 1.9.6 is the algebra over k described by

$$1_{A\otimes H} = 1_A \otimes 1_H \quad \text{and} \quad (a \otimes h)(a' \otimes h') = a(h_{(1)}\cdot a') \otimes h_{(2)}h'$$

for $a, b \in A$ and $h, k \in H$.

Now let B also be an algebra of ${}_H{}^H\mathcal{QYB}$. We define a product on $A \otimes B$ by $m_{A\otimes B} = (m_A \otimes m_B)(1_A \otimes \varsigma_{B,A} \otimes 1_B)$, or equivalently

$$(a \otimes b)(c \otimes d) = a(b^{(1)}\cdot c) \otimes b^{<2>}d$$

for $a, c \in A$ and $b, d \in B$. It is a straightforward exercise to see that $A \otimes B$ is an algebra of ${}_H{}^H\mathcal{QYB}$ with this product with unity $1_A \otimes 1_B$.

Definition 3.12.2 *Let C be a coalgebra of ${}_H{}^H\mathcal{QYB}$. Then C is a* cocommutative coalgebra *of ${}_H{}^H\mathcal{QYB}$ if $\Delta = \varsigma_{C,C}\Delta$, or equivalently*

$$c_{(1)} \otimes c_{(2)} = ({c_{(1)}}^{(1)}\cdot c_{(2)}) \otimes {c_{(1)}}^{<2>}$$

for $c \in C$.

The cosmash product of $C \otimes H$ of Exercise 1.9.13 is defined in this context by

$$\epsilon(c \otimes h) = \epsilon(c)\epsilon(h)$$

and

$$\Delta(c \otimes h) = (c_{(1)} \otimes {c_{(2)}}^{(1)}h_{(1)}) \otimes ({c_{(2)}}^{<2>} \otimes h_{(2)})$$

for all $c \in C$ and $h \in H$.

Now suppose that A is both an algebra and a coalgebra of ${}_H{}^H\mathcal{QYB}$, and thus is an algebra and a coalgebra over k. Then the coalgebra structure morphisms Δ and ϵ are algebra morphisms if and only if the algebra structure morphisms m and η are coalgebra morphisms. Thus a bialgebra of ${}_H{}^H\mathcal{QYB}$ is a 7-tuple $(A, \mu, \rho, m, \eta, \Delta, \epsilon)$, where (A, μ, ρ, m, η) is an algebra of ${}_H{}^H\mathcal{QYB}$ and $(A, \mu, \rho, \Delta, \epsilon)$ is a coalgebra of ${}_H{}^H\mathcal{QYB}$ such that Δ and ϵ are algebra morphisms (or equivalently m and η are coalgebra morphisms). In particular we require

$$\Delta(ab) = a_{(1)}({a_{(2)}}^{(1)}\cdot b_{(1)}) \otimes {a_{(2)}}^{<2>}b_{(2)} \quad \text{and} \quad \Delta(1) = 1 \otimes 1$$

for all $a, b \in A$ in case A is a bialgebra of ${}_H{}^H\mathcal{QYB}$. A rather lengthy calculation shows that the vector space $A \otimes H$ with the smash product algebra structure and cosmash product coalgebra structure is a bialgebra over k. Observe that

$$H \xrightarrow{\iota} A \otimes H \quad \text{and} \quad A \otimes H \xrightarrow{\pi} H$$

defined by $\iota(h) = 1 \otimes h$ and $\pi(a \otimes h) = \epsilon(h)a$ for all $h \in H$ and $a \in A$ are bialgebra maps which satisfy $\pi\iota = 1_H$. Identifying H with $\mathrm{Im}\iota$ we see that π is a projection onto the sub-bialgebra H of $H \otimes A$. When H is a Hopf algebra, all bialgebras which contain H as a sub-bialgebra and have a projection onto it have the structure of a smash product and cosmash product as above [Radford, 1985]. Also see this reference in connection with the following exercises.

Exercise 3.12.1 Suppose that $H = k[G]$ is the group algebra of a finite group over the field k. Show that all bialgebras B of ${}_H{}^H\mathcal{QYB}$ are described as follows: B is an algebra and a coalgebra over k with a representation $\pi : G \longrightarrow \mathrm{End}(B)$ and a subspace decomposition $B = \oplus_{g\in G} B_g$ such that:

a) $\pi(g)$ is an algebra and a coalgebra automorphism of B for all $g \in G$,

b) $1 \in B_1$ and $B_g B_h \subseteq B_{gh}$ for all $g, h \in G$,

c) $\epsilon(B_g) = (0)$ unless $g = 1$ and $\Delta(B_g) \subseteq \sum_{\ell h = g} B_\ell \otimes B_h$, and

d) $h{\cdot}B_g \subseteq B_{hgh^{-1}}$ for all $g, h \in G$.

Exercise 3.12.2 Let G be a cyclic group of order 2 or 3 and $H = k[G]$. Find all the bialgebras of ${}_H{}^H\mathcal{QYB}$ of dimension at most 4. Compute the corresponding smash biproducts.

Exercise 3.12.3 Let $H = k[G]$, where G is a group of order 4, and $B = k[\mathcal{G}]$ be the group algebra of the cyclic group $\mathcal{G}$ of order 3.

a) Find all ways that B can be realized as an object of ${}_H{}^H\mathcal{QYB}$ in such a manner that B with its k-coalgebra structure is in fact a bialgebra of ${}_H{}^H\mathcal{QYB}$.

b) Work out the coalgebra structure of the 12-dimensional smash biproducts arising from part a).

4 MORE ON THE BIALGEBRA ASSOCIATED TO THE QUANTUM YANG–BAXTER EQUATION

Many ideas in the subject of quantum groups are motivated by the material presented in Chapter 2. Now that the reader has had the opportunity to become familiar with that material, we will re-examine and extend it more formally.

4.1 Module–Comodule Compatibility Revisited

Let A be a bialgebra over the field k and consider a triple (M, μ, ρ), where (M, μ) is a left A-module and (M, ρ) is a right A-comodule. We further assume that M is finite-dimensional. In this section we will closely re-examine the connection between when $R_{(\mu,\rho)} : M \otimes M \longrightarrow M \otimes M$ defined by (2.12)

$$R_{(\mu,\rho)}(m \otimes n) = m^{<1>} \otimes m^{(2)}{\cdot}n$$

for all $m, n \in M$ is a solution to the QYBE and the compatibility condition (2.17)

$$a_{(1)}{\cdot}m^{<1>} \otimes a_{(2)}m^{(2)} = (a_{(2)}{\cdot}m)^{<1>} \otimes (a_{(2)}{\cdot}m)^{(2)}a_{(1)}$$

holds for all $a \in A$ and $m \in M$. Our calculations will lead to a basis free description of the FRT construction.

Choose a basis $\mathcal{B} = \{m_1, \ldots, m_r\}$ for M. For $1 \le j \le r$ write

$$\rho(m_j) = m_i \otimes t^i_j \tag{4.1}$$

where $t^i_j \in A$. Then the set of t^i_j's satisfies the comatrix identities (1.2) and (1.3). Recall there is unique minimal subspace $A(\rho)$ of A such that $\rho(M) \subseteq M \otimes A(\rho)$. In this case $A(\rho)$ is the subcoalgebra of A spanned by the t^i_j's. See Proposition 1.4.3 for details.

Now let $\mathcal{R} = \{R^{k,\ell}_{i,j}\}_{1 \le i,j,k,\ell \le r} \subseteq k$ and define a linear transformation $R : M \otimes M \longrightarrow M \otimes M$ by

$$R(m_i \otimes m_j) = R^{k,\ell}_{i,j} m_k \otimes m_\ell. \tag{4.2}$$

Since $R_{(\mu,\rho)}(m_i \otimes m_j) = m_k \otimes t^k_i{\cdot}m_j$ it follows that

$$R = R_{(\mu,\rho)} \quad \text{if and only if} \quad t^k_i{\cdot}m_j = R^{k,\ell}_{i,j} m_\ell$$

for all $1 \le i, j, k \le r$.

Now assume that $R = R_{(\mu,\rho)}$. The reader can check that

$$R_{(1,2)}R_{(1,3)}R_{(2,3)}(m_i \otimes m_j \otimes m_k) = m_s \otimes m_t \otimes (R^{s,t}_{p,q} t^p_i t^q_j){\cdot}m_k$$

and

$$R_{(2,3)}R_{(1,3)}R_{(1,2)}(m_i \otimes m_j \otimes m_k) = m_s \otimes m_t \otimes (R^{p,q}_{i,j} t^t_q t^s_p){\cdot}m_k$$

for $1 \le i, j, k, \le r$. For $\mathcal{E} = \{t^i_j\}_{1 \le i,j \le r} \subseteq A$ which satisfies the comatrix identities we define

$$d_{\mathcal{E},\mathcal{R}}(i,j,s,t) = R^{s,t}_{p,q} t^p_i t^q_j - R^{p,q}_{i,j} t^t_q t^s_p \tag{4.3}$$

for $1 \le i, j, s, t \le r$. When $\mathcal{E}$ is determined by a comodule basis as in (4.1) we will frequently write $d_{\mathcal{E},\mathcal{R}} = d_{\rho,\mathcal{B},\mathcal{R}}$. We deduce from the two equations preceding (4.3) that

$$(R_{(1,2)}R_{(1,3)}R_{(2,3)} - R_{(2,3)}R_{(1,3)}R_{(1,2)})(m_i \otimes m_j \otimes m_k)$$
$$= m_s \otimes m_t \otimes d_{\rho,\mathcal{B},\mathcal{R}}(i,j,s,t){\cdot}m_k$$

for $1 \le i, j, k \le r$. Let $a = t^s_i$ and $m = m_j$. Then

$$\begin{aligned} a_{(1)}{\cdot}m^{<1>} \otimes a_{(2)}m^{(2)} &= t^s_p{\cdot}m_q \otimes t^p_i t^q_j \\ &= m_t \otimes R^{s,t}_{p,q} t^p_i t^q_j \end{aligned}$$

and

$$\begin{aligned}(a_{(2)}\cdot m)^{<1>} \otimes (a_{(2)}\cdot m)^{(2)}a_{(1)} &= (t^p_i\cdot m_j)^{<1>} \otimes (t^p_i\cdot m_j)^{(2)}t^s_p \\ &= R^{p,q}_{i,j}m^{<1>}_q \otimes m^{(2)}_q t^s_p \\ &= m_t \otimes R^{p,q}_{i,j}t^t_q t^s_p.\end{aligned}$$

Therefore

$$a_{(1)}\cdot m^{<1>} \otimes a_{(2)}m^{(2)} - (a_{(2)}\cdot m)^{<1>} \otimes (a_{(2)}\cdot m)^{(2)}a_{(1)}$$
$$= \sum_{t=1}^{r} m_t \otimes d_{\rho,\mathcal{B},\mathcal{R}}(i,j,s,t).$$

The $d_{\rho,\mathcal{B},\mathcal{R}}(i,j,s,t)$'s behave well with respect to the coalgebra structure of A. For assume that $\mathcal{E} = \{t^i_j\}_{1\leq i,j\leq r} \subseteq A$ satisfies the comatrix identities. Then

$$\epsilon(d_{\mathcal{E},\mathcal{R}}(i,j,s,t)) = 0 \tag{4.4}$$

and

$$\Delta(d_{\mathcal{E},\mathcal{R}}(i,j,s,t)) = d_{\mathcal{E},\mathcal{R}}(u,v,s,t) \otimes t^u_i t^v_j + t^t_u t^s_v \otimes d(i,j,v,u) \tag{4.5}$$

for all $1 \leq i,j,s,t \leq r$ with no restrictions on $\mathcal{R}$ in fact. Observe that the set of all $a \in A$ such that compatibility (2.17) holds for all $m \in M$ is a subalgebra of A. To summarize the highlights:

Proposition 4.1.1 *Let A be a bialgebra over the field k and M be a finite-dimensional vector space over k with a left A-module structure (M,μ) and a right A-comodule structure (M,ρ). Suppose that $\mathcal{B} = \{m_1,\ldots,m_r\}$ is a basis for M and $t^i_j \in A$ for $1 \leq i,j \leq r$ is determined by $\rho(m_j) = m_i \otimes t^i_j$. Let $\mathcal{R} = \{R^{k,\ell}_{i,j}\}_{1\leq i,j,k,\ell\leq r} \subseteq k$. Then:*

a) $R_{(\mu,\rho)}(m_i\otimes m_j) = R^{k,\ell}_{i,j}m_k\otimes m_\ell$ *for all* $1 \leq i,j \leq r$ *if and only if* $t^k_i\cdot m_j = R^{k,\ell}_{i,j}m_\ell$ *for all* $1 \leq i,j,k \leq r$.

Assume that $R_{(\mu,\rho)}(m_i \otimes m_j) = R^{k,\ell}_{i,j}m_k \otimes m_\ell$ for all $1 \leq i,j \leq r$ and let $\mathcal{I}$ be the span of the $d_{\rho,\mathcal{B},\mathcal{R}}(i,j,s,t)$'s.

b) *Then $\mathcal{I}$ is a coideal of A.*

c) *$R_{(\mu,\rho)}$ satisfies the QYBE if and only if $\mathcal{I} \subseteq \operatorname{ann}_A(M)$.*

d) *If (M,μ,ρ) is a left QYB A-module then $\mathcal{I} = (0)$.*

e) *Suppose that A is generated as an algebra by $A(\rho)$. If $\mathcal{I} = (0)$ then (M,μ,ρ) is a left QYB A-module.*

∎

We note that Proposition 4.1.1 is a minor variation of [Radford, 1993c, Proposition 3].

Parts c) – e) of the proposition suggest the notion of M-reduced bialgebra.

Definition 4.1.1 *Suppose that A is any bialgebra over the field k and M is a left A-module. Then A is M*-reduced *if* (0) *is the only coideal of A contained in* $\mathrm{ann}_A(M)$.

The following corollary is [Radford, 1994b, Proposition 2] for $R^{\tau}_{(\mu,\rho)}$.

Corollary 4.1.1 *Suppose that A is a bialgebra over the field k and M is a finite-dimensional vector space over k with a left A-module structure (M,μ) and a right A-comodule structure (M,ρ). Assume further that A is M-reduced and that A is generated as an algebra by $A(\rho)$. Then the following are equivalent:*

a) *$R_{(\mu,\rho)}$ is a solution to the QYBE.*

b) *(M,μ,ρ) is a left QYB A-module.*

∎

Let A be any bialgebra over the field k and suppose that M is a left A-module. Let I be the sum of all coideals contained in $\mathrm{ann}_A(M)$. Then I is a coideal of A. Since the ideal of A generated by a coideal of A is a coideal of A, and since $\mathrm{ann}_A(M)$ is an ideal of A, it follows that I is an ideal of A. Therefore I is a bi-ideal of A. For a description of I in terms of the wedge product see part b) of Corollary 1.2.2.

Let $\widetilde{A} = A/I$ and $\pi : A \longrightarrow \widetilde{A}$ be the projection. Since I is an ideal of A and $I \subseteq \mathrm{ann}_A(M)$ it follows that M has a (unique) left $\widetilde{A}$-module structure such that $\pi(a)\cdot m = a\cdot m$ for all $a \in A$ and $m \in M$. Since π is a coalgebra map it follows that $\widetilde{A}$ is M-reduced. See Exercise 4.1.2.

Now suppose that M is a finite-dimensional vector space over k and $R : M\otimes M \longrightarrow M \otimes M$ is a solution to the QYBE. Let A be a bialgebra over k and suppose that $R = R_{(\mu,\rho)}$ where (M,μ) is a left A-module and (M,ρ) is right A-comodule. We can replace A by $\widetilde{B}$, where B is a certain sub-bialgebra of A, replace (M,μ) by a left $\widetilde{B}$-module $(M,\widetilde{\mu})$, and replace (M,ρ) by a right $\widetilde{B}$-comodule $(M,\widetilde{\rho})$ so that $R = R_{(\widetilde{\mu},\widetilde{\rho})}$ and $(M,\widetilde{\mu},\widetilde{\rho})$ is a left QYB $\widetilde{B}$-module as we now explain.

Let B be the subalgebra of A generated by $A(\rho)$. Since $A(\rho)$ is a subcoalgebra of A it follows that B is a sub-bialgebra of A. Regard M as a left B-module by restriction of A-module action to B. Since $\rho(M) \subseteq M \otimes B$ we may think of (M,ρ) as a right B-comodule. Notice that $R = R_{(\mu,\rho)}$ with these modified actions.

Let $\pi : B \longrightarrow \widetilde{B}$ be the projection. Let $(M,\widetilde{\mu})$ denote the left $\widetilde{A}$-module structure on M defined by $\pi(a)\cdot m = a\cdot m$ for $a \in A$. Since π is a coalgebra map $(M,\widetilde{\rho})$ is a right $\widetilde{B}$-comodule where $\widetilde{\rho} = (1_M \otimes \pi)\rho$. It is easy to see that $R = R_{(\mu,\rho)} = R_{(\widetilde{\mu},\widetilde{\rho})}$. By Corollary 4.1.1 it follows that $(M, \widetilde{\mu}, \widetilde{\rho})$ is a left QYB $\widetilde{B}$-module.

At this point we recall the computer calculation described in Section 2.11.2. We recall the parameterized class of solutions with matrix

$$\begin{pmatrix} 1 & 0 & 0 & 0 & 0 & 0 & 0 & 0 & 0 \\ 0 & 1 & 0 & 0 & 0 & 0 & 0 & 0 & 0 \\ 0 & 0 & 1 & 0 & 0 & 0 & 0 & 0 & 0 \\ 0 & 0 & 0 & -e & 0 & 0 & -f & 0 & 0 \\ 0 & 0 & 0 & 0 & e & 0 & 0 & f & 0 \\ 0 & 0 & 0 & 0 & 0 & e & 0 & 0 & f \\ 0 & 0 & 0 & -g & 0 & 0 & e-1 & 0 & 0 \\ 0 & 0 & 0 & 0 & g & 0 & 0 & 1-e & 0 \\ 0 & 0 & 0 & 0 & 0 & g & 0 & 0 & 1-e \end{pmatrix}$$

where e, f, and g are arbitrary constants. Let $R = R(e, f, g)$ denote the corresponding QYBE solution and let $\{t_1^1, t_2^1, \ldots, t_3^3\}$ be the algebra generators of $A(R)$ as usual. Also let $M = \mathbb{C}^3$ and give M the usual ${}_{A(R)}\mathcal{QYB}$ structure.

There are quite a few relations (which we generated using computer algebra) for $A(R)$. Let I denote the maximal coideal contained in the $A(R)$-annihilator of M.

It is straightforward to calculate that t_i^j, for $i \neq j$, acts on M trivially as does $t_2^2 - t_3^3$. Note that if

$$Y = \{t_i^j \mid i \neq j\}$$

then the subspace of $A(R)$ spanned by Y is a coideal. But note that

$$\Delta(t_2^2 - t_3^3) = t_1^2 \otimes t_2^1 + t_2^2 \otimes t_2^2 + t_2^3 \otimes t_2^3 - t_1^3 \otimes t_3^1 - t_2^3 \otimes t_3^2 - t_3^3 \otimes t_3^3$$

and

$$t_2^2 \otimes t_2^2 - t_3^3 \otimes t_3^3 = (t_2^2 - t_3^3) \otimes t_2^2 + t_3^3 \otimes (t_3^3 - t_2^2).$$

Therefore the subspace of $A(R)$ spanned by the union $Z = Y \cup \{t_2^2 - t_3^3 \otimes t_3^3\}$ is a coideal of $A(R)$. Since it is contained in the annihilator of M, it lies in I.

We row reduced the FRT-relations generated above and found the new relation

$$-t_3^3 t_1^1 - t_3^2 t_1^1 + t_1^1 t_2^2.$$

Let $\pi : A(R) \longrightarrow \widetilde{A(R)} = A(R)/I$ be the projection. The above calculations show that $\pi(t_i^j) = 0$ for $i \neq j$ and therefore $\pi(t_i^i)$ is grouplike. Furthermore, we have that $\pi(t_2^2) = \pi(t_3^3)$ and $xy = yx$ where $x = \pi(t_1^1)$ and $y = \pi(t_2^2)$.

Examining the full set of (row reduced) relations, we have seen that there are no further reductions. Therefore $\widetilde{A}(R)$ is isomorphic to the monoid ring

$$k{<}x, y{>}/(xy - yx)$$

of the free commutative monoid on two variables x and y. See Section 8.2.

Throughout the following exercises A and B are bialgebras over the field k.

Exercise 4.1.1 Suppose that H is a commutative cocommutative Hopf algebra over the field k and suppose that M has a left H-module structure (M, μ) and a right H-comodule structure (M, ρ). Show that (M, μ, ρ) is a left QYB H-module if and only if

$$\rho(a{\cdot}m) = a{\cdot}m^{<1>} \otimes m^{(2)}$$

for all $a \in H$ and $m \in M$. [Hint: By Lemma 3.5.1 (M, μ, ρ) is a left QYB H-module if and only if $\rho(a{\cdot}m) = a_{(2)}{\cdot}m^{<1>} \otimes a_{(3)}m^{(2)}\varsigma(a_{(1)})$ for all $a \in H$ and $m \in M$ when H^{op} is a Hopf algebra with antipode ς.]

Exercise 4.1.2 Suppose that M is a left A-module. Show that $\widetilde{A}$ is M-reduced. [Hint: The preimage of a coideal under a coalgebra map is a coideal.]

Exercise 4.1.3 Let $f : A \longrightarrow B$ be a bialgebra map which is onto and consider a triple (M, μ, ρ), where (M, μ) is a left A-module, (M, ρ) is a right A-comodule and M is finite-dimensional. Suppose that $\ker f \subseteq \mathrm{ann}_A(M)$. Let (M, μ_f) be the left B-module structure defined by $f(a){\cdot}m = a{\cdot}m$ for all $a \in A$ and $m \in M$, and let (M, ρ_f) be the right B-comodule structure defined by $\rho_f(m) = m^{<1>} \otimes f(m^{(2)})$ for all $m \in M$.

a) Show that $R_{(\mu,\rho)} = R_{(\mu_f,\rho_f)}$

b) Set $R = R_{(\mu,\rho)}$, choose a basis $\mathcal{B} = \{m_1, \ldots, m_r\}$ for M, and write $R(m_i \otimes m_j) = R^{k,\ell}_{i,j} m_k \otimes m_\ell$ for $1 \leq i, j \leq r$. Show that

$$f(d_{\rho,\mathcal{B},\mathcal{R}}(i, j, s, t)) = d_{\rho_f,\mathcal{B},\mathcal{R}}(i, j, s, t)$$

for all $1 \leq i, j, s, t \leq r$.

Exercise 4.1.4 Let $A = k[X]$ be the algebra of polynomials in indeterminant X over k with bialgebra structure determined by $\Delta(X) = X \otimes 1 + 1 \otimes X$. Assume that the characteristic of k is 0.

a) Show that the association

$$(M, S, T) \longmapsto (M, \mu_S, \rho_T),$$

where

i) M is a finite-dimensional vector space over k,

ii) $S, T \in \mathrm{End}(M)$ are commuting operators, T is nilpotent,

iii) (M, μ_S) is the left A-module structure defined on M determined by $X{\cdot}m = S(m)$ for all $m \in M$, and

iv) (M, ρ_T) is the right A-comodule structure defined by

$$\rho_T(m) = \sum_{\ell=0}^{\mathrm{Dim}M} T^\ell(m) \otimes \frac{X^\ell}{\ell!}$$

for all $m \in M$,

is a bijective correspondence between the set of triples satisfying i) – iv) and the set of left QYB $k[X]$-modules. [Hint: $k[X]$ can be identified with a dense subalgebra of A^*. See Exercises 1.4.8 and 1.4.14.]

b) Let (M, μ_S, ρ_T) satisfy i) – iv) above and set $R = R_{(\mu_S, \rho_T)}$. Show that

$$R(m \otimes n) = \sum_{\ell=0}^{\mathrm{Dim} M} \frac{1}{\ell!} T^\ell(m) \otimes S^\ell(n)$$

for $m, n \in M$.

Exercise 4.1.5 Generalize Exercise 4.1.4 to $A = k[X_1, \ldots, X_n]$, where $\Delta(X_i) = X_i \otimes 1 + 1 \otimes X_i$ for all $1 \leq i \leq n$. [Hint: The operators S and T are replaced by commuting families of operators.]

Exercise 4.1.6 Suppose that M is a left A-module and that A is M-reduced. Let $g, g' \in G(A)$ and $\pi : A \longrightarrow \mathrm{End}(M)$ be the representation of A afforded by M. Show that:

a) $\pi(g) = \pi(g')$ implies $g = g'$.

b) If $x \in P_{g,g'}(A)$ then $\pi(x) = 0$ implies $x = 0$.

Exercise 4.1.7 Let $H = k[X]$ be the Hopf algebra of Exercise 4.1.4. Let M be a left H-module and $\pi : H \longrightarrow \mathrm{End}(M)$ be the representation afforded by M. Show that H is M-reduced if and only if $\pi(X) \neq 0$.

Exercise 4.1.8 Let M be a finite-dimensional left A-module and let (M, ρ) be the (unique) right A^o-comodule structure on M such that $a{\cdot}m = m^{<1>^o} <m^{(2)^o}, a>$ for all $a \in A$ and $m \in M$. See Exercise 1.4.9

a) Show that the following are equivalent:

  i) A is M-reduced.

  ii) $A^o(\rho)$ generates a dense subalgebra of A^*.

b) Suppose that A is finite-dimensional. Show that A is M-reduced if and only if $A^*(\rho)$ generates A^* as an algebra.

Exercise 4.1.9 Reconsider Proposition 4.1.1.

a) Investigate possible generalizations to the one-parameter case.

b) Do the same for the two-parameter case.

4.2 A Basis-Free Description of the FRT Construction

Let M be a finite-dimensional vector space over the field k and suppose that $R : M \otimes M \longrightarrow M \otimes M$ is a solution to the QYBE. Choose a basis $B = \{m_1, \ldots, m_r\}$ for M and write $R(m_i \otimes m_j) = R^{k,\ell}_{i,j} m_k \otimes m_\ell$ where $\mathcal{R} = \{R^{k,\ell}_{i,j}\}$ is the set of B-coordinates of R. Now let $C = C_r(k)$ and define a right C-comodule structure on (M, ρ) on M by the rule $\rho(m_j) = m_i \otimes t^i_j$ for $1 \leq j \leq r$, where $\{t^i_j\}_{1 \leq i,j \leq r}$ is a standard basis for C.

Let $(\jmath, T(C))$ be the free bialgebra on the coalgebra C, which we have noted is the free algebra on the vector space C when the coalgebra structure is ignored. Now $\jmath$ is a one-one coalgebra map. To simplify the subsequent discussion we will identify C with $\jmath(C)$ and thus assume that C is a subcoalgebra of $T(C)$. In particular we may regard (M, ρ) as a right $T(C)$-comodule.

We have observed in Section 2.7.1 that M has a left $T(C)$-module structure (M, μ) determined by $t^j_i {\cdot} m_k = R^{k,\ell}_{i,j} m_\ell$. Let $\Pi : T(C) \longrightarrow \mathrm{End}(M)$ be the corresponding representation. We also observed that $R = R_{(\mu,\rho)}$.

Let $\mathcal{I}$ be the span of the $d_{\rho,B,\mathcal{R}}(i,j,s,t)$'s where $1 \leq i,j,s,t \leq r$. Then $\mathcal{I}$ is a coideal of $T(C)$ which satisfies $\mathcal{I} \subseteq \mathrm{ann}_{T(C)}(M)$ by parts b) and c) of Proposition 4.1.1. Note that the $I = \mathrm{Ker}\Pi$ defined in Section 2.7.1 is clearly the ideal generated by $\mathcal{I}$ which proves again that I is a bi-ideal of $T(C)$. Furthermore it is clear that $I \subseteq \mathrm{ann}_{T(C)}(M)$. The projection $\pi : T(C) \longrightarrow A(R) = T(C)/I$ is a bialgebra map which satisfies $\mathrm{Ker}\pi \subseteq \mathrm{ann}_{T(C)}(M)$. Consider the right $A(R)$-comodule structure (M, ρ_π) defined by $\rho_\pi = (1_M \otimes \pi)\rho$ and the left $A(R)$-module structure (M, μ_π) defined by $\mu_\pi(\pi \otimes 1_M) = \mu$. Then $R_{(\mu,\rho)} = R_{(\mu_\pi,\rho_\pi)}$ by part a) of Exercise 4.1.3. Using part b) of the same we calculate

$$d_{\rho_\pi,B,\mathcal{R}}(i,j,s,t) = \pi(d_{\rho,B,\mathcal{R}}(i,j,s,t)) \in \pi(\mathcal{I}) = (0).$$

Therefore (M, μ_π, ρ_π) is a left QYB $A(R)$-module by part e) of Proposition 4.1.1. We have shown part a) of the following, which is [Radford, 1993c, Theorem 2] for R^τ.

Theorem 4.2.1 *Suppose that M is a finite-dimensional vector space over the field k and that $R : M \otimes M \longrightarrow M \otimes M$ is a solution to the QYBE. Then the bialgebra $A(R)$ satisfies the following properties:*

a) *There exists a left QYB $A(R)$-module structure (M, μ, ρ) on M such that $R = R_{(\mu,\rho)}$.*

b) *Suppose that A is a bialgebra over the field k and that (M, μ', ρ') is a left QYB A-module structure on M such that $R = R_{(\mu',\rho')}$. Then there is a bialgebra map $F : A(R) \longrightarrow A$ uniquely defined by $(1_M \otimes F)\rho = \rho'$. Furthermore $\mu = \mu'(F \otimes 1_M)$*

Proof: We need only to establish part b). Continuing the discussion preceding the statement of the theorem we note that the left QYB $A(R)$-module structure (M, μ_π, ρ_π) satisfies the requirements for part a).

Suppose that the hypothesis for part b) is satisfied. Let $\rho'(m_j) = m_i \otimes a^i_j$. Then $\{a^i_j\}_{1\leq i,j\leq r}$ satisfies the comatrix identities and the linear map $f : C \longrightarrow A$ defined by $f(t^i_j) = a^i_j$ is a coalgebra map. Let $\mathcal{F} : T(C) \longrightarrow A$ be the bialgebra map which uniquely extends f. Since $\mathcal{F}(t^i_j) = f(t^i_j) = a^i_j$ it follows that

$$\mathcal{F}(d_{\rho,\mathcal{B},\mathcal{R}}(i,j,s,t)) = d_{\rho',\mathcal{B},\mathcal{R}}(i,j,s,t) = 0.$$

The last equation follows by part d) of Proposition 4.1.1 since (M, μ', ρ') is a left QYB A-module. Therefore $\mathcal{F}(I) = (0)$, and consequently there exists a bialgebra map $F : A(R) \longrightarrow A$ uniquely determined by $F\pi = \mathcal{F}$. Since $F(\pi(t^i_j)) = \mathcal{F}(t^i_j) = a^i_j$ we have that $(1_M \otimes F)\rho_\pi = \rho'$. Since $F(\pi(t^i_j)) = a^i_j$ for all $1 \leq i,j \leq r$ and $\pi(C)$ generates $A(R)$ as an algebra, the uniqueness statement of part b) follows. By part a) of Proposition 4.1.1 $t^i_j{\cdot}m_k = a^i_j{\cdot}m_k$. Thus $t^i_j{\cdot}m_k = F(\pi(t^i_j)){\cdot}m_k$ for all $1 \leq i,j,k \leq r$. Therefore $a{\cdot}m = F(a){\cdot}m$ for all $a \in A(R)$ and $m \in M$ since the $\pi(t^i_j)$'s generate $A(R)$ as an algebra and the m_k's span M. This last equation gives the statement $\mu = \mu'(F \otimes 1_M)$ of part b). ∎

Let $R : M \otimes M \longrightarrow M \otimes M$ be a solution to the QYBE where M is a finite-dimensional vector space over k, and suppose that (M, μ, ρ) is the left QYB $A(R)$-module of part a) of Theorem 4.2.1. Let I be the bi-ideal of $A(R)$ which is the unique coideal of $A(R)$ which is maximal with respect to the property that $I \subseteq \mathrm{ann}_{A(R)}(M)$.

Definition 4.2.1 $\widetilde{A(R)} = A(R)/I$ *is the* reduced FRT construction.

By Exercise 4.1.2 $\widetilde{A(R)}$ is M-reduced. Now the projection $\pi : A(R) \longrightarrow \widetilde{A(R)}$ is a bialgebra map. Observe that $\mathrm{Ker}\pi = I \subseteq \mathrm{ann}_{A(R)}(M)$. Thus by Exercise 4.1.3 (M, μ_π, ρ_π) is a left QYB $\widetilde{A(R)}$-module, where $\rho_\pi = (1_M \otimes \pi)\rho$ and $\mu_\pi(\pi \otimes 1_M) = \mu$, such that $R = R_{(\mu,\rho)} = R_{(\mu_\pi,\rho_\pi)}$. We have shown part a) of the following which is a slight improvement of [Radford, 1994b, Theorem 2] for R^τ.

Theorem 4.2.2 *Suppose that M is a finite-dimensional vector space over the field k and that $R : M \otimes M \longrightarrow M \otimes M$ is a solution to the QYBE. Then the bialgebra $\widetilde{A(R)}$ satisfies the following properties:*

a) *There exists a left QYB $\widetilde{A(R)}$-module structure (M, μ, ρ) on M such that $\widetilde{A(R)}$ is M-reduced and $R = R_{(\mu,\rho)}$.*

b) *Suppose that A is a bialgebra over the field k and (M, μ', ρ') is a left QYB A-module structure on M such A is M-reduced and $R = R_{(\mu',\rho')}$. There is a bialgebra map $F : \widetilde{A(R)} \longrightarrow A$ uniquely defined by $(1_M \otimes F)\rho = \rho'$. Furthermore $\mu = \mu'(F \otimes 1_M)$,*

i) *F is one-one, and*

ii) *F is an isomorphism when $A(\rho')$ generates A as an algebra.*

Proof: We will continue the discussion preceding the statement of the theorem. The QYB $\widetilde{A(R)}$-module required for part a) is satisfied with (M, μ_π, ρ_π).

Assume that the hypothesis if part b) is satisfied. Let (M, μ, ρ) be the left QYB $A(R)$-module structure of part a) of Theorem 4.2.1. By this theorem there is a bialgebra map $\mathcal{F} : A(R) \longrightarrow A$ satisfying $(1_M \otimes \mathcal{F})\rho = \rho'$ and $\mu'(\mathcal{F} \otimes 1_M) = \mu$. Now $\mathcal{F}(I)$ is a coideal of A since I is a coideal of $A(R)$ and $\mathcal{F}$ is a coalgebra map. Since $\mathcal{F}(I) \cdot M = I \cdot M = (0)$ it follows that $\mathcal{F}(I) = (0)$ since A is M-reduced. Consequently there is a bialgebra map $F : \widetilde{A(R)} \longrightarrow A$ uniquely determined by $\mathcal{F} = F\pi$. Thus

$$(1_M \otimes F)\rho_\pi = (1_M \otimes F\pi)\rho = (1_M \otimes \mathcal{F})\rho = \rho'.$$

Since $\widetilde{A(R)}(\rho_\pi)$ generates $\widetilde{A(R)}$ as an algebra, the uniqueness statement for F follows. Observe that

$$\mu'(F\pi \otimes 1_M) = \mu'(\mathcal{F} \otimes 1_M) = \mu = \mu_\pi(\pi \otimes 1_M).$$

Since $\mathrm{Im}\pi = \widetilde{A(R)}$ it follows that $\mu'(F \otimes 1_M) = \mu_\pi$.

We will next show that F is one-one. Since $F(a) \cdot m = a \cdot m$ for all $a \in \widetilde{A(R)}$ and $m \in M$ it follows that $\mathrm{Ker}F \subseteq \mathrm{ann}_{\widetilde{A(R)}}(M)$. But $\mathrm{Ker}F$ is a coideal of $\widetilde{A(R)}$. Therefore $\mathrm{Ker}F = (0)$ since $A(R)$ is M-reduced. Therefore F is one-one. Since $F(\widetilde{A(R)}(\rho_\pi)) = A(\rho')$ it follows that F is onto whenever $A(\rho')$ generates A as an algebra. This concludes our proof. ∎

Exercise 4.2.1 Let A be a bialgebra over the field k and suppose that (M, ρ) is a finite-dimensional right A-comodule. Let (M, μ_ρ) be the left A^o-module structure on M defined by

$$a^o \cdot m = a^o \rightharpoonup m = m^{<1>} <a^o, m^{(2)}>$$

for $a^o \in A$ and $m \in M$; that is the rational left A^*-module action on M restricted to A^o. Show that the following are equivalent:

a) A^o is M-reduced.

b) $\iota(A(\rho))$ generates a dense subalgebra of $(A^o)^*$, where $\iota : A \longrightarrow (A^o)^*$ is defined by $\iota(a)(a^o) = a^o(a)$ for $a \in A$ and $a^o \in A^o$.

(Thus if A is generated as an algebra by $A(\rho)$ then A^o is M-reduced.)

4.3 Opposites and Co-Opposites of the FRT Construction as FRT Constructions

Our main goal in this section is to show that the bialgebras obtained from twisting the algebraic structures of the FRT construction $A(R)$ are themselves FRT constructions. As a result the QYB operators R^τ and R^T of Section 2.3 are cast in an interesting light. We will uncover a very important connection between $A(R)^{op}$ and $A(R)^o$, and we will discuss close connections between $\widetilde{A(R)}$ and $\widetilde{A(R^\tau)}^o$.

Our secondary goal in this section is to describe a necessary and sufficient condition for $A(R) \simeq A(R')$. This we do first. To this end we need to examine the algebraic structure of $A(R)$ in some detail.

Proposition 4.3.1 *Suppose that M is an r-dimensional vector space over the field k and let $R : M \otimes M \longrightarrow M \otimes M$ be a solution to the QYBE. Then:*

a) *$A(R)$ contains a unique smallest subcoalgebra C which generates $A(R)$ as an algebra.*

b) $C \simeq \mathrm{C}_r(k)$.

c) *Set $A(R)_{(0)} = k1$ and $A(R)_{(n)} = C \cdots C$ (n factors) for $n \geq 1$. Then:*

   (i) *$A(R)_{(n)}$ is a subcoalgebra of $A(R)$ for each $n \geq 0$,*

   (ii) *$A(R) = \oplus_{n=0}^{\infty} A(R)_{(n)}$, and*

   (iii) *$A(R)_{(m)}A(R)_{(n)} \subseteq A(R)_{(m+n)}$ for all $m, n \geq 0$.*

Proof: Let $B = \{m_1, \dots, m_r\}$ be a basis for M. Let $\mathcal{R} = \{R_{i,j}^{k,\ell}\}$ be the B-coordinates for R. We apply Exercise 4.3.1 to the tensor bialgebra $T(\mathrm{C}_r(k))$ of the coalgebra $\mathrm{C}_r(k)$, where $T(\mathrm{C}_r(k))_{(0)} = k1$ and $T(\mathrm{C}_r(k))_{(n)} = \mathrm{C}_r(k) \otimes \cdots \otimes \mathrm{C}_r(k)$ (n factors) for $n \geq 1$, and the coideal $\mathcal{I} \subseteq T(\mathrm{C}_r(k))_{(2)}$ of $T(\mathrm{C}_r(k))$ which is the span of the $d_{\mathcal{E},\mathcal{R}}(i, j, s, t)$'s where $\mathcal{E} = \{t_j^i\}_{1 \leq i,j, \leq r}$ is a standard basis for $\mathrm{C}_r(k)$. Thus part c) follows with $A(R)_{(n)} = \pi(T(\mathrm{C}_r(k))_{(n)})$, where $\pi : T(\mathrm{C}_r(k)) \longrightarrow T(\mathrm{C}_r(k))/I = A(R)$ is the projection.

We note that $A(R)_{(0)} = k1$ and $C = A(R)_{(1)} \simeq \mathrm{C}_r(k)$. To complete the proof we need only show that if D is a subcoalgebra of $A(R)$ which generates $A(R)$ as an algebra then $C \subseteq D$. Now

$$D = D \cap (\oplus_{n=0}^{\infty} A(R)_{(n)}) = \oplus_{n=0}^{\infty}(D \cap A(R)_{(n)})$$

by part b) of Lemma 1.7.1. If $D \cap A(R)_{(1)} = (0)$ then the subalgebra of $A(R)$ which D generates does not contain C. Thus $D \cap A(R)_{(1)} = D \cap C \neq (0)$. Since $C \simeq \mathrm{C}_r(k)$ is simple, necessarily $D \supseteq C$. ∎

Let M be a finite-dimensional vector space over k with basis $B = \{m_1, \ldots, m_r\}$ and suppose that $R : M \otimes M \longrightarrow M \otimes M$ is a solution to the QYBE. Let $\mathcal{R} = \{R_{i,j}^{k,\ell}\}$ be the B-coordinates of R. If C is a coalgebra over k and $\mathcal{E} = \{t_j^i\}_{1 \leq i,j \leq r} \subseteq C$ satisfies the comatrix identities then we set

$$\delta_{\mathcal{E},\mathcal{R}}(i,j,s,t) = R_{p,q}^{s,t} t_i^p \otimes t_j^q - R_{i,j}^{p,q} t_q^t \otimes t_p^s \in C \otimes C$$

for $1 \leq i,j,k,\ell \leq r$. Let $\mathcal{I}_{\mathcal{E},\mathcal{R}}$ be the span of these elements. Note that if $f : C \longrightarrow D$ is a coalgebra map then $f(\mathcal{E}) = \{d_j^i\}_{1 \leq i,j \leq r} \subseteq D$ satisfies the comatrix identities, where $d_j^i = f(c_j^i)$. Regarding C as a subcoalgebra of the tensor bialgebra $T(C)$ on C we deduce from (4.4) and (4.5) that $\mathcal{I}_{\mathcal{E},\mathcal{R}}$ is a coideal of $C \otimes C$.

Theorem 4.3.1 *Suppose that M and M′ are finite-dimensional vector spaces over the field k, $R : M \otimes M \longrightarrow M \otimes M$ and $R' : M' \otimes M' \longrightarrow M' \otimes M'$ are solutions to the QYBE. Then the following are equivalent:*

a) $A(R) \simeq A(R')$ *as bialgebras.*

b) $\mathrm{Dim}M = \mathrm{Dim}M' = r$ *and for any standard basis* $\mathcal{E} = \{t_j^i\}_{1 \leq i,j \leq r}$ *for* $\mathrm{C}_r(k)$ *and for any bases B for M and B′ for M′ there exits a coalgebra automorphism* $f : \mathrm{C}_r(k) \longrightarrow \mathrm{C}_r(k)$ *such that*

$$\mathcal{I}_{\mathcal{E},\mathcal{R}'} = \mathcal{I}_{f(\mathcal{E}),\mathcal{R}},$$

where $\mathcal{R}$ is the set of B-coordinates for R and $\mathcal{R}'$ is the set of B′-coordinates for R′.

c) $\mathrm{Dim}M = \mathrm{Dim}M' = r$ *and for some standard basis* $\mathcal{E} = \{t_j^i\}_{1 \leq i,j \leq r}$ *for* $\mathrm{C}_r(k)$ *and for some basis B for M and for some basis B′ for M′ there exits a coalgebra automporhism* $f : \mathrm{C}_r(k) \longrightarrow \mathrm{C}_r(k)$ *such that*

$$\mathcal{I}_{\mathcal{E},\mathcal{R}'} = \mathcal{I}_{f(\mathcal{E}),\mathcal{R}},$$

where $\mathcal{R}$ is the set of B-coordinates for R and $\mathcal{R}'$ is the set of B′-coordinates for R′.

Proof: Part b) clearly implies part c). We will first show that part c) implies part a). Assume the hypothesis of part c). Let $T(\mathrm{C}_r(k))$ be the tensor bialgebra on the coalgebra $\mathrm{C}_r(k)$. Now $A(R') = T(\mathrm{C}_r(k))/I'$, where I' is the ideal of $T(\mathrm{C}_r(k))$ generated by $\mathcal{I}_{\mathcal{E},\mathcal{R}'}$, where $\mathcal{E} = \{t_j^i\}_{1 \leq i,j \leq r}$ is a standard basis for $\mathrm{C}_r(k)$. Since $f(\mathcal{E}) = \{f(t_j^i)\}_{1 \leq i,j \leq r}$ is a basis for $\mathrm{C}_r(k)$ which satisfies the comatrix identities, $A(R) = T(\mathrm{C}_r(k))/I$ where I is the ideal of $T(\mathrm{C}_r(k))$ generated by $\mathcal{I}_{\mathcal{E},\mathcal{R}}$. By the universal mapping property of the tensor bialgebra $T(\mathrm{C}_r(k))$ on $\mathrm{C}_r(k)$ it follows that the coalgebra automorphism $f : \mathrm{C}_r(k) \longrightarrow \mathrm{C}_r(k)$ extends uniquely to a bialgebra automorphism $\mathcal{F} : T(\mathrm{C}_r(k)) \longrightarrow T(\mathrm{C}_r(k))$. Since

$$\mathcal{F}(\mathcal{I}_{\mathcal{E},\mathcal{R}}) = \mathcal{I}_{f(\mathcal{E}),\mathcal{R}} = \mathcal{I}_{\mathcal{E},\mathcal{R}'}$$

it follows that $\mathcal{F}(I) = I'$. Therefore $\mathcal{F}$ gives rise to an isomorphism of bialgebras

$$A(R) = T(\mathrm{C}_r(k))/I \xrightarrow{F} T(\mathrm{C}_r(k))/I' = A(R')$$

uniquely defined by $F\pi = \pi'\mathcal{F}$, where

$$T(\mathrm{C}_r(k)) \xrightarrow{\pi} T(\mathrm{C}_r(k))/I$$

and

$$T(\mathrm{C}_r(k)) \xrightarrow{\pi'} T(\mathrm{C}_r(k))/I'$$

are the projections. We have shown that part c) implies part a).

To see that part a) implies part b), let $F : A(R) \longrightarrow A(R')$ be an isomorphism of bialgebras, $r = \mathrm{Dim}M$, and $r' = \mathrm{Dim}M'$. By parts a) and b) of Proposition 4.3.1 it follows that $A(R)$ and $A(R')$ are generated as algebras by unique smallest subcoalgebras C and C' respectively. Since F and F^{-1} are onto bialgebra maps we conclude that $F(C) \supseteq C'$ and $F^{-1}(C') \supseteq C$. Therefore $F(C) = C'$ and $F^{-1}(C') = C$. As $C \simeq \mathrm{C}_r(k)$ and $C' \simeq \mathrm{C}_{r'}(k)$ we conclude that $\mathrm{Dim}M = r = r' = \mathrm{Dim}M'$.

Now let B and B' be any bases for M and M' respectively. Let $\mathcal{R}$ be the set of B-coordinates for R and $\mathcal{R}'$ be the set of B'-coordinates for M'. By virtue of Theorem 4.2.1 we may assume that $A(R) = T(\mathrm{C}_r(k))/I$ and $A(R') = T(\mathrm{C}_r(k))/I'$, where I is the ideal of $T(\mathrm{C}_r(k))$ generated by $\mathcal{I}_{\mathcal{E},\mathcal{R}}$ and I' is the ideal of $T(\mathrm{C}_r(k))$ generated by $\mathcal{I}_{\mathcal{E},\mathcal{R}'}$. Let $\pi : T(\mathrm{C}_r(k)) \longrightarrow T(\mathrm{C}_r(k))/I$ and $\pi' : T(\mathrm{C}_r(k)) \longrightarrow T(\mathrm{C}_r(k))/I'$ be the projections. Then $C = \pi(\mathrm{C}_r(k))$ and $C' = \pi'(\mathrm{C}_r(k))$ are the smallest generating subcoalgebras described in Proposition 4.3.1. Thus the restrictions of π and π' to $\mathrm{C}_r(k)$ are isomorphisms $\pi_r : \mathrm{C}_r(k) \longrightarrow C$ and $\pi'_r : \mathrm{C}_r(k) \longrightarrow C'$.

We have shown that $F(C) = C'$. By the universal mapping property of the tensor bialgebra $T(\mathrm{C}_r(k))$ on $\mathrm{C}_r(k)$ the coalgebra automorphism $f : \mathrm{C}_r(k) \longrightarrow \mathrm{C}_r(k)$ uniquely determined by $\pi'_r f = F\pi_r$ extends uniquely to a bialgebra automorphism $\mathcal{F} : T(\mathrm{C}_r(k)) \longrightarrow T(\mathrm{C}_r(k))$. Observe that $\pi'\mathcal{F} = F\pi$ since both $\pi'\mathcal{F}$ and $F\pi$ are algebra maps which agree on the algebra generating set $\mathrm{C}_r(k)$. Therefore

$$\pi'\mathcal{F}(\mathcal{I}_{\mathcal{E},\mathcal{R}}) = F(\pi(\mathcal{I}_{\mathcal{E},\mathcal{R}})) = (0).$$

Since $\pi\mathcal{F}^{-1} = F^{-1}\pi'$ we compute in the same manner

$$\pi(\mathcal{F}^{-1}(\mathcal{I}_{\mathcal{E},\mathcal{R}'})) = F^{-1}(\pi'(\mathcal{I}_{\mathcal{E},\mathcal{R}'})) = (0).$$

Therefore

$$\mathcal{F}(\mathcal{I}_{\mathcal{E},\mathcal{R}}) \subseteq I' \cap (\mathrm{C}_r(k) \otimes \mathrm{C}_r(k)) = \mathcal{I}_{\mathcal{E},\mathcal{R}'}$$

and

$$\mathcal{F}^{-1}(\mathcal{I}_{\mathcal{E},\mathcal{R}'}) \subseteq I \cap (\mathrm{C}_r(k) \otimes \mathrm{C}_r(k)) = \mathcal{I}_{\mathcal{E},\mathcal{R}}$$

which means $\mathcal{F}(\mathcal{I}_{\mathcal{E},\mathcal{R}}) = \mathcal{I}_{\mathcal{E},\mathcal{R}'}$. Since $\mathcal{F}(\mathcal{I}_{\mathcal{E},\mathcal{R}}) = \mathcal{I}_{f(\mathcal{E}),\mathcal{R}}$ we have $\mathcal{I}_{f(\mathcal{E}),\mathcal{R}} = \mathcal{I}_{\mathcal{E},\mathcal{R}'}$. Therefore part a) implies part b), and the proof of the theorem is complete. ∎

In more concrete terms, what it means for $\mathcal{I}_{f(\mathcal{E}),\mathcal{R}} = \mathcal{I}_{\mathcal{E},\mathcal{R}'}$ is that the span of the differences

$$\delta_{f(\mathcal{E}),\mathcal{R}}(i,j,s,t) = R^{s,t}_{p,q} f(t^p_i) \otimes f(t^q_j) - R^{p,q}_{i,j} f(t^t_q) \otimes f(t^s_p) \in \mathrm{C}_r(k) \otimes \mathrm{C}_r(k),$$

where $1 \leq i,j,s,t \leq r$, is the span of the differences

$$\delta_{\mathcal{E},\mathcal{R}'}(i,j,s,t) = R'^{s,t}_{p,q} t^p_i \otimes t^q_j - R'^{p,q}_{i,j} t^t_q \otimes t^s_p \in \mathrm{C}_r(k) \otimes \mathrm{C}_r(k),$$

where $1 \leq i,j,s,t \leq r$.

We now consider the bialgebra structures obtained by twisting the algebra structures of $A(R)$. These are FRT constructions in their own right.

Let M be a finite-dimensional vector space over the field k and suppose that $R : M \otimes M \longrightarrow M \otimes M$ is a solution to the QYBE. Let $B = \{m_1, \ldots, m_r\}$ be a basis for M and $\mathcal{R} = \{R^{k,\ell}_{i,j}\}$ be the B-coordinates of R. We define

$$(R^{op})^{k,\ell}_{i,j} = R^{\ell,k}_{j,i}, \tag{4.6}$$

$$(R^{cop})^{k,\ell}_{i,j} = R^{j,i}_{\ell,k}, \tag{4.7}$$

and

$$(R^{op\ cop})^{k,\ell}_{i,j} = R^{i,j}_{k,\ell} \tag{4.8}$$

for $1 \leq i,j,k,\ell \leq r$. We let $\mathcal{R}^{op}$, $\mathcal{R}^{cop}$ and $\mathcal{R}^{op\ cop}$ denote the sets of scalars described by (4.6) – (4.8) respectively.

Recall from Section 2.3 that R gives rise to solutions $R^\tau : M \otimes M \longrightarrow M \otimes M$ and $R^T : M^* \otimes M^* \longrightarrow M^* \otimes M^*$ to the QYBE, where $R^\tau = \tau_{M,M} R \tau_{M,M}$. For the basis B note that the coordinates of R^τ are given by

$$(R^\tau)^{k,\ell}_{i,j} = R^{\ell,k}_{j,i} = (R^{op})^{k,\ell}_{i,j} \tag{4.9}$$

and for the basis

$$\mathcal{B} = \{m^1, \ldots, m^r\}$$

for M^* dual to B note that the $\mathcal{B}$-coordinates of R^T are given by

$$(R^T)^{k,\ell}_{i,j} = R^{i,j}_{k,\ell} = (R^{op\ cop})^{k,\ell}_{i,j}. \tag{4.10}$$

Thus the $\mathcal{B}$-coordinates for $(R^T)^\tau$ are given by

$$(R^{T\ \tau})^{k,\ell}_{i,j} = R^{j,i}_{\ell,k} = (R^{cop})^{k,\ell}_{i,j}. \tag{4.11}$$

Proposition 4.3.2 *Suppose that M is a finite-dimensional vector space over the field k and that $R : M \otimes M \longrightarrow M \otimes M$ is a solution to the QYBE. Then there are bialgebra isomorphisms*

a) $A(R)^{op} \simeq A(R^{op}) = A(R^{\tau})$,

b) $A(R)^{cop} \simeq A(R^{cop}) = A(R^{T\,\tau})$, *and*

c) $A(R)^{op\ cop} \simeq A(R^{op\ cop}) = A(R^{T})$.

Proof: The fact that $A(R^{op}) = A(R^{\tau}), A(R^{op\ cop}) = A(R^{T})$, and $A(R^{cop}) = A(R^{T\,\tau})$ follow from (4.9) – (4.11) respectively. Thus to prove the proposition we need only to establish the isomorphisms.

Choose a basis $B = \{m_1, \ldots, m_r\}$ for M and let $\mathcal{R} = \{R^{k,\ell}_{i,j}\}$ be the B-coordinates for R. Let $\mathcal{E} = \{t^i_j\}_{1 \le i,j \le r}$ be a standard basis for $\mathrm{C}_r(k)$.

To show part a), we first note the that identity map $f : \mathrm{C}_r(k) \longrightarrow \mathrm{C}_r(k)$ extends uniquely to a bialgebra isomorphism $\mathcal{F} : T(\mathrm{C}_r(k)) \longrightarrow T(\mathrm{C}_r(k))^{op}$. Observe that

$$\mathcal{F}(d_{\mathcal{E},\mathcal{R}^{op}}(i,j,s,t)) = d_{\mathcal{E},\mathcal{R}}(j,i,t,s)$$

for all $1 \le i,j,s,t \le r$. Consequently if I is the ideal of $T(\mathrm{C}_r(k))$ generated by $\mathcal{I}_{\mathcal{E},\mathcal{R}^{op}}$ then $\mathcal{F}(I)$ is the ideal of $T(\mathrm{C}_r(k))$ generated by $\mathcal{I}_{\mathcal{E},\mathcal{R}}$. Thus $\mathcal{F}$ determines an isomorphism of bialgebras

$$A(R^{op}) = T(\mathrm{C}_r(k))/I \longrightarrow T(\mathrm{C}_r(k))^{op}/\mathcal{F}(I) = A(R)^{op}.$$

and part a) follows.

To show part b), consider the coalgebra isomorphism $f : \mathrm{C}_r(k) \longrightarrow \mathrm{C}_r(k)^{cop}$ defined by $f(t^i_j) = t^j_i$ and its unique extension to a bialgebra isomorphism $\mathcal{F} : T(\mathrm{C}_r(k)) \longrightarrow T(\mathrm{C}_r(k))^{cop}$. Since

$$\mathcal{F}(d_{\mathcal{E},\mathcal{R}^{cop}}(i,j,s,t)) = -d_{\mathcal{E},\mathcal{R}}(t,s,j,i)$$

for all $1 \le i,j,s,t \le r$, we conclude, using the argument for part a), that $\mathcal{F}$ induces an isomorphism $A(R^{cop}) \simeq A(R)^{cop}$.

To show part c) we note that the map f of part b) extends uniquely to a bialgebra isomorphism $\mathcal{F} : T(\mathrm{C}_r(k)) \longrightarrow T(\mathrm{C}_r(k))^{op\ cop}$. Since

$$\mathcal{F}(d_{\mathcal{E},\mathcal{R}^{op\ cop}}(i,j,s,t)) = -d_{\mathcal{E},\mathcal{R}}(s,t,i,j)$$

for all $1 \le i,j,k,\ell \le r$ it follows, using the argument for part b), that $\mathcal{F}$ induces an isomorphism of bialgebras $A(R^{op\ cop}) \simeq A(R)^{op\ cop}$. This completes our proof. ∎

Let M be a vector space over k with basis $B = \{m_1, \ldots, m_r\}$ and suppose that $R : M \otimes M \longrightarrow M \otimes M$ is a solution to the QYBE. Suppose that A is a bialgebra over k and (M, μ, ρ) is a left QYB A-module with associated QYBE solution R.

Recall the left A-module (M, μ) gives rise to a right A^o-comodule structure (M, ρ_μ) determined by

$$a{\cdot}m = m^{<1>^o} {<}m^{(2)^o}, a{>}$$

for $a \in A$ and $m \in M$, where we write $\rho_\mu(m) = m^{<1>^o} \otimes m^{(2)^o} \in M \otimes A^o$, and that the right A-comodule structure (M, ρ) gives rise to a (rational) left A^o-module structure (M, μ_ρ) described by

$$a^o{\cdot}m = m^{<1>} {<}a^o, m^{(2)}{>}$$

for all $a^o \in A^o$ and $m \in M$. The main result of Section 3.6 is that (M, μ_ρ, ρ_μ) is a left QYB A^o-module and that R^τ is the associated QYBE solution.

Theorem 4.3.2 *Let M be a finite-dimensional vector space over the field k and suppose that $R : M \otimes M \longrightarrow M \otimes M$ is a solution to the QYBE. Let (M, μ, ρ) be a left QYB $A(R)$-module such that R is the associated QYBE solution. Then:*

a) *There is a bialgebra map $F : A(R^\tau) \longrightarrow A(R)^o$ uniquely determined by $(1_M \otimes F)\rho_{(\tau)} = \rho_\mu$, where $(M, \mu_{(\tau)}, \rho_{(\tau)})$ is a left QYB $A(R^\tau)$-module structure with associated QYBE solution R^τ.*

b) Ker$F$ *is the largest coideal contained in* $\mathrm{ann}_{A(R^\tau)}M$.

c) *$\widetilde{A(R^\tau)} \simeq (A(R)^o(\rho_\mu))$ as bialgebras, where the isomorphism arises from the one-one bialgebra map*

$$\widetilde{A(R^\tau)} = A(R^\tau)/\mathrm{Ker}F \xrightarrow{f} A(R)^o$$

defined by $f(a + \mathrm{Ker}F) = F(a)$ for $a \in A(R^\tau)$.

Proof: Part c) follows from parts a) and b) directly. Part a) follows from part b) of Theorem 4.2.1. To see part b) we note that part b) of Theorem 4.2.1 also gives $\mu_{(\tau)} = \mu_\rho(F \otimes 1_M)$. Therefore

$$a{\cdot}m = F(a){\rightharpoonup}m = m^{<1>}{<}F(a), m^{(2)}{>}$$

for all $a \in A(R^\tau)$ and $m \in M$. We deduce from this equation that Ker$F \subseteq \mathrm{ann}_{A(R^\tau)}M$. Since F is a coalgebra map Ker$F$ is a coideal of $A(R^\tau)$. Conversely, let I be a coideal of $A(R^\tau)$ contained in $\mathrm{ann}_{A(R^\tau)}M$. Then $(0) = I{\cdot}M = F(I){\rightharpoonup}M$ means that $F(I) \subseteq \mathrm{ann}_{A(R)^o}M$. By Exercise 4.2.1 it follows that $A(R)^o$ is (M, μ_ρ)-reduced. Now $F(I)$ is a coideal of $A(R)^o$ since F is a coalgebra map and I is a coideal of $A(R^\tau)$. Therefore $F(I) = (0)$, which is what was needed to complete the proof of part b). ∎

The previous theorem and the details of its proof have important implications for $A(R)$ and $\widetilde{A(R)}$. These will be explored in the exercises. We note that part a) along with part a) of Proposition 4.3.2 give the existence of a bialgebra map

$$A(R)^{op} \simeq A(R^\tau) \longrightarrow A(R)^o$$

which the reader is encouraged to examine.

Exercise 4.3.1 Let A be a bialgebra over the field k and suppose that

i) $A = \oplus_{n=0}^{\infty} A_{(n)}$, where

ii) $A_{(n)}$ is a subcoalgebra of A for $n \geq 0$, and

iii) $A_{(m)}A_{(n)} \subseteq A_{(m+n)}$ for all $m, n \geq 0$.

Show that:

a) $1 \in A_{(0)}$.

b) If A has an antipode then $A = A_{(0)}$.

Let $\mathcal{I}$ be a coideal of A and suppose that $\mathcal{I} \subseteq A_{(N)}$ for some $N > 0$. Let I be the ideal of A generated by $\mathcal{I}$ and $\pi : A \longrightarrow \mathcal{A}$ be the projection onto the bialgebra $\mathcal{A} = A/I$.

c) Show that (i)–(iii) are satisfied for $\mathcal{A}$, where $\mathcal{A}_{(n)} = \pi(A_{(n)})$ for $n \geq 0$.

d) Show that $\pi_n : A_{(n)} \longrightarrow \mathcal{A}_{(n)}$ is an isomorphism of coalgebras for $0 \leq n < N$, where $\pi_n(a) = \pi(a)$ for $a \in A_{(n)}$.

For the remainder of these exercises M is a finite-dimensional vector space over k and $R : M \otimes M \longrightarrow M \otimes M$ is a solution to the QYBE.

Exercise 4.3.2 Show that $A(R)$ has an antipode if and only if $M = (0)$.

Exercise 4.3.3 Show that $A(R) \simeq A(\alpha R)$ for non-zero $\alpha \in k$. (Note the R and αR are not similar operators in general.)

Exercise 4.3.4 Suppose that A is a bialgebra over k and that (M, μ, ρ) is a left QYB A-module such that R is the associated QYBE. If A is generated by $A(\rho)$ as an algebra, show that $\widetilde{A(R^\tau)} \simeq \mathcal{A}$ as bialgebras, where $\mathcal{A}$ is the subalgebra of A^o generated by $A^o(\rho_\mu)$.

Exercise 4.3.5 Show that $\widetilde{A(R^\tau)}$ can be identified with a sub-bialgebra of $\widetilde{A(R)}^o$ which is a dense subspace of $\widetilde{A(R)}^*$.

Exercise 4.3.6 Let $B = \{m_1, \ldots, m_r\}$ be a basis for M and let $\mathcal{R} = \{R_{i,j}^{k,\ell}\}$ be the B-coordinates of R. Suppose that $F : A(R^\tau) \longrightarrow A(R)^o$ is the bialgebra map of part a) of Theorem 4.3.2.

a) Show that there is a standard basis $\{t^i_j\}_{1\leq i,j\leq r}$ for the smallest subcoalgebra $C \simeq C_r(k)$ of $A(R^\tau)$ which generates $A(R^\tau)$ as an algebra such that

$$F(t^k_i)(t^\ell_j) = R^{\ell,k}_{j,i}$$

for all $1 \leq i, j, k, \ell \leq r$.

b) Suppose that $f : A(R)^{op} \longrightarrow A(R)^o$ is the composite of the bialgebra maps of part a) of Proposition 4.3.2 and part a) of Theorem 4.3.2 respectively. Show that there is a standard basis $\{t^i_j\}_{1\leq i,j\leq r}$ for the smallest subcoalgebra $C \simeq C_r(k)$ of $A(R)$ which generates $A(R)$ as an algebra such that

$$f(t^k_i)(t^\ell_j) = R^{k,\ell}_{i,j}$$

for all $1 \leq i, j, k, \ell \leq r$. This map was defined for arbitrary finite-dimensional Hopf algebras in [Majid, 1990b, p. 12].

4.4 Necessary and Sufficient Conditions for $\widetilde{A(R)}$ to be a Pointed Bialgebra

In this section we find necessary and sufficient conditions for $\widetilde{A(R)}$ to be a pointed bialgebra and for $\widetilde{A(R)}$ to be a pointed Hopf algebra in terms of the operator R.

Suppose that M is a finite-dimensional vector space over the field k.

Definition 4.4.1 *A* flag of subspaces *for M is a sequence of subspaces* $(0) = M_0 \subseteq M_1 \subseteq M_2 \subseteq \ldots \subseteq M_r = M$ *such that* $\text{Dim}M_i = i$ *for all* $1 \leq i \leq r$.

Definition 4.4.2 *If M is a comodule then a* flag of subcomodules *for M is a flag of subspaces for M such that each term is a subcomodule of M.*

Theorem 4.4.1 *Let M be a finite-dimensional vector space over the field k and suppose that $R : M \otimes M \longrightarrow M \otimes M$ is a solution to the QYBE. Then the following are equivalent:*

a) *$\widetilde{A(R)}$ is a pointed bialgebra.*

b) *There exists a flag of subspaces $(0) = M_0 \subseteq M_1 \subseteq M_2 \subseteq \cdots \subseteq M_r = M$ for M such that $R(M_i \otimes M) \subseteq M_i \otimes M$ for all $1 \leq i \leq r$.*

Proof: Set $A = \widetilde{A(R)}$ and let (M, μ, ρ) be any left QYB A-module such that $R = R_{(\mu,\rho)}$, A is M-reduced, and $A(\rho)$ generates A as an algebra. If N is a subcomodule of M then $R(N \otimes M) \subseteq N \otimes M$. Thus to show part a) implies part b) it suffices to find a flag of subcomodules for M.

Assume that A is pointed and pick a composition series $(0) = M_0 \subseteq M_1 \subseteq M_2 \subseteq \cdots \subseteq M_r = M$ for the comodule (M, ρ). Then by definition M_i/M_{i-1} is a simple right A-comodule for $1 \leq i \leq r$. Since A is pointed all simple right A-comodules are one-dimensional by Exercise 1.4.4. We have showed that part a) implies part b).

To show part b) implies part a) we let $(0) = M_0 \subseteq M_1 \subseteq \cdots \subseteq M_r = M$ be a flag of subspaces for M such that $R(M_i \otimes M) \subseteq M_i \otimes M$ for all $1 \leq i \leq r$. Choose a basis $\{m_1, \ldots, m_r\}$ for M so that $\{m_1, \ldots, m_i\}$ is a basis for M_i for all $1 \leq i \leq r$. Let $\{t^i_j\}_{1 \leq i,j \leq r} \subseteq A$ be defined by $\rho(m_j) = m_i \otimes t^i_j$. Then

$$R(m_i \otimes m) = m_\ell \otimes t^\ell_i{\cdot}m \in M_i \otimes M$$

means that $t^\ell_i{\cdot}M = (0)$ whenever $\ell > i$. Thus $t^\ell_i \in \mathrm{ann}_A(M)$ whenever $\ell > i$. The span $\mathcal{I}$ of the t^ℓ_i's, where $1 \leq i < \ell \leq r$ is a coideal of A. But $\mathcal{I} \subseteq \mathrm{ann}_A(M)$ means that $\mathcal{I} = (0)$ since $\mathcal{I}$ is a coideal and A is M-reduced. Therefore $t^\ell_i = 0$ when $i < \ell$. The calculation

$$\Delta(t^i_i) = \sum_{\ell=1}^{r} t^i_\ell \otimes t^\ell_i = t^i_i \otimes t^i_i$$

for fixed $1 \leq i \leq r$ shows that $t^1_1, \ldots, t^r_r$ are grouplike elements of A.

Let $C = A(\rho)$, which is the span of the t^ℓ_i's where $\ell \leq i$. For $m \geq 0$ let $C_{(m)}$ be the span of the t^ℓ_i's such that $i - \ell \leq m$. Then it easy to check that $C_{(0)}, C_{(1)}, C_{(2)} \cdots$ is a filtration of C. Therefore $C_0 \subseteq C_{(0)}$ by Proposition 1.7.2. But $C_{(0)}$ is the span of grouplike elements $t^1_1, \ldots, t^r_r$. Therefore $C_0 = C_{(0)}$. Now A is generated as an algebra by $C = A(\rho)$. Therefore A_0 is contained in the subalgebra of A generated by C_0 by Corollary 1.7.2. Let S be the multiplicative semigroup of A generated by $t^1_1, \ldots, t^r_r$. Then $S \subseteq G(A)$ and the subalgebra of A which C_0 generates is $k[S]$ since $G(A)$ is linearly independent. Therefore $A_0 = k[S]$. We have shown that A is pointed and $G(A) = S$. Thus part b) implies part a). ∎

The reader should compare the preceding theorem with [Radford, 1994b, Theorem 3]. It is interesting to look at Theorem 4.4.1 in terms of matrix representations of R. Assume that $A = \widetilde{A(R)}$ is pointed and let $(0) = M_0 \subseteq M_1 \subseteq M_2 \subseteq \ldots \subseteq M_r = M$ be a flag of subspaces for M such that $R(M_i \otimes M) \subseteq M_i \otimes M$ for all $0 \leq i \leq r$. Let $B = \{m_1, \ldots, m_r\}$ be a basis for M such that $\{m_1, \ldots, m_i\}$ is a basis for M_i for all $1 \leq i \leq r$ and let $t^i_j \in A$ be defined for B as in the proof of Theorem 4.4.1. Now for any endomorphism $T \in \mathrm{End}(M \otimes M)$ there are endomorphisms $T^i_j \in \mathrm{End}(M)$ uniquely determined by

$$T(m_j \otimes m) = m_i \otimes T^i_j(m)$$

for all $1 \leq i \leq r$ and $m \in M$. Order the basis $\mathcal{B} = \{m_i \otimes m_j\}_{1 \leq i,j \leq r}$ for $M \otimes M$ lexicographically, reading left to right, and set $A^i_j = [T^i_j]_B$. Then

$$[T]_{\mathcal{B}} = \begin{pmatrix} A^1_1 & \cdots & A^1_r \\ \vdots & & \vdots \\ A^r_1 & \cdots & A^r_r \end{pmatrix}.$$

For $T = R$ we note that $A^i_j = 0$ whenever $i > j$ since $R(M_i \otimes M) \subseteq M_i \otimes M$. Thus the $r \times r$ array

$$[R]_{\mathcal{B}} = \begin{pmatrix} A^1_1 & \cdots & A^1_r \\ & \ddots & \vdots \\ & & A^r_r \end{pmatrix}$$

is upper triangular. Since A is pointed, it follows by Corollary 1.7.5 and part a) of Proposition 1.5.1 that A is a Hopf algebra if and only if $G(A) = S$ is a group. Now let $\pi : A \longrightarrow \text{End}(M)$ be the representation afforded by (M, μ) and let $\mathcal{S}$ be the multiplicative semigroup of $\text{End}(M)$ generated by $\pi(t^1_1), \ldots, \pi(t^r_r)$. Since A is M-reduced, by Exercise 4.1.6 the restriction $\pi|_S : S \longrightarrow \mathcal{S}$ is an isomorphism of semigroups. Since $R(m_j \otimes m) = \sum_{i=1}^r m_i \otimes t^i_j{\cdot}m$ we conclude that

$$\pi(t^i_i) = R^i_i$$

for $1 \leq i \leq r$. Lastly, suppose that $\mathcal{B}' = \{m'_1, \ldots, m'_r\}$ is a basis for M such that $[R]_{\mathcal{B}'} = (A'^i_j)$, where $A'^i_j \in \text{M}_r(k)$ and $A'^i_j = 0$ whenever $i > j$. Then $(0) = M'_0 \subseteq M'_1 \subseteq \ldots \subseteq M'_r = M$ is a flag of subspaces for M, where M'_i is the span of $\{m'_1, \ldots, m'_i\}$, and $R(M'_i \otimes M) \subseteq M'_i \otimes M$ for all $1 \leq i \leq r$. We have proved:

Theorem 4.4.2 *Suppose that M is an r-dimensional vector space over the field k and that $R : M \otimes M \longrightarrow M \otimes M$ is a solution to the QYBE.*

a) *The following are equivalent:*

  i) *$\widetilde{A(R)}$ is pointed.*

  ii) *There exists a matrix $\mathcal{B}$ for M such that the array*

$$[R]_{\mathcal{B}} = \begin{pmatrix} A^1_1 & \cdots & A^1_r \\ & \ddots & \vdots \\ & & A^r_r \end{pmatrix}$$

  is upper triangular in form, where $A^i_j \in \text{M}_r(k)$.

b) *Suppose that there exists a basis $\mathcal{B}$ for M such that $[R]_{\mathcal{B}}$ has the form described in part a.ii). Then $\widetilde{A(R)}$ is a Hopf algebra if and only if the multiplicative semigroup of $\text{M}_r(k)$ generated by $A^1_1, \ldots, A^r_r$ is a group.*

∎

We shall use the ideas of this section to study the solutions to the QYBE when $\text{Dim}M = 2$ in Sections 8.2 and 8.3.

Exercise 4.4.1 Let M be a vector space over the field k with basis $B = \{m_1, \ldots, m_r\}$ and let $\omega_{i,j} \in k$ for $1 \leq i, j \leq r$. Define $R : M \otimes M \longrightarrow M \otimes M$ by $R(m_i \otimes m_j) = \omega_{i,j} m_i \otimes m_j$ for all $1 \leq i, j \leq r$.

a) Show that R is a solution to the QYBE.

b) Show that the array

$$[R]_{\mathcal{B}} = \begin{pmatrix} A_1^1 & & \\ & \ddots & \\ & & A_r^r \end{pmatrix}$$

is diagonal, and that $A_i^i \in \mathrm{M}_r(k)$ is the diagonal matrix

$$A_i^i = \begin{pmatrix} \omega_{i,1} & & \\ & \ddots & \\ & & \omega_{i,r} \end{pmatrix}.$$

(Thus $\widetilde{A(R)}$ is pointed by Theorem 4.4.2.)

c) Show that $G(\widetilde{A(R)})$ is commutative.

d) Show that $\widetilde{A(R)}$ is a Hopf algebra if and only if there are positive integers $n_1, \ldots, n_r$ such that

$$\omega_{i,1}^{n_1} \cdots \omega_{i,r}^{n_r} = 1$$

for all $1 \leq i \leq r$.

e) Show that if $R^n = I$ for some $n > 0$ then $\widetilde{A(R)}$ is a Hopf algebra.

f) Find examples of invertible R with infinite order

  i) such that $\widetilde{A(R)}$ is a Hopf algebra,

  ii) such that $\widetilde{A(R)}$ is not a Hopf algebra.

Exercise 4.4.2 Suppose that A is a bialgebra over the field k and (M, μ, ρ) is a finite-dimensional left QYB A-module. Let $(0) = M_0 \subseteq M_1 \subseteq \ldots \subseteq M_r = M$ be subspaces of M. Let $d(i) = \mathrm{Dim} M_i$ for $1 \leq i \leq r$ and let $B = \{m_1, \ldots, m_r\}$ be a basis for M such that $\{m_1, \ldots, m_{d(i)}\}$ is a basis for M_i for $1 \leq i \leq r$.

a) Suppose that $(0) = M_0 \subseteq M_1 \subseteq \ldots \subseteq M_r = M$ is a composition series for the comodule (M, ρ).

  i) Show that $R(M_i \otimes M) \subseteq M_i \otimes M$ for all $1 \leq i \leq r$.

  ii) Describe $[R]_{\mathcal{B}}$ in terms of block arrangements.

b) Suppose that $(0) = M_0 \subseteq M_1 \subseteq \ldots \subseteq M_r = M$ is a composition series for the module (M, μ).

  i) Show that $R(M \otimes M_i) \subseteq M \otimes M_i$ for all $1 \leq i \leq r$.

ii) Describe $[R]_{\mathcal{B}}$ in terms of block arrangements.

Exercise 4.4.3 Suppose that A is a bialgebra over k and that (M, μ, ρ) is a finite-dimensional left QYB A-module. Suppose further that A is M-reduced and N is a subspace of M. Set $R = R_{(\mu,\rho)}$. Show that the following are equivalent:

a) $R(N \otimes M) \subseteq N \otimes M$.

b) N is an A-subcomodule of M.

Exercise 4.4.4 Suppose that A is a bialgebra over k and that (M, μ, ρ) is a finite-dimensional left QYB A-module. Suppose further that A is generated as an algebra by $A(\rho)$ and N is a subspace of M. Set $R = R_{(\mu,\rho)}$. Show that the following are equivalent:

a) $R(M \otimes N) \subseteq M \otimes N$.

b) N is an A-submodule of M.

Exercise 4.4.5 Let M be a finite-dimensional vector space over k. Suppose that $S, T \in \mathrm{End}(M)$ are non-zero commuting operators, and T is nilpotent. Recall from Exercise 4.1.4 that $R : M \otimes M \longrightarrow M \otimes M$ defined by

$$R(m \otimes n) = \sum_{\ell=0}^{\mathrm{Dim}M} \frac{1}{\ell!} T^{\ell}(m) \otimes S^{\ell}(n)$$

for $m, n \in M$ is a solution to the QYBE. Show that $\widetilde{A(R)} = k[X]$, the polynomial algebra in indeterminant X over k where $\Delta(X) = X \otimes 1 + 1 \otimes X$.

5 THE FUNDAMENTAL EXAMPLE OF A QUANTUM GROUP

Throughout this chapter k is an algebraically closed field of characteristic zero. With the exception of Section 5.1.1, we follow [Lambe and Radford, 1993, Section 9]

5.1 Review of the Special Linear Group

Definition 5.1.1 *For a given positive integer n the Lie group $\mathrm{SL}(n, k)$ is the multiplicative group of all invertible $n \times n$ matrices of $\mathrm{M}_n(k)$ with determinant 1.*

Thus as a set

$$\mathrm{SL}(2, k) = \left\{ \left[\begin{array}{cc} a & b \\ c & d \end{array} \right] \,\middle|\, ad - cb = 1 \right\}. \tag{5.1}$$

We can think of $\mathrm{SL}(2, k)$ as an algebraic variety [Shafarevich, 1974] of dimension 3 in k^4.

5.1.1 The Coordinate Ring of the Special Linear Group

Let X be a subset of affine n-space k^n. Then the set of all functions $\mathrm{Map}(X, k)$ from X to k is an algebra over k under pointwise operations. For fixed $x = (\alpha_1, \ldots, \alpha_n) \in X$ the substitution map $s_x : k[t_1, \ldots, t_n] \longrightarrow k$ defined by $s_x(f) = f(\alpha_1, \ldots, \alpha_n)$ is an algebra homomorphism from the polynomial algebra in n commuting indeterminants $t_1, \ldots, t_n$ over k to the field k. Set $f(x) = f(\alpha_1, \ldots, \alpha_n)$. Observe that the map $\varsigma : k[t_1, \ldots, t_n] \longrightarrow \mathrm{Map}(X, k)$ defined by $\varsigma(f)(x) = s_x(f) = f(x)$ for all $x \in X$ is an algebra homomorphism.

Definition 5.1.2 *Let X be a subset of k^n. The subalgebra* $\mathrm{Im}\varsigma$ *of* $\mathrm{Map}(X, k)$ *is the* coordinate ring of X, *or the* ring of polynomial functions on X, *and is denoted by* $k[X]$.

Suppose that $X = k^n$. Then ς is one-one since k is infinite. In this case we have an identification of $k[t_1, \ldots, t_n]$ with the polynomial functions of X and will thus refer to the polynomial ring $k[t_1, \ldots, t_n]$ as the ring of polynomial functions of $X = k^n$.

Suppose that $X \subseteq k^n$. Let $I(X)$ be the subset of all polynomials $f \in k[t_1, \ldots, t_n]$ such that $f(x) = 0$ for all $x \in X$. Then $I(X) = \mathrm{Ker}\varsigma$. Therefore we have an identification of k-algebras

$$k[X] = k[t_1, \ldots, t_n]/I(X). \tag{5.2}$$

Notice for $x \in X$ the map $\hat{x} : k[X] \longrightarrow k$ defined by $\hat{x}(f) = f(x)$ for all $f \in k[X]$ is an algebra homomorphism. Observe that the map

$$X \xrightarrow{\ \iota\ } \mathrm{Alg}(k[X], k). \tag{5.3}$$

defined by $\iota(x) = \hat{x}$ for all $x \in X$ is one-one.

Definition 5.1.3 *A subset X of k^n is an* affine algebraic variety *if ι is onto.*

In particular an affine variety X can be recovered from its coordinate ring $k[X]$. Thus we may think of affine algebraic varieties as certain algebras over k. This is the correct point of view for quantum groups.

If, in addition, the algebra $k[X]$ has a coproduct such that the resulting structure is a Hopf algebra then (5.3) determines a group structure on X since $\mathrm{Alg}(k[X], k) = G(k[X]^o)$ is a group by Theorem 1.6.1 and part e) of Proposition 1.6.1.

Definition 5.1.4 *An affine algebraic variety X is an* affine algebraic group *if its coordinate ring $k[X]$ is a Hopf algebra.*

Thus we may think of affine algebraic groups as certain Hopf algebras. This is the correct point of view for quantum groups.

In more familiar terms, X is an affine algebraic variety if and only if X is the set of all zeros in k^n of some ideal of the polynomial functions $k[t_1, \ldots, t_n]$ of k^n. See the first exercise below.

We sketch a proof that $\mathrm{SL}(2, k)$ is an affine algebraic group. In the exercises below the reader is lead through more of the details.

Consider the free commutative bialgebra $A = S(\mathrm{C}_2(k))$ on the comatrix coalgebra $\mathrm{C}_2(k)$. As an algebra $A = k[e^1_1, e^1_2, e^2_1, e^2_2]$ is the algebra of polynomials in commuting indeterminants which we may assume form a standard basis for $\mathrm{C}_2(k)$. See Exercise 1.5.11. Let I be the ideal of A generated by $e^1_1e^2_2 - e^1_2e^2_1 - 1$. Then it is not hard to see that I is a bi-ideal of A. Therefore $H = A/I$ is a bialgebra over k which in fact has an antipode. By virtue of [Shafarevich, 1974, p. 17] it follows that $H = k[\mathrm{SL}(2, k)]$.

Let $a = e^1_1, b = e^1_2, c = e^2_1$, and $d = e^2_2$. Then

$$k[\mathrm{SL}(2, k)] = k[a, b, c, d]/(ad - bc - 1). \tag{5.4}$$

For the connection between Hopf algebras and affine algebraic varieties, see [Waterhouse, 1979].

Exercise 5.1.1 Let $A = k[t_1, \ldots, t_n]$ be the ring of polynomial functions of k^n. For a subset X of k^n define

$$X^{\perp} = \{f \in A \mid f(X) = (0)\}$$

and for a subset I of A define

$$I^{\perp} = \{x \in k^n \mid I(x) = (0)\}.$$

a) Show that $X^{\perp}$ is an ideal of A for all $X \subseteq k^n$.

b) If X and Y are both subsets of k^n or are both subsets of A, show the following:

   i) $X \subseteq X^{\perp\perp}$.

   ii) If $X \subseteq Y$ then $Y^{\perp} \subseteq X^{\perp}$.

   iii) $X^{\perp} = X^{\perp\perp\perp}$.

Definition 5.1.5 *A subset X of k^n is* closed *if $X = X^{\perp\perp}$.*

c) Show that $X \subseteq k^n$ is closed if and only if $X = I^{\perp}$ for some ideal I of A (in which case $X = I(X)^{\perp}$).

d) Show that $X \in k^n$ is an affine algebraic variety if and only if X is a closed subset of k^n. [Hint: Use (5.2) and note that if $\eta : A \longrightarrow k$ is an algebra map then there exists a unique $x \in k^n$ such that $\eta(f) = f(x)$ for all $f \in A$.]

The reader should compare the notion of closed of this exercise with that of Section A.4.2.

Exercise 5.1.2 Let $A = S(\mathrm{C}_r(k))$ be the free commutative bialgebra on the coalgebra $\mathrm{C}_r(k)$ and suppose that $\{e^i_j\}_{1 \leq i,j \leq r}$ is a standard basis for $\mathrm{C}_r(k)$.

a) Show that the map $G(A^o) \longmapsto \mathrm{M}_r(k)$ defined by

$$\eta \longrightarrow \begin{pmatrix} \eta(e^1_1) & \cdots & \eta(e^1_r) \\ \vdots & & \vdots \\ \eta(e^r_1) & \cdots & \eta(e^r_r) \end{pmatrix}$$

is an isomorphism of multiplicative semigroups.

b) Show that A is the coordinate ring of the affine algebraic variety $X = \mathrm{M}_r(k)$.

Exercise 5.1.3 Let $A = S(\mathrm{C}_2(k)) = k[e^1_1, e^1_2, e^2_1, e^2_2]$ be the free commutative bialgebra on a standard basis for $\mathrm{C}_2(k)$.

a) Let I be the ideal of A generated by $e^1_1e^2_2 - e^1_2e^2_1 - 1$. Show that I is a bi-ideal of A (and thus $H = A/I$ is a bialgebra).

b) Show that the linear map $\varsigma : \mathrm{C}_2(k) \longrightarrow A$ determined by

$$\varsigma(e^1_1) = e^2_2, \qquad \varsigma(e^1_2) = -e^1_2, \qquad \varsigma(e^2_1) = -e^2_1 \qquad \text{and} \qquad \varsigma(e^2_2) = e^1_1$$

gives rise to an algebra map $s : H \longrightarrow H$ which is an antipode for H.

c) Show that the map $G(H^o) \longrightarrow \mathrm{SL}(2, k)$ given by

$$\eta \longmapsto \begin{pmatrix} \eta(e^1_1) & \eta(e^1_2) \\ \eta(e^2_1) & \eta(e^2_2) \end{pmatrix}$$

is a well-defined map of multiplicative groups.

5.1.2 The Lie Algebra of the Special Linear Group

Recall from Section 1.5 that an associative algebra A over k has a Lie algebra structure given by $[a, b] = ab - ba$ for all $a, b \in A$. When $A = \mathrm{M}_n(k)$ the associated Lie algebra is $\mathrm{gl}(n, k)$. The set of all $n \times n$ matrices in $\mathrm{M}_n(k)$ with trace 0 is the Lie subalgebra $\mathrm{sl}(n, k)$. Thus as a set

$$\mathrm{sl}(2, k) = \left\{ \begin{bmatrix} a & b \\ c & d \end{bmatrix} \;\middle|\; a + d = 0 \right\}.$$

Observe that $\mathrm{sl}(2, k)$ has linear basis

$$e_1 = \begin{bmatrix} 0 & 1 \\ 0 & 0 \end{bmatrix}, \quad e_2 = \begin{bmatrix} 0 & 0 \\ 1 & 0 \end{bmatrix}, \quad \text{and} \quad h = \begin{bmatrix} 1 & 0 \\ 0 & -1 \end{bmatrix}. \tag{5.5}$$

Since $h = [e_1, e_2] = e_1e_2 - e_2e_1$ it follows that $\mathrm{sl}(2, k)$ is generated as a Lie algebra by $\{e_1, e_2\}$. The structure constants are derived from the relations

$$[e_1, e_2] = h, \quad [e_1, h] = -2e_1 \quad \text{and} \quad [e_2, h] = 2e_2.$$

Suppose that $A = k[X]$ is the coordinate ring of an affine algebraic variety X. Then the subspace $P_\epsilon(A^o)$ of primitive elements of A^o is the space of ϵ:ϵ derivations of A; that is the set of all functionals $f : A \longrightarrow k$ which satisfy

$$f(ab) = \epsilon(a)f(b) + f(a)\epsilon(b)$$

for all $a, b \in A$. See Exercises 1.1.3, 1.3.1, and 1.3.2. When X is an affine algebraic group A is a Hopf algebra. In this case $P_\epsilon(A^o)$ is a Lie sub-algebra of the Lie algebra derived from the associative algebra A^o by part d) of Proposition 1.5.1 and part b) of Proposition 1.5.2.

Definition 5.1.6 *Suppose that X is an affine algebraic group and $A = k[X]$ is its coordinate ring. Then* $\mathrm{Lie}(X) = P_\epsilon(A^o)$ *is the* Lie algebra of X.

As one might suspect

$$\mathrm{Lie}(\mathrm{SL}(2, k)) = \mathrm{sl}(2, k).$$

See Exercise 5.1.6 below.

Exercise 5.1.4 Let A be an algebra over the field k and suppose $\eta, \xi \in G(A^o)$. Let $f, g : A \longrightarrow k$ be η:ξ-derivations. (See Exercises 1.3.1 and 1.3.2.)

a) Show that $\mathrm{Ker} f$ is a subalgebra of A.

b) Suppose that V is a subspace of A and $V \subseteq \mathrm{Ker}\eta, \mathrm{Ker}\xi$, and $\mathrm{Ker} f$. Show that $f(I) = (0)$, where I is the ideal of A generated by V.

c) Suppose that V is a subspace of A which generates A as an algebra. Show that $f = g$ if and only if $f|_V = g|_V$.

Exercise 5.1.5 Suppose that C is a coalgebra over k. Show that the correspondence

$$f \longmapsto f|_C$$

defines a Lie algebra isomorphism

$$P_\epsilon(T(C)) \simeq C^*$$

from the space of ϵ:ϵ derivations of the tensor bialgebra of C over k to the dual algebra of the coalgebra C. [Hint: any ϵ:ϵ derivation of $T(C)$ is determined by its restriction to C. Show that any functional on C determines an ϵ:ϵ derivation of $T(C)$.]

Exercise 5.1.6 Show that $\mathrm{Lie}(\mathrm{SL}(2, k)) \simeq \mathrm{sl}(2, k)$. [Hint: First observe that the coordinate ring is a quotient of the form $k[\mathrm{SL}(2, k)] = T(\mathrm{C}_2(k))/I$ for some bi-ideal I of $T(\mathrm{C}_2(k))$. Then show that $P_\epsilon(T(\mathrm{C}_2(k))/I)$ is identified with the ϵ:ϵ derivations of $T(\mathrm{C}_2(k))$ which vanish on $e^1_1e^2_2 - e^1_2e^2_1 - 1$.]

5.1.3 Irreducible Representations of the Lie Algebra of the Special Linear Group

Consider the derivations

$$E_1 = x\frac{\partial}{\partial y}, \quad E_2 = y\frac{\partial}{\partial x} \tag{5.6}$$

of the polynomial algebra $A = k[x,y]$. Observe that $H = [E_1, E_2] = x\frac{\partial}{\partial x} - y\frac{\partial}{\partial y}$. Let V be the span of x and y. Since V generates A as an algebra, any two derivations of A are equal if and only if they agree on V. Let $\mathcal{L}$ be the Lie subalgebra of $\mathrm{Der}(A)$ generated by E_1 and E_2. Then $\mathcal{L}(V) \subseteq V$. Thus $\mathcal{L}$ is isomorphic to a Lie subalgebra of $\mathrm{sl}(2,k)$. An easy computation shows that:

Lemma 5.1.1 *Suppose that $A = k[x,y]$ is the algebra of polynomials in commuting indeterminants x and y over k. Suppose that $\mathcal{L}$ is the Lie subalgebra of* $\mathrm{Der}(A)$ *described above. Then* $\mathrm{sl}(2,k) \simeq \mathcal{L}$. *More precisely there is a one-one map of Lie algebras*

$$\mathrm{sl}(2,k) \xrightarrow{\rho} \mathrm{Der}(A)$$

determined by $\rho(e_i) = E_i$ for $1 \le i \le 2$, where $\{e_1, e_2\}$, is the basis for $\mathrm{sl}(2,k)$ *described in Section 5.1.2.*

The proof of this is left as an exercise. ∎

The representation $\rho : \mathrm{sl}(2,k) \longrightarrow \mathrm{Der}(A)$ described in the preceding lemma accounts for all of the finite-dimensional irreducible representations of $\mathrm{sl}(2,k)$. For $n \ge 0$ let A_n be the subspace of A generated by polynomials of homogeneous degree n. Then $\mathcal{L}(A_n) \subseteq A_n$ for all $n \ge 0$. Regard A as a left $\mathcal{L}$-module under function evaluation and regard A as a left $\mathrm{sl}(2,k)$-module by pullback along ρ. Note that $\mathrm{Dim}A_n = n+1$ for all $n \ge 0$. Since there is exactly one finite-dimensional irreducible $\mathrm{sl}(2,k)$-module for every positive integer we have:

Theorem 5.1.1 *The finite-dimensional irreducible representations of* $\mathrm{sl}(2,k)$ *are realized by the* $\mathrm{sl}(2,k)$*-submodules A_n of A where A is given the pullback* $\mathrm{sl}(2,k)$*-module structure described above.* ∎

See e.g. [Jacobson, 1962], [Serre, 1992]. This theorem is covered by the special case $q = 1$ in a more general result developed in the next sections.

By Section 1.5 the module action of $\mathrm{sl}(2,k)$ on A derived from the representation $\rho : \mathrm{sl}(2,k) \longrightarrow \mathrm{Der}(A)$ above gives rise to a left $\mathcal{U}(\mathrm{sl}(2,k))$-module algebra structure on the polynomial algebra A, where $\mathcal{U}(\mathrm{sl}(2,k))$ is the universal enveloping algebra of the Lie algebra $\mathrm{sl}(2,k)$. Recall that the enveloping algebra is a Hopf algebra. In the following sections of this chapter we will present an example that generalizes all of what we have said so far for $\mathrm{SL}(2,k)$ and its associated Lie algebra $\mathrm{sl}(2,k)$.

5.2 Derivations and (Co)Algebra Actions Revisited

Suppose that H is a bialgebra over the field k and that A is a left H-module algebra. In this section we show that the action of skew primitives gives rise to skew derivations. We will also show how comodule algebra actions and give rise to module algebra actions.

The reader is referred to Section 1.9 for a general discussion of module and comodule algebras. Module algebras and comodule algebras are important for understanding our presentation of a *quantum calculus* in Section 5.6.

Definition 5.2.1 *Suppose that A is an algebra over the field k. Let $\gamma, \gamma' \in \mathrm{End}(A)$ be algebra endomorphisms and $\alpha \in \mathrm{End}(A)$. We say that α is a $\gamma{:}\gamma'$-*derivation *of A if*

$$\alpha(ab) = \gamma(a)\alpha(b) + \alpha(a)\gamma'(b) \tag{5.7}$$

for all $a, b \in A$.

We refer to $\gamma{:}\gamma'$-derivations more informally as skew derivations. Specializing the discussion from [Sweedler, 1969, pp. 139–140] we have [Lambe and Radford, 1993, Proposition 9.1.1]:

Proposition 5.2.1 *Suppose that H is a bialgebra over a field k. Let A be a left H-module algebra over k and let $\pi : H \longrightarrow \mathrm{End}(A)$ be the induced representation. Then:*

a) *If $g \in G(H)$ and $\gamma = \pi(g)$ then γ is an algebra endomorphism of A.*

b) *If $\ell \in H$ is a primitive and $\alpha = \pi(\ell)$ then α is a derivation of A.*

c) *Suppose that $\ell \in H$ is a $g{:}g'$-skew primitive. Set $\alpha = \pi(\ell), \gamma = \pi(g)$, and $\gamma' = \pi(g')$. Then α is a $\gamma{:}\gamma'$-derivation of A.*

Proof: By reformulating the module action in terms of the representation π we have

$$\pi(h)(1) = h{\cdot}1 = \epsilon(h)1$$

and

$$\begin{aligned}\pi(h)(ab) &= h{\cdot}(ab)\\ &= (h_{(1)}{\cdot}a)(h_{(2)}{\cdot}b)\\ &= (\pi(h_{(1)})(a))(\pi(h_{(2)})(b))\end{aligned}$$

for $h \in H$ and $a, b \in A$. The proposition follows immediately from this calculation. ∎

We will need the following relationship between comodule and module algebra actions for later use.

Proposition 5.2.2 *Suppose that H is a bialgebra over a field k and let A be a right H-comodule algebra. Suppose that (A, μ) is the resulting rational left H^*-module structure on A. Then the restriction of the rational action to the subalgebra H^o of H^* gives A the structure of a left H^o-module algebra.*

Proof: Let $a, b \in A$ and $h^o \in H^o$. Notice that

$$h^o{\cdot}1 = 1{<}h^o, 1{>} = \epsilon_{H^o}(h^o)1.$$

Since the comodule structure map is multiplicative,

$$\begin{aligned} h^o{\cdot}(ab) &= (ab)^{<1>}{<}h^o, (ab)^{(2)}{>} \\ &= a^{<1>}b^{<1>}{<}h^o, a^{(2)}b^{(2)}{>} \\ &= a^{<1>}b^{<1>}{<}h^o_{(1)}, a^{(2)}{>}{<}h^o_{(2)}, b^{(2)}{>} \\ &= (h^o_{(1)}{\cdot}a)(h^o_{(2)}{\cdot}b) \end{aligned}$$

follows by definition of comultiplication in H^o. ∎

5.3 A Hopf Algebra Closely Related to the Coordinate Ring of the Special Linear Group

Let $C = C_2(k)$. Choose a standard basis for C and denote its elements by $a = e^1_1, b = e^1_2, c = e^2_1$ and $d = e^2_2$. Then

$$\begin{aligned} \Delta(a) &= a \otimes a + b \otimes c, & \Delta(b) &= a \otimes b + b \otimes d, \\ \Delta(c) &= c \otimes a + d \otimes c, & \Delta(d) &= c \otimes b + d \otimes d \end{aligned}$$

and

$$\begin{aligned} \epsilon(a) &= 1, & \epsilon(d) &= 1, \\ \epsilon(b) &= 0, & \epsilon(c) &= 0. \end{aligned}$$

We encode these formulas in the *grouplike* formalism

$$\Delta \begin{bmatrix} a & b \\ c & d \end{bmatrix} = \begin{bmatrix} a & b \\ c & d \end{bmatrix} \otimes \begin{bmatrix} a & b \\ c & d \end{bmatrix}$$

and

$$\epsilon \begin{bmatrix} a & b \\ c & d \end{bmatrix} = I.$$

Let M be the right C-comodule with basis $\{x, y\}$ over k whose comodule structure is given by

$$\begin{aligned} \rho(x) &= x \otimes a + y \otimes c, \\ \rho(y) &= x \otimes b + y \otimes d. \end{aligned}$$

We use the formalism above and write

$$\rho\begin{bmatrix} x & y \end{bmatrix} = \begin{bmatrix} x & y \end{bmatrix} \otimes \begin{bmatrix} a & b \\ c & d \end{bmatrix}. \tag{5.8}$$

The coalgebra structure on C extends uniquely to a bialgebra structure on the tensor algebra $T(C)$, and the comodule structure on M extends uniquely to a right $T(C)$-comodule algebra structure $(T(M), \rho)$ on $T(M)$. As the reader has probably noticed, we are using ρ in two different ways in the same context, which is a slight abuse of notation we will continue. Now suppose that $q \in k$ is not zero. Consider the quotient $H_q = T(C)/I$, where I is the ideal of $T(C)$ generated by the relations

$$ba = qab, \quad ca = qac, \quad db = qbd, \quad dc = qcd, \quad bc = cb,$$

and

$$da - qbc = 1, \quad ad - q^{-1}bc = 1.$$

These relations determine a coideal of $T(C)$; thus I is a bi-ideal of $T(C)$. Therefore H_q is a bialgebra, which is in fact a Hopf algebra with antipode given by

$$s(a) = d, \quad s(b) = -qb, \quad s(c) = -q^{-1}c, \quad s(d) = a.$$

By abuse of notation we use lower case letters denote cosets as well. Notice that

$$H_q = k[\mathrm{SL}(2,k)] \quad \text{when} \quad q = 1.$$

Remark 5.3.1 Since H_q is the coordinate ring $k[\mathrm{SL}(2,k)]$ of the special linear group $\mathrm{SL}(2,k)$ over k when $q = 1$, we write

$$H_q = k[\mathrm{SL}_q(2,k)]$$

for $q \in k\backslash 0$.

Now the projection $\pi : T(C) \longrightarrow T(C)/I = H_q$ is a map of bialgebras. Since π is a coalgebra map, pushout along π gives $T(M)$ a right H_q-comodule structure, which is a comodule algebra structure since π is an algebra map. Since $\rho(yx - qxy) = (yx - qxy) \otimes (da - qcb)$ and ρ is an algebra map, the ideal $(yx - qxy)$ of $T(M)$ is a sub-comodule of $T(M)$. Thus $k[x,y]_q = T(M)/(yx - qxy)$ has the structure of a right H_q-comodule algebra $(k[x,y]_q, \rho)$.

5.4 Grouplikes and Skew Primitives of the Dual of the Coordinate Ring of the Special Linear Group

Let (A, m, η) be any algebra over k and let (A^o, Δ, ϵ) be its dual coalgebra. For $\ell \in A^o$ recall that $\Delta(\ell) = \ell_{(1)} \otimes \ell_{(2)} \in A^o \otimes A^o$ is determined by $\ell(ab) = \ell_{(1)}(a)\ell_{(2)}(b)$ for

$a, b \in A$ and $G(A^o) = \mathrm{Alg}(A, k)$. Let $\xi, \eta \in G(A^o)$ and suppose that $\ell \in A^o$ is a $\eta{:}\xi$ skew primitive. Then by definition $\Delta(\ell) = \eta \otimes \ell + \ell \otimes \xi$, or

$$\ell(ab) = \eta(a)\ell(b) + \ell(a)\xi(b) \tag{5.9}$$

for all $a, b \in A$.

We continue the discussion of the previous section. First we characterize the group $G(k[\mathrm{SL}_q(2, k)]^o)$ for all non-zero $q \in k$. When A is a bialgebra recall that A^o is a bialgebra with multiplication determined by

$$(a^o b^o)(c) = a^o(c_{(1)})b^o(c_{(2)}) \tag{5.10}$$

for all $a^o, b^o \in A^o$ and $c \in A$. If $q = 1$ then $k[\mathrm{SL}_q(2, k)] = k[\mathrm{SL}(2, k)]$. By part b) of Exercise 5.1.3 the map defined by

$$\begin{array}{rcl} G(k[\mathrm{SL}(2, k)]^o) & \longrightarrow & \mathrm{SL}(2, k) \\ \eta & \longmapsto & \begin{pmatrix} \eta(a) & \eta(b) \\ \eta(c) & \eta(d) \end{pmatrix} \end{array}$$

is a group isomorphism.

Now suppose that $q \neq 1$. We first consider the case $q = -1$. Let $\eta \in G(k[\mathrm{SL}_q(2, k)]^o)$. The algebra defining relations for $k[\mathrm{SL}_{-1}(2, k)]$ imply that $\eta(a) = 0$, in which case $\eta(d) = 0$ and $\eta(b)\eta(c) = 1$, or $\eta(b) = 0$, in which case $\eta(c) = 0$ and $\eta(a)\eta(d) = 1$. Let k^* denote the group of units of k and

$$\mathbb{Z}_2 \xrightarrow{\sigma} \mathrm{Aut}(k^*)$$

be the group homomorphism determined by $\sigma(1) = (\ )^{-1}$. It follows that the map $G(k[\mathrm{SL}_{-1}(2, k)]) \longrightarrow k^* \times_\sigma \mathbb{Z}_2$ defined by

$$\eta \longmapsto (\eta(a) + \eta(b), \delta_{\eta(a),0})$$

is a group isomorphism. The case when q is not a root of unity is of most interest to us. Using (5.10) it follows that:

Lemma 5.4.1 *Suppose that k is a field and that $q \in k$ is not zero and satisfies $q^2 \neq 1$. Then:*

a) *If $\eta \in G(k[\mathrm{SL}_q(2, k)]^o)$, then $\eta(a)\eta(d) = 1$ and $\eta(b) = 0 = \eta(c)$.*

b) *The map $G(k[\mathrm{SL}_q(2, k)]^o) \longrightarrow k^*$ defined by $\eta \longmapsto \eta(a)$ is an isomorphism of groups.* ∎

The preceding lemma is [Lambe and Radford, 1993, Lemma 9.3.3].

Now assume that $q \in k$ is not zero, and suppose $\eta, \xi \in G(k[\mathrm{SL}_q(2,k)]^o)$ vanish on b and c. Suppose that $\ell \in k[\mathrm{SL}_q(2,k)]^o$ is a $\eta{:}\xi$ skew primitive and let $I = \mathrm{Ker}\xi \cap \mathrm{Ker}\eta$. Then I is an ideal of $k[\mathrm{SL}_q(2,k)]$, and $\ell(I^2) = (0)$ by (5.9). Since $J = (b^2, bc, c^2) \subseteq I^2$, it follows that $\ell(J) = (0)$. It is easy to see that the algebra $B = k[\mathrm{SL}_q(2,k)]/J$ has a linear basis of monomials $a^n, a^n b$ and $a^n c$, where $n \in \mathbb{Z}$. Let $\pi : A \longrightarrow B$ be the projection. Since π is an onto algebra map, the restriction of the transpose

$$B^o \xrightarrow{\pi^o} k[\mathrm{SL}_q(2,k)]^o$$

is a one-one coalgebra map. Since $\ell(J) = (0)$ it follows that $\ell \in B^o$, where we identify B^o with $\mathrm{Im}\pi^o$. Thus the grouplike elements η, ξ which vanish on b, c and the $\eta{:}\xi$-skew primitives of $k[\mathrm{SL}_q(2,k)]^o$ lie in the subcoalgebra B^o.

Let $\eta, \xi \in G(B^o)$ and suppose that $\ell \in B^o$ is a $\eta{:}\xi$ skew primitive. Suppose that $u = b$ or c. Observe that $\ell(a^n u) = \eta(a)^n \ell(u)$, and the relation $ua = qau$ implies that $\ell(u)(\xi(a) - q\eta(a)) = 0$. By induction

$$\ell(a^n) = \frac{\eta(a)^n - \xi(a)^n}{\eta(a) - \xi(a)}\ell(a)$$

for $n \in \mathbb{Z}$ if $\xi \neq \eta$, and

$$\ell(a^n) = n\xi(a)^{n-1}\ell(a)$$

for $n \in \mathbb{Z}$ if $\xi = \eta$. This means ℓ is determined by the values $\ell(a), \ell(b)$ and $\ell(c)$. An element $\ell \in B^o$ satisfying these conditions is easily seen to be a $\eta{:}\xi$ skew primitive. To summarize [Lambe and Radford, 1993, Lemma 9.3.4]:

Lemma 5.4.2 *Suppose that k is a field and $q \in k$ is not zero. Let $\eta, \xi \in k[\mathrm{SL}_q(2,k)]^o$ be grouplike elements which vanish on b and c. Then:*

a) *If $\ell \in k[\mathrm{SL}_q(2,k)]^o$ is a $\eta{:}\xi$ skew primitive, then $\ell(b)(\xi(a) - q\eta(a)) = 0 = \ell(c)(\xi(a) - q\eta(a))$.*

b) *Conversely, if $\mathcal{A}, \mathcal{B}, \mathcal{C} \in k$ and $\mathcal{B}(\xi(a) - q\eta(a)) = 0 = \mathcal{C}(\xi(a) - q\eta(a))$, then there is a unique $\eta{:}\xi$ skew primitive $\ell \in k[\mathrm{SL}_q(2,k)]^o$ such that $\ell(a) = \mathcal{A}$, $\ell(b) = \mathcal{B}$ and $\ell(c) = \mathcal{C}$.* ∎

5.5 Embedding the Universal Enveloping Algebra of the Special Linear Group into the Dual Coordinate Ring of the Special Linear Group

It is well-known that the Hopf algebra $\mathcal{U}(\mathrm{sl}(2,k))$ is realized as a sub-Hopf algebra of $k[\mathrm{SL}(2,k)]^o$. Specifically $\mathcal{U}(\mathrm{sl}(2,k))$ can be identified with the connected component of the identity of $k[\mathrm{SL}(2,k)]^o$. See the exercises below.

The purpose of this section is to examine this embedding in detail. We continue to use the notation and conventions of Section 5.4.

As noted in Remark 5.3.1, $k[\mathrm{SL}(2,k)] = k[\mathrm{SL}_q(2,k)]$ when $q = 1$. In addition, when $q = 1$, the algebra $k[x,y]_q$ is just the ordinary commutative polynomial algebra $k[x,y]$ over k. We will examine the results of Section 5.4 in this case.

Let $L = P_\epsilon(k[\mathrm{SL}(2,k)]^o)$ be the subspace of primitives of $k[\mathrm{SL}(2,k)]^o$. Then $\mathrm{Dim}L = 3$ and there are primitives ℓ_a, ℓ_b and $\ell_c \in L$ determined by $\ell_u(v) = \delta_{uv}$ for $u, v \in \{a, b, c\}$ by part b) of Lemma 5.4.2. Clearly ℓ_a, ℓ_b and ℓ_c form a basis for L. Using part b) of Lemma 5.4.2 again, it follows that $[\ell_b, \ell_c] = \ell_b\ell_c - \ell_c\ell_b = \ell_a$, $[\ell_a, \ell_b] = 2\ell_b$ and $[\ell_a, \ell_c] = -2\ell_c$. Consequently $L \simeq \mathrm{sl}(2,k)$.

Recall from Section 5.3 that $k[x,y]$ has a right $k[\mathrm{SL}(2,k)]$-comodule algebra structure $(k[x,y], \rho)$, where ρ is determined by (5.8). Let $(k[x,y], \mu_\rho)$ be the induced left rational $k[\mathrm{SL}(2,k)]^*$-module structure on $k[x,y]$ restricted to $k[\mathrm{SL}(2,k)]^o$. Then $(k[x,y], \mu_\rho)$ is a left $k[\mathrm{SL}(2,k)]^o$-module algebra by Proposition 5.2.2. Let

$$k[\mathrm{SL}(2,k)]^o \xrightarrow{\ \pi\ } \mathrm{End}(k[x,y])$$

be the induced representation. Then by definition:

$$\begin{array}{rclcrcl} \ell_b{\cdot}x &=& 0, && \ell_c{\cdot}x &=& y \\ \ell_b{\cdot}y &=& x, && \ell_c{\cdot}y &=& 0. \end{array} \tag{5.11}$$

For $u = a, b$ or c set $D_u = \pi(\ell_u)$. Since ℓ_u is a primitive, and $(k[x,y], \mu_\rho)$ is a left $k[\mathrm{SL}_q(2,k)]^o$-module algebra, it follows that D_u is a derivation of $k[x,y]$ by part b) of Proposition 5.2.2. Now $x\frac{\partial}{\partial y}$ and $y\frac{\partial}{\partial x}$ are also derivations of $k[x,y]$. Since two derivations are the same they agree on algebra generators, it follows that

$$D_b = x\frac{\partial}{\partial y}, \quad D_c = y\frac{\partial}{\partial x},$$

and

$$D_a = [D_b, D_c] = x\frac{\partial}{\partial x} - y\frac{\partial}{\partial y}.$$

We have recaptured the results of Section 5.1.3 in a natural way. Indeed, a proof of the fact that $\mathcal{U}(\mathrm{sl}(2,k))$ can be embedded in $k[\mathrm{SL}(2,k)]^o$ as a sub-Hopf algebra can be made easily by showing that formally different "monomials" of the type $D_a^r D_b^m D_c^n$ are linearly independent, where $0 \leq r, m, n$. This technique is adapted in the next section to show that the quantized enveloping algebra $\mathcal{U}_q'(\mathrm{sl}(2,k))$ can be embedded into $k[\mathrm{SL}_q(2,k)]^o$.

Exercise 5.5.1 Suppose that C is a coalgebra over the field k.

a) Let D be a subcoalgebra of the coradical C_0 of C. Show that there is a unique subcoalgebra $\mathcal{C}$ of C maximal with respect to the property that $\mathcal{C}_0 = D$.

b) Show that $\mathcal{C} = \cup_{n=0}^{\infty}(\wedge^n D)$.

c) Suppose that C is a bialgebra over k and that D is a sub-bialgebra of C. Show that $\mathcal{C}$ is a sub-bialgebra of C.

d) Suppose that C is a Hopf algebra over k and that D is a sub-Hopf algebra of C. Show that $\mathcal{C}$ is a sub-Hopf algebra of C.

Definition 5.5.1 *Let C be a Hopf algebra and $D = k1$. Then the sub-Hopf algebra $\mathcal{C} = \cup_{n=0}^{\infty}(\wedge^n D)$ of C is the* connected component of the identity.

In particular the connected component of the identity is a pointed irreducible subcoalgebra of C.

Exercise 5.5.2 Show that the subalgebra of $k[\mathrm{SL}(2,k)]^o$ generated by the primitive elements ℓ_a and ℓ_b is a sub-Hopf algebra isomorphic to $\mathcal{U}(\mathrm{sl}(2,k))$. [Hint: The main problem is to show that the formally different monomials $\ell_a^r \ell_b^m \ell_c^m$ form a linearly independent set. Independence will follow once it is shown that the formally different monomials $D_a^r D_b^m D_c^n$ form a linearly independent set.]

5.6 Quantum Analogs of the Universal Enveloping Algebra $\mathcal{U}(\mathbf{sl}(2,\mathbf{k}))$

We will consider two quantum analogs of the universal enveloping algebra $\mathcal{U}(\mathrm{sl}(2,k))$ denoted by $\mathcal{U}_q(\mathrm{sl}(2,k))$ and $\mathcal{U}'_q(\mathrm{sl}(2,k))$. The latter $\mathcal{U}'_q(\mathrm{sl}(2,k))$ is a sub-Hopf algebra of the former $\mathcal{U}_q(\mathrm{sl}(2,k))$. In [Koorwinder, 1990] it is noted that an identification of $\mathcal{U}_q(\mathrm{sl}(2,k))$ in $k[\mathrm{SL}_q(2,k)]^o$ is found in [Soibelman and Vaksman, 1988] when $k = \mathbb{C}$ is the field of complex numbers. The reverse, namely, the identification of $k[\mathrm{SL}_q(2,k)]$ in $\mathcal{U}'_q(\mathrm{sl}(2,k))^o$ is given in [Takeuchi, 1992]. In fact [Takeuchi, 1992] characterizes the entire Hopf algebra dual $\mathcal{U}'_q(\mathrm{sl}(2,k))^o$ in terms of $k[\mathrm{SL}_q(2,k)]$. Also noted in this paper is that [Drinfel'd, 1987] gives quite natural Hopf algebra maps

$$\mathcal{U}_q(\mathrm{sl}(n,k)) \longrightarrow k[\mathrm{SL}_q(n,k)]^o \quad \text{and} \quad k[\mathrm{SL}_q(n,k)] \longrightarrow \mathcal{U}_q(\mathrm{sl}(n,k))^o$$

which are adjoint to each other. Related to these notions of quantum calculus is [Montgomery and Smith, 1990] in which all possible module algebra actions of $\mathcal{U}(\mathrm{sl}(2,k))$ on the polynomial algebra $\mathbb{C}[x]$ are determined and a natural module algebra action of $\mathcal{U}(\mathrm{sl}(2,k))$ and a sub-Hopf algebra (quite similar to $\mathcal{U}'_q(\mathrm{sl}(2,k))$ mentioned above) on the quantum affine plane $k[x,y]_q$ are given.

We treat the quantum calculus for $\mathcal{U}_q(\mathrm{sl}(2,k))$ using only basic results from the theory of Hopf algebras. We also work over an arbitrary field. From this point of view the embedding of $\mathcal{U}_q(\mathrm{sl}(2,k))$ into $k[\mathrm{SL}_q(2,k)]^o$ follows by very elementary reasons. The calculus for $\mathcal{U}_q(\mathrm{sl}(2,k))$ is seen to arise from a basic $k[\mathrm{SL}_q(2,k)]$ structure on the quantum affine plane $k[x,y]_q$. We use results on maximal weights in describing the irreducible representations of $\mathcal{U}(\mathrm{sl}(2,k))$ from the perspective of quantum calculus [Rosso, 1988].

Let k be any field, and suppose that $q \in k$ is a square which is neither zero nor a root of unity. Let $\mathcal{U}_q(\mathrm{sl}(2,k))$ be generated as an algebra by symbols k, k^{-1}, e and f subject to the relations

$$kk^{-1} = 1 = k^{-1}k, \quad kek^{-1} = qe, \quad kfk^{-1} = q^{-1}f,$$

and

$$[e, f] = \frac{k^2 - k^{-2}}{q - q^{-1}}.$$

The coalgebra structure of $\mathcal{U}_q(\mathrm{sl}(2,k))$ is determined by

$$\Delta(k) = k \otimes k, \quad \Delta(k^{-1}) = k^{-1} \otimes k^{-1},$$

and

$$\Delta(e) = e \otimes k^{-1} + k \otimes e, \quad \Delta(f) = f \otimes k^{-1} + k \otimes f;$$

that is k, k^{-1} are grouplike and e, f are $k{:}k^{-1}$ skew primitives. By Lemma 1.1.1 the set of grouplike elements of a coalgebra is linearly independent. Thus by Corollaries 1.7.3 and 1.7.5 it follows that $\mathcal{U}_q(\mathrm{sl}(2,k))$ has an antipode.

The possibilities for embedding $\mathcal{U}_q(\mathrm{sl}(2,k))$ into $k[\mathrm{SL}_q(2,k)]^o$ are quite limited. Suppose that $\ell \in k[\mathrm{SL}_q(2,k)]^o$ is a $\eta{:}\xi$ skew primitive, where $\xi \neq \eta$, and $\ell(b) = 0 = \ell(c)$. Since $\xi - \eta$ is also a $\eta{:}\xi$ skew primitive, we see that

$$\ell = \frac{\ell(a)}{\xi(a) - \eta(a)}(\xi - \eta)$$

by using part b) of Lemma 5.4.2. To find an embedding of

$$\mathcal{U}_q(\mathrm{sl}(2,k))$$

into

$$k[\mathrm{SL}_q(2,k)]^o,$$

let $k = \eta$ and $\xi = \eta^{-1} = k^{-1}$. We need to find $k : k^{-1}$ skew primitives ℓ such that $\ell(b) \neq 0$ or $\ell(c) \neq 0$. But by part a) of Lemma 5.4.2, $k(a)^{-1} - qk(a) = 0$. Thus,

$$k(a)^2 = q^{-1}.$$

By part b) of Lemma 5.4.2 again, there are $k{:}k^{-1}$ skew primitives $e, f \in k[\mathrm{SL}_q(2,k)]^o$ determined by

$$\begin{array}{rclrclrcl} e(a) & = & 0, & e(b) & = & 0, & e(c) & = & \alpha, \\ f(a) & = & 0, & f(b) & = & \alpha^{-1}, & f(c) & = & 0, \end{array}$$

where $\alpha \in k$ and is not zero. Let $(k[\mathrm{SL}_q(2,k)]^o, \Delta, \epsilon)$ be the dual coalgebra of the algebra $k[\mathrm{SL}_q(2,k)]$. Then

$$\Delta(k) = k \otimes k, \quad \Delta(k^{-1}) = k^{-1} \otimes k^{-1},$$

and

$$\Delta(e) = e \otimes k^{-1} + k \otimes e, \quad \Delta(f) = f \otimes k^{-1} + k \otimes f.$$

Since e and f are $k{:}k^{-1}$ skew primitives, it follows that kek^{-1} and kfk^{-1} are also. Note that $k\ell k^{-1}(a) = \ell(a), k\ell k^{-1}(b) = q^{-1}\ell(b)$, and that $k\ell k^{-1}(c) = q\ell(c)$ for $\ell \in k[\mathrm{SL}_q(2,k)]^o$. By part b) of (5.4.2) we have

$$kek^{-1} = qe \quad \text{and} \quad kfk^{-1} = q^{-1}f.$$

From the last two equations we deduce $[e,f] = ef - fe$ is a $k^2{:}k^{-2}$ skew primitive. Since $q^2 \neq 1$ it follows that $k^2 \neq k^{-2}$. By part b) of Lemma 5.4.2 again we have that

$$[e,f] = \frac{k^2 - k^{-2}}{q - q^{-1}}.$$

Let $(k[x,y]_q, \rho)$ be the right $k[\mathrm{SL}_q(2,k)]$-comodule algebra structure described in Section 5.3 and $(k[x,y]_q, \mu)$ be the induced rational left $k[\mathrm{SL}_q(2,k)]^*$-module action restricted to $k[\mathrm{SL}_q(2,k)]^o$. Then by (5.2.2) we have that $(k[x,y]_q, \mu)$ is a left $k[\mathrm{SL}_q(2,k)]^o$-module algebra. Note that

$$\begin{aligned} e{\cdot}x &= \alpha y, & f{\cdot}x &= 0, \\ e{\cdot}y &= 0, & f{\cdot}y &= \alpha^{-1}x, \end{aligned} \tag{5.12}$$

and

$$k{\cdot}x = k(a)x, \quad k{\cdot}y = k(a)^{-1}y,$$

where $k(a)^2 = q^{-1}$. When $\alpha = q$, x and y are reversed, and q is replaced by q^2, the module algebra action is that of [Montgomery and Smith, 1990].

Let $\widetilde{\mathcal{U}_q}$ be the subalgebra of $k[\mathrm{SL}_q(2,k)]^o$ generated by k, k^{-1}, e and f. To show that $\widetilde{\mathcal{U}_q} = \mathcal{U}_q(\mathrm{sl}(2,k))$, we need to show that the monomials of the form $k^\ell e^m f^n$, where $\ell \in \mathbb{Z}$ and $m, n \geq 0$, are linearly independent. To this end note that the $k[\mathrm{SL}_q(2,k)]^*$-module action on the quantum plane distinguishes formally different monomials. To see this, observe that $e{\cdot}(uv) = (e{\cdot}u)(k^{-1}{\cdot}v) + (f{\cdot}u)(e{\cdot}v)$ and $k{\cdot}(uv) = (k{\cdot}u)(k{\cdot}v)$ for $u, v \in k[x,y]_q$. Let $\beta = k(a)$. By induction

$$\begin{aligned} k{\cdot}(x^n y^m) &= \beta^{n-m} x^n y^m \quad \text{for } n, m \geq 0, \\ e{\cdot}y^m &= 0 \quad \text{for } m \geq 0, \end{aligned}$$

and

$$e{\cdot}x^n = \alpha\beta^{-3(n-1)}\left(\frac{\beta^{4n} - 1}{\beta^4 - 1}\right)x^{n-1}y \quad \text{for } n \geq 1.$$

Since $\beta^2 = q^{-1}$ and $e{\cdot}(x^n y^m) = (e{\cdot}x^n)(k^{-1}{\cdot}y^m)$ we have

$$e{\cdot}(x^n y^m) = \alpha\beta^{m-n+1}[n]_q x^{n-1}y^{m+1} \quad \text{for } n, m \geq 0,$$

where

$$[0]_q = 0 \tag{5.13}$$

and

$$[n]_q = \frac{q^n - q^{-n}}{q - q^{-1}} = q^{n-1}(\frac{q^{-2n} - 1}{q^{-2} - 1}) = q^{-(n-1)}(\frac{q^{2n} - 1}{q^2 - 1}) \tag{5.14}$$

for $n \geq 1$. Compare these coefficients with the q-binomial coefficients of Section 6.5.1.

Similarly $f{\cdot}x^n = 0$ for $n \geq 0$ and

$$f{\cdot}y^m = \alpha^{-1}\beta^{3(m-1)}(\frac{\beta^{-4m} - 1}{\beta^{-4} - 1})y^{m-1}x$$

for $m \geq 1$. Since $f{\cdot}(x^n y^m) = (k{\cdot}x^n)(f{\cdot}y^m)$ it now follows

$$f{\cdot}(x^n y^m) = \alpha^{-1}\beta^{n-m+1}[m]_q x^{n+1}y^{m-1} \quad \text{for all } n, m \geq 0.$$

A variation of $\mathcal{U}_q(\mathrm{sl}(2,k))$ is used in [Takeuchi, 1992]. We denote it here by $\mathcal{U}'_q(\mathrm{sl}(2,k))$. It is defined to be the following sub-Hopf algebra of $\mathcal{U}_q(\mathrm{sl}(2,k))$ generated by E, F, and K where $K = k^{-2}, F = ke$ and $E = fk^{-1}$. It is easy to to establish the algebra defining relations

$$KK^{-1} = 1 = K^{-1}K, \quad KEK^{-1} = q^2E, \quad KFK^{-1} = q^{-2}F \quad \text{and}$$

$$[E, F] = \frac{K - K^{-1}}{q - q^{-1}}.$$

It is clear that

$$\Delta(K) = K \otimes K, \quad \Delta(K^{-1}) = k^{-1} \otimes K^{-1},$$

and

$$\Delta(E) = E \otimes K + 1 \otimes E, \quad \Delta(F) = F \otimes 1 + K^{-1} \otimes F,$$

which determines the coalgebra structure of $\mathcal{U}'_q(\mathrm{sl}(2,k))$. Set $\alpha = k(a)$. Then

$$E{\cdot}x = 0, \quad F{\cdot}x = y,$$

$$E{\cdot}y = x, \quad F{\cdot}y = 0,$$

and

$$K{\cdot}x = qx, \quad K{\cdot}y = q^{-1}y,$$

which determine a $\mathcal{U}_q'(\mathrm{sl}(2,k))$-module algebra action on the quantum plane $k[x,y]_q$. Let $\pi : \mathcal{U}_q'(\mathrm{sl}(2,k)) \longrightarrow \mathrm{End}(k[x,y]_q)$ be the induced representation.

It is very natural to try to define a $\mathcal{U}_q'(\mathrm{sl}(2,k))$-module algebra action on $k[x,y]$, extending the action of E, F and K on x and y described above. However, the calculations $E\cdot(xy) = x^2$ and $E\cdot(yx) = qx^2$ are inconsistent with $xy = yx$. To remove this inconsistency, we can deform the multiplication of $k[x,y]$ by setting $yx = qxy$. Remarkably, this leads to a module algebra action involving $\mathcal{U}_q(\mathrm{sl}(2,k))$ that parallels the ordinary calculus for $\mathcal{U}(\mathrm{sl}(2,k))$ described in Section 5.1.3.

Using our work above, we find that the action of E, F, and K on monomials has the following description:

$$E\cdot(x^n y^m) = [m]_q x^{n+1} y^{m-1}, \tag{5.15}$$

$$F\cdot(x^n y^m) = [n]_q x^{n-1} y^{m+1}, \tag{5.16}$$

$$K\cdot(x^n y^m) = q^{n-m} x^n y^m, \tag{5.17}$$

for $n, m \geq 0$.

Let $\kappa : k[x] \longrightarrow k[x]$ be the algebra automorphism determined by $\kappa(x) = qx$, and let

$$k[x] \xrightarrow{\frac{\partial}{\partial x}_q} k[x]$$

be the linear map defined by

$$\frac{\partial}{\partial x}_q (x^n) = [n]_q x^{n-1}$$

for $n \geq 0$. Then $\frac{\partial}{\partial x}_q$ is easily seen to be a $\kappa{:}\kappa^{-1}$-derivation. Since two $\kappa{:}\kappa^{-1}$-derivation are the same if the agree on x, it follows that the $\kappa{:}\kappa^{-1}$-derivations of $k[x]$ are the $(\ell(u))(\frac{\partial}{\partial x}_q)$'s, where $\ell(u) : k[x] \longrightarrow k[x]$ is left multiplication by $u \in k[x]$.

Notice that the skew derivations determined by left multiplication by E and F are naturally expressible in terms of $\frac{\partial}{\partial x}_q$ and $\frac{\partial}{\partial y}_q$, namely

$$E\cdot(x^n y^m) = x^{n+1}(\frac{\partial}{\partial y}_q (y^m)) \quad \text{and} \quad F\cdot(x^n y^m) = (\frac{\partial}{\partial x}_q (x^n)) y^{m+1}$$

for $m, n \geq 0$. The description of the basic finite-dimensional simple $\mathcal{U}_q'(\mathrm{sl}(2,k))$-modules in [Takeuchi, 1992] can now be given. Fix $n \geq 0$ and set $u_i = x^{n-i}y^i$ for $0 \leq i \leq n$. Then $\{u_0, \ldots, u_n\}$ is a basis for M_n. Using the highest weight theory developed in [Rosso, 1988], we see that M_n is an irreducible $\mathcal{U}_q'(\mathrm{sl}(2,k))$-module and any finite-dimensional irreducible representation of $\mathcal{U}_q'(\mathrm{sl}(2,k))$ is equivalent to M_n for some n. We have have recaptured the theorems on irreducible representations of $\mathcal{U}_q'(\mathrm{sl}(2,k))$ from [Rosso, 1988] and [Takeuchi, 1992].

Theorem 5.6.1 *Suppose that the field k has characteristic* 0 *and let*

$$\mathcal{U}_q'(\mathrm{sl}(2,k)) \xrightarrow{\pi} \mathrm{End}(k[x,y]_q)$$

be the representation defined above. Let $M_n \subseteq k[x,y]_q$ *be the submodule of homogeneous polynomials of degree* n. *Then* M_n *is an irreducible* $\mathcal{U}_q'(\mathrm{sl}(2,k))$*-module, and any finite-dimensional irreducible* $\mathcal{U}_q'(\mathrm{sl}(2,k))$*-module is equivalent to* M_n *for some* $n \geq 0$. ∎

By virtue of (5.15), (5.16), and (5.17) the action of $\mathcal{U}_q'(\mathrm{sl}(2,k))$ on M_n has a very simple description using the basis $\{u_0, \ldots, u_n\}$. Identifying operators and their matrices with respect to this basis, we have

$$\pi(E) = \begin{pmatrix} 0 & [1]_q & 0 & \cdots & 0 \\ 0 & 0 & [2]_q & \cdots & 0 \\ \vdots & \vdots & \vdots & \vdots & \vdots \\ 0 & 0 & 0 & 0 & [n]_q \\ 0 & 0 & 0 & 0 & 0 \end{pmatrix},$$

$$\pi(F) = \begin{pmatrix} 0 & 0 & 0 & 0 & 0 \\ [n]_q & 0 & 0 & \cdots & 0 \\ 0 & [n-1]_q & 0 & \cdots & 0 \\ \vdots & \vdots & \vdots & \vdots & \vdots \\ 0 & 0 & 0 & [1]_q & 0 \end{pmatrix},$$

and

$$\pi(K) = \begin{pmatrix} q^n & 0 & 0 & 0 & 0 \\ 0 & q^{n-2} & 0 & \cdots & 0 \\ & & \ddots & & \\ 0 & \cdots & 0 & q^{2-n} & 0 \\ 0 & 0 & 0 & 0 & q^{-n} \end{pmatrix}.$$

6 QUASITRIANGULAR ALGEBRAS, BIALGEBRAS, HOPF ALGEBRAS AND THE QUANTUM DOUBLE

Quasitriangular algebras give rise to solutions to the QYBE through their representations. Quasitriangular Hopf algebras have been very widely studied. See [Drinfel'd, 1987], [Drinfel'd, 1990], [Majid, 1990b], [Majid, 1991a], and [Radford, 1993b] for example. One of the most important examples of a quasitriangular Hopf algebra is the quantum double [Drinfel'd, 1987, p. 816]. In this chapter our focus will be on finite-dimensional objects. Every finite-dimensional Hopf algebra over a field is a sub-Hopf algebra of a quantum double in a natural way. In particular there is a close connection between finite-dimensional Hopf algebras and solutions to the QYBE. Our treatment of quasitriangular Hopf algebras and the quantum double follows [Radford, 1993b].

6.1 Quasitriangular Algebras

One approach to solving the QYBE is through representations of special kinds of algebras A over the field k. These algebras possess an element $R = \sum_{i=1}^{r} a_i \otimes b_i \in A \otimes A$ such that for any left A-module M the linear map $R_M : M \otimes M \longrightarrow M \otimes M$

defined by

$$R_M(m \otimes n) = \sum_{i=1}^{r} a_i{\cdot}m \otimes b_i{\cdot}n \tag{6.1}$$

for all $m, n \in M$ is a solution to the QYBE. Observe that R_M is a solution to the QYBE for all left A-modules M if and only if R_A is a solution. The latter is the case if and only if

$$\sum_{i,j,k=1}^{r} a_i a_j \otimes b_i a_k \otimes b_j b_k = \sum_{i,j,k=1}^{r} a_j a_i \otimes a_k b_i \otimes b_k b_j. \tag{6.2}$$

Let $R_{(1,2)} = \sum_{i=1}^{r} a_i \otimes b_i \otimes 1$, $R_{(1,3)} = \sum_{i=1}^{r} a_i \otimes 1 \otimes b_i$, and $R_{(2,3)} = \sum_{i=1}^{r} 1 \otimes a_i \otimes b_i$. Then (6.2) is rewritten in the more familiar form

$$R_{(1,2)}R_{(1,3)}R_{(2,3)} = R_{(2,3)}R_{(1,3)}R_{(1,2)}. \tag{6.3}$$

This is consistent with the notation used in Section 2.1.

Definition 6.1.1 *Let A be an algebra over the field k and $R \in A \otimes A$. The pair (A, R) is a* quasitriangular algebra *if (6.2) holds, or equivalent if (6.3) holds.*

Definition 6.1.2 *Let (A, R) and (A', R') be quasitriangular algebras over k. Then a* map of quasitriangular algebras *$f : (A, R) \longrightarrow (A', R')$ is a map of algebras $f : A \longrightarrow A'$ such that $(f \otimes f)(R) = R'$.*

A very extensive class of finite-dimensional Hopf algebras H have have a distinguished element $R \in H \otimes H$ resulting in a quasitriangular algebra structure (H, R), namely the finite-dimensional quasitriangular Hopf algebras. Finite-dimensional quantum doubles belong to this class. There are very close connections between quantum doubles and quasitriangular Hopf algebras in general. Solutions to the QYBE arising from distinguished elements R of finite-dimensional quasitriangular Hopf algebras according to (6.1) arise from finite-dimensional quantum doubles. For this reason modules for the quantum double will be of great interest to us. Also see [Majid, 1991a].

6.2 Quasitriangular Structures Arising from Integrals

This section is based on [Lambe and Radford, 1993, Section 8.3]. Suppose that H is a finite-dimensional Hopf algebra over k. We show that integrals for H equip H with a quantum algebra structure in certain cases. Perhaps this not too surprising since integrals are at the foundation of the structure theory for finite-dimensional Hopf algebras. We begin this section by discussing some basic facts about integrals.

Definition 6.2.1 *A* left *(respectively* right*)* integral *for a Hopf algebra H is an element $\Lambda \in H$ such that $a\Lambda = \epsilon(a)\Lambda$ (respectively $\Lambda a = \epsilon(a)\Lambda$) for all $a \in H$.*

Left integrals (and also right integrals) form an ideal of H, which is one-dimensional by [Sweedler, 1969, Corollary 5.1.6].

Definition 6.2.2 *A Hopf algebra H is* unimodular *if the ideal of left integrals for H and the ideal of right integrals for H are the same. In this case left, and hence right, integrals are* two-sided integrals.

Suppose that Λ is a non-zero left or right integral for H. Then H is semisimple as an algebra if and only if $\epsilon(\Lambda) \neq 0$ by [Sweedler, 1969, Theorem 5.1.8]. An easy exercise for the reader is to show that semisimple implies unimodular.

Let $\Lambda \in H$ be a left integral for H and let $a \in H$. Then

$$\Lambda_{(1)} \otimes a\Lambda_{(2)} = s(a)\Lambda_{(1)} \otimes \Lambda_{(2)}. \tag{6.4}$$

This equation follows from the calculation

$$\begin{aligned} \Lambda_{(1)} \otimes a\Lambda_{(2)} &= \epsilon(a_{(1)})\Lambda_{(1)} \otimes a_{(2)}\Lambda_{(2)} \\ &= s(a_{(1)})a_{(2)}\Lambda_{(1)} \otimes a_{(3)}\Lambda_{(2)} \\ &= s(a_{(1)})(a_{(2)}\Lambda)_{(1)} \otimes (a_{(2)}\Lambda)_{(2)} \\ &= s(a_{(1)})\epsilon(a_{(2)})\Lambda_{(1)} \otimes \Lambda_{(2)} \\ &= s(a)\Lambda_{(1)} \otimes \Lambda_{(2)}. \end{aligned}$$

In a similar manner one can show that

$$\Lambda'_{(1)}a \otimes \Lambda'_{(2)} = \Lambda'_{(1)} \otimes \Lambda'_{(2)}s(a) \tag{6.5}$$

where Λ' is a right integral for H.

By [Larson and Radford, 1988, Theorem 4] the following theorem holds when H is semisimple as an algebra and the characteristic of k is 0. It also holds when H is commutative. The following is [Lambe and Radford, 1993, Theorem 8.3.3].

Theorem 6.2.1 *Suppose that H is a finite-dimensional unimodular Hopf algebra with antipode s over the field k such that $s^2 = 1_H$. Let Λ be a non-zero two-sided integral for H and set $R = \Lambda_{(1)} \otimes s(\Lambda_{(2)})$. Let R_H be defined by (6.1). Then:*

a) *R_H is a solution to the QYB.*

b) *R_H is invertible if and only if $H = k$.*

Proof: Write $\Lambda = \sum_{i=1}^{r} a_i \otimes b_i$. Then $R = \sum_{i=1}^{r} a_i \otimes s(b_i)$ by definition. Part a) is equivalent to

$$\sum_{i,j,k=1}^{r} a_i a_j \otimes s(b_i)a_k \otimes s(b_j)s(b_k) = \sum_{i,j,k=1}^{r} a_j a_i \otimes a_k s(b_i) \otimes s(b_k)s(b_j). \tag{6.6}$$

Since Λ is a right integral for H we use (6.5) to rewrite the left hand side of (6.6)

$$\begin{aligned}\sum_{i,j,k=1}^{r} a_i a_j \otimes s(b_i)a_k \otimes s(b_j)s(b_k) &= \sum_{i,j,k=1}^{r} a_i a_j \otimes s(b_i)a_k \otimes s(b_k b_j)\\ &= \sum_{i,j,k=1}^{r} a_i a_j \otimes s(b_i)a_k s^{-1}(b_j) \otimes s(b_k).\end{aligned}$$

Since Λ is a left integral for H we use (6.4) to rewrite the right hand side of (6.6)

$$\begin{aligned}\sum_{i,j,k=1}^{r} a_j a_i \otimes a_k s(b_i) \otimes s(b_k)s(b_j) &= \sum_{i,j,k=1}^{r} a_j a_i \otimes a_k s(b_i) \otimes s(b_j b_k)\\ &= \sum_{i,j,k=1}^{r} a_j a_i \otimes s(b_j)a_k s(b_i) \otimes s(b_k)\end{aligned}$$

Now $s = s^{-1}$ since $s^2 = I$. Therefore (6.6) is established and part a) follows.

To show part b) we first of all suppose that R_H is invertible. Then $\sum_{i=1}^{r} \Lambda_{(1)}c_i \otimes s(\Lambda_{(2)})d_i = 1 \otimes 1$ for some $\nu = \sum_{i=1}^{n} c_i \otimes d_i \in H \otimes H$. Since Λ is a right integral for H it follows that $\sum_{i=1}^{n} \Lambda\Lambda_{(1)}c_i \otimes s(\Lambda_{(2)})d_i = \Lambda \otimes (s(\Lambda)(\sum_{i=1}^{n} d_i \epsilon(c_i)))$. Therefore $\Lambda \otimes s(\Lambda)a = \Lambda \otimes 1$ for some $a \in H$. This means $s(\Lambda)$ has a right inverse. Consequently Λ has a left inverse since s is bijective. Now Λ generates a one-dimensional ideal of H; thus $H = k$. Since the converse is clear we are done. ∎

6.3 Quasitriangular Bialgebras and Quasitriangular Hopf Algebras

Definition 6.3.1 *A* quasitriangular bialgebra *(respectively* quasitriangular Hopf algebra*) over the field k is a pair (A, R), where A is a bialgebra (respectively Hopf algebra) over k and $R = \sum_{i=1}^{r} a_i \otimes b_i \in A \otimes A$ satisfies the following:*

(QT.1) $\sum_{i=1}^{r} \Delta(a_i) \otimes b_i = \sum_{i,j=1}^{r} a_i \otimes a_j \otimes b_i b_j$,

(QT.2) $\sum_{i=1}^{r} \epsilon(a_i)b_i = 1$,

(QT.3) $\sum_{i=1}^{r} a_i \otimes \Delta^{cop}(b_i) = \sum_{i,j=1}^{r} a_i a_j \otimes b_i \otimes b_j$,

(QT.4) $\sum_{i=1}^{r} a_i \epsilon(b_i) = 1$, *and*

(QT.5) $(\Delta^{cop}(a))R = R(\Delta(a))$ *for all $a \in A$.*

Notice that (QT.1) and (QT.2) imply that R is invertible when A is a Hopf algebra with antipode s. For

$$\sum_{i,j=1}^{r} s(a_i)a_j \otimes b_i b_j = \sum_{i=1}^{r} s(a_{i(1)})a_{i(2)} \otimes b_i = \sum_{i=1}^{r} \epsilon(a_i)1 \otimes b_i$$

by (QT.1) and

$$\sum_{i=1}^{r} \epsilon(a_i)1 \otimes b_i = 1 \otimes (\sum_{i=1}^{r} \epsilon(a_i)b_i) = 1 \otimes 1$$

by (QT.2). Likewise $\sum_{i,j=1}^{r} a_i s(a_j) \otimes b_i b_j = 1 \otimes 1$. We have shown that R has an inverse and $R^{-1} = \sum_{i=1}^{r} s(a_i) \otimes b_i = (s \otimes I)(R)$. The definition of quasitriangular Hopf algebra we make here, which is used in [Radford, 1992, Section 2], is easily seen to be equivalent to the definition [Drinfel'd, 1987, p. 811] in the category of finite-dimensional vector spaces over a field.

Definition 6.3.2 *A finite-dimensional bialgebra A over the field k* admits a quasitriangular structure *if (A, R) is quasitriangular for some $R \in A \otimes A$.*

A cocommutative bialgebra A has the quasitriangular structure $(A, 1 \otimes 1)$. By (QT.5) a commutative quasitriangular Hopf algebra is necessarily cocommutative.

Definition 6.3.3 *A* morphism $f : (A, R) \longrightarrow (A', R')$ *of* quasitriangular bialgebras *(respectively* quasitriangular Hopf algebras*) over k is a bialgebra map $f : A \longrightarrow A'$ such that $R' = (f \otimes f)(R)$.*

There appears to be an asymmetry in the definition. There really is none. Recall from part c) of Lemma 1.6.1 that bialgebra maps of Hopf algebras are automatically Hopf algebra maps.

Conditions (QT.1)-(QT.5) are encoded in the existence of certain bialgebra maps. Let V be a vector space over the field k and suppose that $R = \sum_{i=1}^{r} a_i \otimes b_i \in V \otimes V$. Define a linear map $f_R : V^* \longrightarrow V$ by

$$f_R(p) = (p \otimes 1_V)(R) = \sum_{i=1}^{r} p(a_i)b_i$$

for all $p \in V^*$.

Proposition 6.3.1 *Suppose that A is a finite-dimensional bialgebra over the field k and $R \in A \otimes A$. Then the following are equivalent:*

a) *(A, R) is a quasitriangular bialgebra.*

b) *$f_R : A^* \longrightarrow A^{cop}$ is a map of bialgebras and*

$$(p_{(1)} \rightharpoonup a) f_R(p_{(2)}) = f_R(p_{(1)})(a \leftharpoonup p_{(2)})$$

for all $p \in A^*$ and $a \in A$.

Proof: Axioms (QT.1)–(QT.2) hold if and only if $f : A^* \longrightarrow A$ is an algebra homomorphism. Axioms (QT.3)–(QT.4) are satisfied if and only if $f : A^* \longrightarrow$

A^{cop} is a coalgebra map. Let $a \in A$. Then $\Delta^{cop}(a)R = R\Delta(a)$ if and only $(p \otimes 1_A)((\Delta^{cop}(a))R) = (p \otimes 1_A)(R(\Delta(a)))$ for all $p \in A^*$. Since

$$\begin{aligned}(p \otimes 1_A)(\Delta^{cop}(a)R) &= \sum_{i=1}^{r} p(a_{(2)}a_i)a_{(1)}b_i \\ &= \sum_{i=1}^{r} p_{(1)}(a_{(2)})p_{(2)}(a_i)a_{(1)}b_i \\ &= (p_{(1)}\rightharpoonup a)f_R(p_{(2)})\end{aligned}$$

and

$$\begin{aligned}(p \otimes 1_A)(R(\Delta(a))) &= \sum_{i=1}^{r} p(a_i a_{(1)})b_i a_{(2)} \\ &= \sum_{i=1}^{r} p_{(1)}(a_i)p_{(2)}(a_{(1)})b_i a_{(2)} \\ &= f_R(p_{(1)})(a\leftharpoonup p_{(2)})\end{aligned}$$

the proposition follows. ▮

Now suppose that (H, R) is a finite-dimensional quasitriangular Hopf algebra. It is observed in [Drinfel'd, 1987, p. 811] that R satisfies (6.3). The fact that R is a solution to the QYBE follows from (QT.1) and (QT.5). To see this we first note that

$$\begin{aligned}R_{(1,2)}R_{(1,3)}R_{(2,3)} &= \sum_{i,j,k=1}^{r} (a_i \otimes b_i \otimes 1)(a_j \otimes 1 \otimes b_j)(1 \otimes a_k \otimes b_k) \\ &= \sum_{i,j,k=1}^{r} a_i a_j \otimes b_i a_k \otimes b_j b_k \\ &= \sum_{i,j=1}^{r} (a_i \otimes b_i \otimes 1)(\Delta(a_j) \otimes b_j)\end{aligned}$$

by (QT.1). On the other hand,

$$\begin{aligned}R_{(2,3)}R_{(1,3)}R_{(1,2)} &= \sum_{i,j,k=1}^{r} (1 \otimes a_k \otimes b_k)(a_j \otimes 1 \otimes b_j)(a_i \otimes b_i \otimes 1) \\ &= \sum_{i,j,k=1}^{r} a_j a_i \otimes a_k b_i \otimes b_k b_j \\ &= \sum_{i,j=1}^{r} (\Delta^{cop}(a_j) \otimes b_j)(a_i \otimes b_i \otimes 1)\end{aligned}$$

by (QT.1) again. Thus $R_{(1,2)}R_{(1,3)}R_{(2,3)} = R_{(2,3)}R_{(1,3)}R_{(1,2)}$ by (QT.5). It turns out the (QT.3) and (QT.5) imply that R is a solution as well. The details are left to the reader to work out. Our calculations have shown:

Lemma 6.3.1 *A quasitriangular bialgebra over the field k is a quasitriangular algebra.*

Let (H, R) be a quasitriangular Hopf algebra over k. Ordinarily $R \notin \mathrm{Im}\Delta$. If $R \in \mathrm{Im}\Delta$ then $R = 1 \otimes 1$ in which case H is cocommutative. To see this, suppose that $R = \Delta(a)$ for some $a \in H$. We use (QT.1) to compute

$$\begin{aligned} a_{(1)} \otimes a_{(2)} \otimes a_{(3)} &= \Delta(a_{(1)}) \otimes a_{(2)} \\ &= a_{(1)} \otimes b_{(1)} \otimes a_{(2)}b_{(2)} \end{aligned}$$

where $a = b$. Therefore

$$\begin{aligned} a &= a_{(1)}a_{(2)}s(a_{(3)}) \\ &= a_{(1)}b_{(1)}s(a_{(2)}b_{(2)}) \\ &= a_{(1)}b_{(1)}s(b_{(2)})s(a_{(2)}) \\ &= a_{(1)}(\epsilon(b)1)s(a_{(2)}) \\ &= \epsilon(a)\epsilon(b)1. \end{aligned}$$

Thus $a = \alpha 1$ for some $\alpha \in k$. Applying (QT.2) gives $a = 1$. Hence $R = \Delta(1) = 1 \otimes 1$. Now by (QT.5) we have $\Delta^{cop} = \Delta$; thus H is cocommutative.

Suppose (A, R) is a quasitriangular bialgebra over k. The bialgebras A^{op}, A^{cop}, and $A^{op\ cop}$ have a quasitriangular structure as well. Let $\widetilde{R} = \sum_{i=1}^{r} b_i \otimes a_i$. Then it is easy to see that $(A^{op}, \widetilde{R})$, $(A^{cop}, \widetilde{R})$, and therefore $(A^{op\ cop}, R)$, are quasitriangular bialgebras.

If $\pi : A \longrightarrow B$ is a bialgebra map which is onto, then $(B, (\pi \otimes \pi)(R))$ is a quasitriangular bialgebra and $\pi : (A, R) \longrightarrow (B, (\pi \otimes \pi)(R))$ is a morphism. Thus quotients of quasitriangular bialgebras admit a quasitriangular structure. Generally sub-bialgebras of A do not admit a quasitriangular structure.

Suppose that (A, R) is a quasitriangular bialgebra over k. If B is a sub-bialgebra of A such that $R \in B \otimes B$ then the pair (B, R) is a quasitriangular bialgebra.

Definition 6.3.4 *Let (A, R) be a quasitriangular bialgebra (respectively Hopf algebra) over the field k and suppose that B is a sub-bialgebra (respectively sub-Hopf algebra) of A such that $R \in B \otimes B$. Then (B, R) is a* sub-quasitriangular bialgebra *(respectively* sub-quasitriangular Hopf algebra*) of (H, R).*

Suppose that (H, R) is a quasitriangular Hopf algebra over k and let s be the antipode of H. Then $R = (s \otimes s)(R)$, which is found in [Majid, 1990b, p. 13]. See also [Radford, 1992, Lemma 2]. Since s is a linear automorphism of H it follows that $s : (H, R) \longrightarrow (H^{op\ cop}, R)$ is an isomorphism of quasitriangular Hopf algebras. As a result $(H^{op}, \widetilde{R})$ and $(H^{cop}, \widetilde{R})$ are isomorphic.

For the remainder of this section we will turn our attention to finite-dimensional quasitriangular Hopf algebras.

Definition 6.3.5 *Let (H, R) be a finite-dimensional quasitriangular Hopf algebra over the field k. Then H_R denotes the smallest sub-Hopf algebra K of H such that $R \in K \otimes K$, and (H, R) is* minimal *if $H = H_R$.*

As far as generating solutions to the QYBE according to (6.1) we may as well assume that quasitriangular Hopf algebras are minimal.

We will show that the underlying Hopf algebra H of a finite-dimensional minimal quasitriangular Hopf algebra (H, R) is the product $H = \mathcal{RL}$ of two sub-Hopf algebras $\mathcal{L}$ and $\mathcal{R}$ which satisfy $\mathcal{LR} = \mathcal{RL}$, $\mathcal{L}^* \simeq \mathcal{R}^{cop}$, and $R \in \mathcal{L} \otimes \mathcal{R}$. One can view these observations as motivation for the quantum double construction of the next section, where $\mathcal{L}$ and $\mathcal{R}$ are separated in the sense that multiplication

$$\mathcal{R} \otimes \mathcal{L} \xrightarrow{m} H$$

is a linear isomorphism.

Let U, V be vector spaces over k and suppose that $R \in U \otimes V$. Consider the subspaces $R_{(\ell)} = (1_U \otimes V^*)(R)$ of U and $R_{(r)} = (U^* \otimes 1_V)(R)$ of V. Suppose that $R \neq 0$ and write $R = \sum_{i=1}^{r} u_i \otimes v_i$ where $r = \mathrm{Rank}R$. Since the u_i's and the v_i's form linearly independent sets it follows that $\{u_1, \ldots, u_r\}$ is a linear basis for $R_{(\ell)}$ and that $\{v_1, \ldots, v_r\}$ is a basis for $R_{(r)}$. We have shown that

$$\mathrm{Dim}R_{(\ell)} = \mathrm{Rank}R = \mathrm{Dim}R_{(r)}. \tag{6.7}$$

To show that $H = \mathcal{RL}$, when (H, R) is a minimal quasitriangular Hopf algebra, we look at other formulations of (QT.5):

$$\sum_{i=1}^{r} a_i c \otimes b_i = \sum_{i=1}^{r} c_{(2)} a_i \otimes c_{(1)} b_i s(c_{(3)}), \tag{6.8}$$

$$\sum_{i=1}^{r} c a_i \otimes b_i = \sum_{i=1}^{r} a_i c_{(2)} \otimes s(c_{(1)}) b_i c_{(3)}, \tag{6.9}$$

$$\sum_{i=1}^{r} a_i \otimes b_i c = \sum_{i=1}^{r} c_{(3)} a_i s^{-1}(c_{(1)}) \otimes c_{(2)} b_i, \tag{6.10}$$

$$\sum_{i=1}^{r} a_i \otimes cb_i = \sum_{i=1}^{r} s^{-1}(c_{(3)})a_i c_{(1)} \otimes b_i c_{(2)} \tag{6.11}$$

for all $c \in H$.

To establish these equations, we first express (QT.5) as

$$\sum_{i=1}^{r} c_{(2)}a_i \otimes c_{(1)}b_i = \sum_{i=1}^{r} a_i c_{(1)} \otimes b_i c_{(2)}$$

for all $c \in A$. Equation 6.8 follows from the calculation

$$\begin{aligned}
\sum_{i=1}^{r} c_{(2)}a_i \otimes c_{(1)}b_i s(c_{(3)}) &= \sum_{i=1}^{r} a_i c_{(1)} \otimes b_i c_{(2)} s(c_{(3)}) \\
&= \sum_{i=1}^{r} a_i c_{(1)} \otimes b_i (\epsilon(c_{(2)})1) \\
&= \sum_{i=1}^{r} a_i c_{(1)}\epsilon(c_{(2)}) \otimes b_i \\
&= \sum_{i=1}^{r} a_i c \otimes b_i
\end{aligned}$$

for all $c \in H$. The other equations are established in a similar manner.

The proof of part e) of the theorem below will be based on two commutation formulas:

$$cf_R(p) = f_R(c_{(1)}{\cdot}p{\cdot}(s^{-1}(c_{(3)})))c_{(2)} \tag{6.12}$$

and

$$f_R(p)c = c_{(2)}f_R((s^{-1}(c_{(1)})){\cdot}p{\cdot}c_{(3)}) \tag{6.13}$$

for all $p \in H^*$ and $c \in H$. To prove the first we use (6.11) and compute

$$\begin{aligned}
cf_R(p) &= \sum_{i=1}^{r} p(a_i)cb_i \\
&= \sum_{i=1}^{r} p(s^{-1}(c_{(3)})a_i c_{(1)})b_i c_{(2)} \\
&= \sum_{i=1}^{r} (c_{(1)}{\cdot}p{\cdot}s^{-1}(c_{(3)}))(a_i)b_i c_{(2)} \\
&= f_R(c_{(1)}{\cdot}p{\cdot}(s^{-1}(c_{(3)})))c_{(2)}.
\end{aligned}$$

To prove the second we use (6.10) and compute

$$
\begin{aligned}
f_R(p)c &= \sum_{i=1}^{r} p(a_i)b_i c \\
&= \sum_{i=1}^{r} p(c_{(3)}a_i s^{-1}(c_{(1)}))c_{(2)}b_i \\
&= \sum_{i=1}^{r} (s^{-1}(c_{(1)}){\cdot}p{\cdot}c_{(3)})(a_i)c_{(2)}b_i \\
&= c_{(2)}f_R((s^{-1}(c_{(1)})){\cdot}p{\cdot}c_{(3)}).
\end{aligned}
$$

The following is a combination of [Radford, 1993b, Proposition 2] and [Radford, 1993b, Theorem 1] in the finite-dimensional case.

Theorem 6.3.1 *Suppose that (H, R) is a finite-dimensional quasitriangular Hopf algebra over the field k. Let $\mathcal{L} = R_{(\ell)}$ and $\mathcal{R} = R_{(r)}$.*

a) *$\mathcal{L}$ and $\mathcal{R}$ are sub-Hopf algebras of H.*

b) $\mathrm{Dim}\mathcal{L} = \mathrm{Rank}R = \mathrm{Dim}\mathcal{R}$

c) $\mathrm{Rank}R \mid \mathrm{Dim}H$.

d) *The map $f : \mathcal{L}^* \longrightarrow \mathcal{R}^{cop}$ defined by $f(p) = (p \otimes 1_H)(R)$ for $p \in \mathcal{L}^*$ is an isomorphism of Hopf algebras.*

e) *If C is any subcoalgebra of A then $\mathcal{L}C = C\mathcal{L}$ and $\mathcal{R}C = C\mathcal{R}$.*

f) $H_R = \mathcal{R}\mathcal{L} = \mathcal{L}\mathcal{R}$.

Proof: Since $f_R : H^* \longrightarrow H^{cop}$ is a Hopf algebra map and $\mathcal{R} = \mathrm{Im} f_R$ it follows that $\mathcal{R}$ is a sub-Hopf algebra of H. Replacing (H, R) by $(H^{op}, \widetilde{R})$ we conclude that $\mathcal{L}$ is a sub-Hopf algebra of H. Thus part a) follows. By (6.7) we have part b). Since H is a free $\mathcal{L}$-module by [Nichols and Zoeller, 1989, Theorem 7], part c) is a consequence of parts a) and b).

To show part d) we first note that $\mathrm{Ker} f_R = \mathcal{L}^{\perp}$. To see this write $R = \sum_{i=1}^{r} a_i \otimes b_i$ where $r = \mathrm{Rank}R$. Since the b_i's form a linearly independent set and the span of the a_i's is $\mathcal{L}$, we conclude that $\mathrm{Ker} f_R = \mathcal{L}^{\perp}$.

Next we observe that $\iota^* : H^* \longrightarrow \mathcal{L}^*$ is an onto Hopf algebra map and $\mathrm{Ker}\iota^* = \mathcal{L}^{\perp} = \mathrm{Ker} f_R$, where $\iota : \mathcal{L} \longrightarrow H$ is the inclusion. Since $\mathrm{Im} f_R = \mathcal{R}$ we may regard $f_R : H^* \longrightarrow \mathcal{R}^{cop}$ as a Hopf algebra map. Therefore there is a Hopf algebra isomorphism $F_R : \mathcal{L}^* \longrightarrow \mathcal{R}^{cop}$ determined by $F_R\iota^* = f_R$. Since $F_R = f$ part d) follows.

To establish part e) we let C be any subcoalgebra of H. It suffices to show that $\mathcal{R}C = C\mathcal{R}$. Assume this is the case. Then replacing (H, R) with $(H^{op}, \widetilde{R})$ it follows that $\mathcal{L}C = C\mathcal{L}$. Now since $\mathcal{R} = \mathrm{Im} f_R$ and C is a subcoalgebra of H, (6.13) and (6.12) show that $\mathcal{R}C \subseteq C\mathcal{R}$ and $C\mathcal{R} \subseteq \mathcal{R}C$. Therefore $\mathcal{R}C = C\mathcal{R}$ and part e) follows.

Lastly, we show part f). By part a) $\mathcal{L}$ and $\mathcal{R}$ are sub-Hopf algebras of H. Since $\mathcal{L}$ and $\mathcal{R}$ are subcoalgebras of H the product $\mathcal{L}\mathcal{R}$ is a subcoalgebra of H. Since $\mathcal{L}$ and $\mathcal{R}$ are subalgebras of H the product $\mathcal{L}\mathcal{R}$ is a subalgebra of H by part e). We have established part f) and thus the proof of the theorem is complete. ▮

Suppose that (H, R) is a finite-dimensional quasitriangular Hopf algebra over k. Then f_R has a counterpart $g_R : H^* \longrightarrow H$ defined by $g_R(p) = (1_H \otimes p)(R)$. We observe that

$$\mathrm{Im} g_R = \mathcal{L} \quad \text{and} \quad g_R = f_R^*$$

where H^{**} and H are identified in the usual way.

We complete this section by discussing the quasitriangular structures admitted by two relatively simple Hopf algebras.

Suppose that G is the cyclic (multiplicative) group of order $n > 0$ and let $H = k[G]$ be the group algebra of G over k. Let $R \in H \otimes H$ and suppose that k has primitive n^{th} root of unity ω. We will determine when (H, R) is quasitriangular.

Since H is commutative and cocommutative, (H, R) is a quasitriangular Hopf algebra if and only if $f_R : H^* \longrightarrow H$ is a Hopf algebra map by Proposition 6.3.1. Write $G = (a)$. Then $G(H^o) = \mathrm{Alg}(H, k)$ is a cyclic group generated by the algebra map $\eta : H \longrightarrow k$ determined by $\eta(a) = \omega$. Since $G(H^o)$ is linearly independent, it follows that $H^* = k[\mathcal{G}]$, where $\mathcal{G} = (\eta)$ is the cyclic group of order n generated by η.

Now let $f : H^* \longrightarrow H$ be a Hopf algebra map. Then $f(\eta)$ must be a grouplike element of H. Therefore $f(\eta) = a^m$ for some $0 \leq m < n$, and such an assignment determines a Hopf algebra map from H^* to H.

Suppose that $f_m : H^* \longrightarrow H$ is the Hopf algebra map determined by $f_m(\eta) = a^m$. We will describe the $R_m \in H \otimes H$ such that $f_m = f_{R_m}$, that is which satisfies $f_m(p) = (p \otimes 1_H)(R_m)$ for $p \in H^*$. Since f_m is an algebra map $f_m(\eta^\ell) = a^{m\ell}$ for $\ell \geq 0$. Let $\{\eta_{(0)}, \ldots, \eta_{(n-1)}\}$ be the basis for H with dual basis $\{\eta^0, \ldots, \eta^{n-1}\}$ for H^*. Then

$$R_m = \sum_{\ell=0}^{n-1} \eta_{(\ell)} \otimes a^{m\ell}.$$

It remains to determine $\eta_{(\ell)}$. We claim that

$$\eta_{(\ell)} = \sum_{i=0}^{n-1} \frac{\omega^{-i\ell}}{n} a^i$$

for $0 \leq \ell < n$. Let $\eta'_{(\ell)}$ denote the sum on the right hand side of the last equation. Then

$$\eta^m(\eta'_{(\ell)}) = \sum_{i=0}^{n-1} \frac{\omega^{(m-\ell)i}}{n}$$

for $0 \leq m < n$. Since an nth root of unity ρ satisfies $\sum_{i=0}^{n-1} \rho^i = 0$ unless $\rho = 1$, it follows that $\eta^m(\eta'_{(\ell)}) = \delta_{m,\ell}$. We have shown the quasitriangular structures which H admits are (H, R_m), where $0 \leq m < n$. These examples are worked out in [Radford, 1992, p. 10].

It can be the case that a finite-dimensional Hopf algebra can admit an infinite number of non-isomorphic quasitriangular structures. Such an example is the 4-dimensional example $\mathrm{T}_{2,-1}(k)$ of Section 6.5.2 in certain cases. See [Radford, 1993b, pp. 295–297].

As an algebra $H = \mathrm{T}_{2,-1}(k)$ is generated by g and x which satisfy the relations

$$g^2 = 1, \quad x^2 = 0, \quad \text{and} \quad xg = -gx.$$

The coalgebra structure of H is determined by

$$\Delta(g) = g \otimes g \quad \text{and} \quad \Delta(x) = x \otimes g + 1 \otimes x,$$

and the set of monomials $\{1, g, x, gx\}$ is a linear basis for H.

We will assume for the remainder of this discussion that the characteristic of k is not 2. Let $\{\overline{1}, \overline{g}, \overline{x}, \overline{gx}\}$ be the dual basis for H^*. Set

$$G = \overline{1} - \overline{g} \quad \text{and} \quad X = \overline{x} + \overline{gx}.$$

A short calculation reveals that

$$G^2 = \epsilon, \quad X^2 = 0, \quad \text{and} \quad XG = -GX,$$

and

$$\Delta(G) = G \otimes G, \quad \Delta(X) = X \otimes G + \epsilon \otimes X,$$

and that $\{\epsilon, G, X, GX\}$ is a linear basis for A^*. In particular there is an isomorphism of Hopf algebras $F : H^* \longrightarrow H$ determined by $F(G) = g$ and $F(X) = x$.

Suppose that $R \in H \otimes H$ affords H a quasitriangular structure (H, R). We will determine the possibilities for $f_R : H^* \longrightarrow H^{cop}$. Since $H^* \simeq H$ as Hopf algebras, the possibilities for f_R are found among the bialgebra maps $f : H \longrightarrow H^{cop}$. A straightforward calculation yields that $f = \mathcal{E}$, where $\mathcal{E}(a) = \epsilon(a)1$ for all $a \in H$, or $f(g) = g$ and $f(x) = \alpha gx$ for some $\alpha \in k$. For each $\alpha \in k$ there is a unique bialgebra map $f_\alpha : H \longrightarrow H^{cop}$ such that

$$f_\alpha(g) = g \quad \text{and} \quad f_\alpha(x) = \alpha gx.$$

Therefore $\mathcal{E}$ and the f_α's constitute all of the bialgebra maps $f : H \longrightarrow H^{cop}$. Notice that f_α is an isomorphism if and only if $\alpha \neq 0$. In particular $H \simeq H^{cop}$, and hence $H \simeq H^{op}$, as Hopf algebras.

For $\alpha \in k$ set

$$R_\alpha = \frac{1}{2}(1 \otimes 1 + 1 \otimes g + g \otimes 1 - g \otimes g) + \frac{\alpha}{2}(x \otimes x + x \otimes gx + gx \otimes gx - gx \otimes x).$$

It is easy to see that

$$(f_\alpha F)(p) = (p \otimes 1_H)(R_\alpha) \tag{6.14}$$

and that $(\mathcal{E}F)(p) = p(1)1$ for $p \in H^*$. Therefore, if (H, R) is quasitriangular, $R = R_\alpha$ for some $\alpha \in k$ or $R = 1 \otimes 1$. But H is not cocommutative; thus $R \neq 1 \otimes 1$.

By (6.14) and the remarks made at the beginning of this section, (QT.1)–(QT.4) are satisfied for $R = R_\alpha$. Therefore R_α is invertible for all $\alpha \in k$. Observe that (QT.5) is satisfied for any R_α since $(\Delta^{cop}(a))R_\alpha = R_\alpha(\Delta(a))$ holds for the algebra generators $a = g, x$. We have shown that:

Proposition 6.3.2 *Suppose that k is a field of characteristic other than 2 and $H = \mathrm{T}_{2,-1}(k)$. Let $\mathcal{G} = 1 + g$. Then the quasitriangular structures on H form a parameterized family $\{(H, R_\alpha)\}_{\alpha \in k}$ where*

$$R_\alpha = \frac{1}{2}(\mathcal{G} \otimes \mathcal{G} - 2g \otimes g) + \frac{\alpha}{2}(\mathcal{G}x \otimes \mathcal{G}x - 2gx \otimes x).$$

∎

We remark that $H = \mathrm{T}_{2,-1}(k)$ admits an infinite number of quasitriangular structures when k is infinite. Note that (H, R_α) is minimal quasitriangular except when $\alpha = 0$.

Lastly, consider the isomorphism classes these different quasitriangular structures fall into. Let $f : (H, R_\alpha) \longrightarrow (H, R_{\alpha'})$ be an isomorphism of quasitriangular Hopf algebras. Then $f = f_1 f_\beta$ for some non-zero $\beta \in k$. Notice that $f(g) = g$ and $f(x) = \beta x$. This means that $\alpha' = \alpha\beta^2$. On the other hand if $\beta \in k$ is not zero, and $\alpha' = \beta^2\alpha$, then $f_1 f_\beta : (H, R_\alpha) \longrightarrow (H, R_{\alpha'})$ is an isomorphism of quasitriangular Hopf algebras. Since $\mathrm{Rank}R_0 = 2$ and $\mathrm{Rank}R_1 = 4$, it follows that $(H, R_0) \not\simeq (H, R_1)$. If k is algebraically closed, then $(H, R_1) \simeq (H, R_\alpha)$ whenever $\alpha \neq 0$. Consequently, in the algebraically closed case, the (H, R_α)'s fall into two isomorphism classes. The (H, R_α)'s may fall into infinitely many isomorphism classes, which happens for example when $k = \mathbb{Q}$ is the field of rational numbers.

Exercise 6.3.1 Let (H, R) be a finite-dimensional quasitriangular Hopf algebra over the field k. Write $R = \sum_{i=1}^r a_i \otimes b_i$, let $u = \sum_{i=1}^r s(b_i)a_i$ and set $c = us(u)$.

a) Show that $R^{-1} = \sum_{i=1}^r a_i \otimes s^{-1}(b_i) = (I \otimes s^{-1})(R)$.

b) Show that $R = (s \otimes s)(R)$.

c) Show that u is invertible, $u^{-1} = \sum_{i=1}^{r} b_i s^2(a_i)$ and that $s^2(a) = uau^{-1}$ for all $a \in A$.

d) Show that $c = s(u)u$ and is in the center of A. (The element c is called the *quantum casimir* element of A.)

See [Drinfel'd, 1990].

Exercise 6.3.2 Let $H = k[G]$ be the group algebra of the cyclic group G of order n discussed in this section.

a) Show that $\eta_{(\ell)}\eta_{(\ell')} = \delta_{\ell,\ell'}\eta_{(\ell)}$ and that $\epsilon = \sum_{\ell=0}^{n-1} \eta_{(\ell)}$.

b) Determine which of the quasitriangular structures (H, R_m) are minimal.

c) Determine the isomorphism classes of the quasitriangular Hopf algebras (H, R_m) for $0 \leq m < n$.

Exercise 6.3.3 Let $H = \mathrm{T}_{2,-1}(k)$ and suppose that the characteristic of k is not 2.

a) Using (6.1), describe all solutions to the QYBE arising from H when $\mathrm{Dim}M = 2$. [Hint: Let $\pi : H \longrightarrow \mathrm{End}(M)$ be a representation. Then $\pi(g)$ is diagonalizable and $\pi(x)$ can be described in terms of the eigenspaces of $\pi(g)$.]

b) Describe all solutions to the QYBE arising from H according to (6.1) when M is finite-dimensional.

Exercise 6.3.4 Suppose that (H, R) is a finite-dimensional minimal quasitriangular Hopf algebra over k. Show that H is commutative if and only if H is cocommutative.

Exercise 6.3.5 Suppose that (H, R) is a finite-dimensional quasitriangular Hopf algebra over k. Let M be a left H-module and $\mathcal{L} = R_{(\ell)}$. Regard M as a left $\mathcal{L}$-module (M, μ) by restriction. Regard M as a left $\mathcal{L}^*$-module by pull-back along the algebra map $f : \mathcal{L}^* \longrightarrow \mathcal{R} \subseteq H$ and let (M, ρ) be the underlying right $\mathcal{L}$-comodule structure for this action.

a) Show that
$$a_{(1)}{\cdot}m^{<1>} \otimes a_{(2)}m^{(2)} = (a_{(2)}{\cdot}m)^{<1>} \otimes (a_{(2)}{\cdot}m)^{(2)}a_{(1)}$$
for all $m \in M$ and $a \in \mathcal{L}$. (Thus (M, μ, ρ) is a left QYB $\mathcal{L}$-module.)

b) Show that the QYBE solution $R_M : M \otimes M \longrightarrow M \otimes M$ described by (6.1) is computed by the formula
$$R_M(m \otimes n) = n^{(2)}{\cdot}m \otimes n^{<1>}$$
for all $m, n \in M$. (Thus $R_M = R^\tau$ where R is the solution associated with (M, μ, ρ).)

Exercise 6.3.6 Suppose that (H, R) is a finite-dimensional quasitriangular Hopf algebra over k. Let $\mathcal{L} = R_{(\ell)}, \mathcal{R} = R_{(r)}$, and write $R = \sum_{i=1}^{r} a_i \otimes b_i$. Suppose that (M, μ) is a left $\mathcal{L}$-module.

a) Show that (M, ρ) is a right $\mathcal{R}^{cop}$-comodule where
$$\rho(m) = \sum_{i=1}^{r} a_i{\cdot}m \otimes b_i$$

for all $m \in M$.

Suppose that (M, μ) is a left H-module, and regard M as a left $\mathcal{L}$-module under restriction. Regard (M, ρ) as a right H^{cop}-comodule, where (M, ρ) is the $\mathcal{R}^{cop}$-comodule described in part a).

b) Show that (M, μ, ρ) is a left QYB H^{cop}-module.

c) Show that $R_M : M \otimes M \longrightarrow M \otimes M$ defined by (6.1) is also described by $R_M(m \otimes n) = m^{<1>} \otimes m^{(2)} . n$ for all $m, n \in M$. (Thus R_M is the solution associated with (M, μ, ρ).)

6.4 The Quantum Double

We are going to motivate the finite-dimensional quantum double, which we also refer to as the Drinfel'd double, from a slightly unconventional point of view. One advantage of this approach is that the algebraic structure of the double is quickly deduced and the relationship between the quantum double and finite-dimensional quasitriangular Hopf algebras in general becomes readily apparent.

It turns out that the quantum double is merely a minimal quasitriangular Hopf algebra (H, R) such that multiplication $m : \mathcal{R} \otimes \mathcal{L} \longrightarrow H$ is a linear isomorphism, where $\mathcal{R} = \mathrm{Im} f_R$ and $\mathcal{L} = \mathrm{Im} g_R$.

In the finite-dimensional case, the coalgebra structure of the quantum double is the tensor product coalgebra structure $\mathcal{L}^{*\ cop} \otimes \mathcal{L}$, but the algebra structure is a twisted tensor product structure [Gugenheim, 1962], [Singer, 1972].

Suppose that (H, R) is a finite-dimensional minimal quasitriangular Hopf algebra. We continue the discussion of the previous section.

Assume that multiplication $m : \mathcal{R} \otimes \mathcal{L} \longrightarrow H$ is a linear isomorphism. Since $f : \mathcal{L}^{*\ cop} \longrightarrow \mathcal{R}$ is an isomorphism of Hopf algebras the composite $F = m(f \otimes 1_{\mathcal{L}})$ defines a linear isomorphism from $D(\mathcal{L}) = \mathcal{L}^{*\ cop} \otimes \mathcal{L}$ to H. Therefore $D(\mathcal{L})$ admits a (unique) quasitriangular Hopf algebra structure $(D(\mathcal{L}), \mathfrak{R})$ such that $F : (D(\mathcal{L}), \mathfrak{R}) \longrightarrow (H, R)$ is an isomorphism of quasitriangular Hopf algebras.

The defining properties of $(D(\mathcal{L}), \mathfrak{R})$ are easily derivable. Since $\mathcal{L}$ is finite-dimensional, the map $\iota : D(\mathcal{L}) \longrightarrow \mathrm{End}(\mathcal{L})$ defined by $\iota(\ell^* \otimes \ell)(a) = \ell^*(a)\ell$ is a linear isomorphism by Proposition A.4.1. Let $C \in \mathcal{L}^* \otimes \mathcal{L}$ be the *canonical element*, that is the element defined by $\iota(C) = 1_{\mathcal{L}}$. Notice that C is realized as follows. Take any linear basis $\{\ell_1, \ldots, \ell_r\}$ for $\mathcal{L}$ and let $\{\ell^1, \ldots, \ell^r\}$ be the dual basis for $\mathcal{L}^*$. Then

$$C = \ell^i \otimes \ell_i. \tag{6.15}$$

By definition

$$<\ell^i, a>\ell_i = a \quad \text{and} \quad <p, \ell_i>\ell^i = p$$

for all $a \in \mathcal{L}$ and $p \in \mathcal{L}^*$. We set $\widetilde{C} = \ell_i \otimes \ell^i$.

Let $\jmath_{\mathcal{L}} : \mathcal{L} \longrightarrow D(\mathcal{L})$ be the map defined by $\jmath_{\mathcal{L}}(\ell) = \epsilon \otimes \ell$ and $\jmath_{\mathcal{L}^*} : \mathcal{L}^* \longrightarrow D(\mathcal{L})$ be the map defined by $\jmath_{\mathcal{L}^*}(\ell^*) = \ell^* \otimes 1$. We will show that the quasitriangular Hopf algebra $(D(\mathcal{L}), \mathfrak{R})$ satisfies the following conditions and is determined by them.

$$\text{As a coalgebra } D(\mathcal{L}) \text{ is the tensor product of } \mathcal{L}^{*\,cop} \text{ and } \mathcal{L}. \tag{6.16}$$

$$\Re = \epsilon \otimes \widetilde{C} \otimes 1 = (\epsilon \otimes \ell_i) \otimes (\ell^i \otimes 1). \tag{6.17}$$

$$\jmath_{\mathcal{L}} \text{ and } \jmath_{\mathcal{L}^*} \text{ are algebra maps, and} \tag{6.18}$$

$$(p \otimes 1)(\epsilon \otimes a) = p \otimes a, \tag{6.19}$$

$$(\epsilon \otimes a)(p \otimes 1) = (a_{(1)} \cdot p \cdot (s^{-1}(a_{(3)}))) \otimes a_{(2)} \tag{6.20}$$

holds for all $a \in \mathcal{L}$ and $p \in \mathcal{L}^*$.

Notice that (6.18)–(6.20) determine the multiplication of $D(\mathcal{L})$ completely. For assuming these equations hold we calculate

$$\begin{aligned}(p \otimes a)(q \otimes b) &= (p \otimes 1)(\epsilon \otimes a)(q \otimes 1)(\epsilon \otimes b)\\ &= (p \otimes 1)(a_{(1)} \cdot q \cdot (s^{-1}(a_{(3)}))) \otimes a_{(2)})(\epsilon \otimes b)\\ &= (p \otimes 1)((a_{(1)} \cdot q \cdot (s^{-1}(a_{(3)}))) \otimes 1)(\epsilon \otimes a_{(2)})(\epsilon \otimes b)\\ &= (p(a_{(1)} \cdot q \cdot (s^{-1}(a_{(3)}))) \otimes 1)(\epsilon \otimes a_{(2)}b)\\ &= p(a_{(1)} \cdot q \cdot (s^{-1}(a_{(3)}))) \otimes a_{(2)}b\end{aligned}$$

and thus

$$(p \otimes a)(q \otimes b) = p(a_{(1)} \cdot q \cdot (s^{-1}(a_{(3)}))) \otimes a_{(2)}b \tag{6.21}$$

for all $p, q \in \mathcal{L}^*$ and $a, b \in \mathcal{L}$.

Now we will show that (6.16) – (6.20) hold. The tensor product coalgebra structure on $D(\mathcal{L})$ is such that F is an isomorphism of coalgebras. There is one coalgebra structure for which this is true. Thus (6.16) follows.

To show (6.17) we write $R = \sum_{i=1}^{r} a_i \otimes b_i$ where $r = \text{Rank}R$. We have noted that the a_i's form a linear basis for $\mathcal{L}$ and that the b_i's form a linear basis for $\mathcal{R}$. Since f is a linear isomorphism we can choose $p_1, \ldots, p_r \in \mathcal{L}^*$ such that $f(p_i) = b_i$. The calculation $b_i = f(p_i) = \sum_{j=1}^{r} p_i(a_j)b_j$ shows that $p_i(a_j) = \delta_{i,j}$. Therefore $\{p_1, \ldots, p_r\}$ is the dual basis for $\{a_1, \ldots, a_r\}$. Now

$$(F \otimes F)(\sum_{i=1}^{r}(\epsilon \otimes a_i) \otimes (p_i \otimes 1)) = \sum_{i=1}^{r} a_i \otimes f(p_i) = \sum_{i=1}^{r} a_i \otimes b_i = R.$$

Since $(F \otimes F)(\Re) = R$ and $F \otimes F$ is one-one it now follows that $\Re = \sum_{i=1}^{r}(\epsilon \otimes a_i) \otimes (p_i \otimes 1)$. We have established (6.17).

Observe that the composites $F\jmath_{\mathcal{L}} = \iota$, where $\iota : \mathcal{L} \longrightarrow A$ is the inclusion, and $F\jmath_{\mathcal{L}^*} = f$ are algebra maps. Therefore $\jmath_{\mathcal{L}} = F^{-1}\iota$ and $\jmath_{\mathcal{L}^*} = F^{-1}f$ are algebra maps, and (6.18) holds. Since F is a one-one algebra map the calculation

$$F((p \otimes 1)(\epsilon \otimes a)) = F(p \otimes 1)F(\epsilon \otimes a) = f(p)a = F(p \otimes a).$$

shows that (6.19) holds. Using the commutation relation (6.12) we calculate

$$\begin{aligned} F((\epsilon \otimes a)(p \otimes 1)) &= F(\epsilon \otimes a)F(p \otimes 1) \\ &= af(p) \\ &= f(a_{(1)}{\cdot}p{\cdot}(s^{-1}(a_{(3)})))a_{(2)} \\ &= F((a_{(1)}{\cdot}p{\cdot}(s^{-1}(a_{(3)}))) \otimes a_{(2)}) \end{aligned}$$

which establishes (6.20).

We have actually shown that $\jmath_{\mathcal{L}} : \mathcal{L} \longrightarrow D(\mathcal{L})$ and $\jmath_{\mathcal{L}^*} : \mathcal{L}^{*\ cop} \longrightarrow D(\mathcal{L})$ are bialgebra maps, hence Hopf algebra maps. Let s be the antipode of $\mathcal{L}$ and S be the antipode of $D(\mathcal{L})$. Now s^{-1} is the antipode of $\mathcal{L}^{op}$, and therefore $s^{*\ -1}$ is the antipode of $\mathcal{L}^{*\ cop}$. Since $S\jmath_{\mathcal{L}} = \jmath_{\mathcal{L}}s$ and $S\jmath_{\mathcal{L}^*} = \jmath_{\mathcal{L}^*}s^{*\ -1}$ we compute for $p \in \mathcal{L}^*$ and $a \in \mathcal{L}$ that

$$\begin{aligned} S(p \otimes a) &= S((p \otimes 1)(\epsilon \otimes a)) \\ &= S(\epsilon \otimes a)S(p \otimes 1) \\ &= S(\jmath_{\mathcal{L}}(a))S(\jmath^*_{\mathcal{L}}(p)) \\ &= \jmath_{\mathcal{L}}(s(a))\jmath_{\mathcal{L}^*}(s^{*\ -1}(p)) \\ &= (\epsilon \otimes s(a))(s^{*\ -1}(p) \otimes 1). \end{aligned}$$

Therefore

$$S(p \otimes a) = (\epsilon \otimes s(a))(s^{*\ -1}(p) \otimes 1). \tag{6.22}$$

Theorem 6.4.1 *Suppose that $\mathcal{L}$ is a finite-dimensional Hopf algebra with antipode s over the field k. Then there exists a quasitriangular Hopf algebra $(D(\mathcal{L}), \mathfrak{R})$ which satisfies (6.16) – (6.20).*

Proof: There are many details to check. We will discuss a portion of them here and relegate the remainder to the reader.

Let $D(\mathcal{L}) = \mathcal{L}^{*\ cop} \otimes \mathcal{L}$ be the tensor product coalgebra. Assume for the moment that the product of (6.21) does in fact give $D(\mathcal{L})$ an associative algebra structure and that Δ and ϵ are algebra maps. Then (6.18)–(6.20) hold. The fact that $\mathfrak{R}$ satisfies (QT.1)–(QT.4) uses (6.18) and is an exercise in algebra of the canonical element $C = \sum_{i=1}^{r} p_i \otimes a_i \in \mathcal{L}^* \otimes \mathcal{L}$. See Exercise 6.4.1.

We will establish (QT.5). Since Δ and Δ^{cop} are algebra maps, the solutions $a \in D(\mathcal{L})$ to the equation $(\Delta^{cop}(a))\mathfrak{R} = \mathfrak{R}(\Delta(a))$ form a subalgebra of $D(\mathcal{L})$. Therefore we may assume that $a = \epsilon \otimes c$ for some $c \in \mathcal{L}$ or $a = p \otimes 1$ for some $p \in \mathcal{L}^*$. In the case of $a = \epsilon \otimes c$ we note that (QT.5) comes down to whether or not

$$\sum_{i=1}^{r} c_{(4)}a_i \otimes c_{(1)} \cdot p_i \cdot (s^{-1}(c_{(3)})) \otimes c_{(2)} = \sum_{i=1}^{r} a_i c_{(1)} \otimes p_i \otimes c_{(2)}. \tag{6.23}$$

Let $d \in \mathcal{L}$ and $\iota : \mathcal{L} \longrightarrow \mathcal{L}^{**}$ be the map defined by $\iota(a)(p) = p(a)$ for $a \in \mathcal{L}$ and $p \in \mathcal{L}^*$. Then applying the operator $1_{\mathcal{L}} \otimes \iota(d) \otimes 1_{\mathcal{L}}$ to the left hand side of (6.23) results in

$$\begin{aligned}
\sum_{i=1}^{r} c_{(4)}a_i(p_i(s^{-1}(c_{(3)}))dc_{(1)}) \otimes c_{(2)} &= c_{(4)}s^{-1}(c_{(3)})dc_{(1)} \otimes c_{(2)} \\
&= \epsilon(c_{(3)}1)dc_{(1)} \otimes c_{(2)} \\
&= dc_{(1)} \otimes c_{(2)}.
\end{aligned}$$

Applying this operator to the right hand side of (6.23) yields the same result. The case $a = p \otimes 1$ follows in a similar manner. Therefore (6.23) is in fact correct.

We will leave the reader with the rather tedious exercise of showing that (6.21) does indeed give $D(\mathcal{L})$ an associative algebra structure and that the coproduct and counit are algebra maps. ∎

Definition 6.4.1 *A finite-dimensional quasitriangular Hopf algebra (H, R) over the field k is a* quantum double, *or a* Drinfel'd double, *if $(H, R) \simeq (D(\mathcal{L}), \mathfrak{R})$ for some finite-dimensional Hopf algebra $\mathcal{L}$ over k.*

Every *minimal* finite-dimensional quasitriangular Hopf algebra is the quotient of a quantum double. The following theorem is [Radford, 1993b, Theorem 2] in the finite-dimensional case.

Theorem 6.4.2 *Suppose that (H, R) is a finite-dimensional quasitriangular Hopf algebra over the field k. Let $\mathcal{L} = R_{(\ell)}$, $\mathcal{R} = R_{(r)}$ and define $f : \mathcal{L}^* \longrightarrow \mathcal{R}$ by $f(p) = (p \otimes 1_{\mathcal{R}})(R)$ for $p \in \mathcal{L}^*$. Then:*

a) *There is a unique morphism $F : (D(\mathcal{L}), \mathfrak{R}) \longrightarrow (H, R)$ of quasitriangular Hopf algebras such that $F\jmath_{\mathcal{L}} = 1_H|_{\mathcal{L}}$ and $F\jmath_{\mathcal{L}^*} = f$.*

b) $\text{Rank}\mathfrak{R} = \text{Rank}R$.

c) $\text{Im}F = H_R$.

Proof: Part b) follows immediately since both ranks are equal to $\text{Dim}\mathcal{L}$. To show part a) we suppose that $F : (D(\mathcal{L}), \mathfrak{R}) \longrightarrow (H, R)$ is a morphism satisfying $F\jmath_{\mathcal{L}} = 1_H|_{\mathcal{L}}$

and $F_{\jmath_{\mathcal{L}^*}} = f$. For $p \in \mathcal{L}^*$ and $a \in \mathcal{L}$ we calculate $F(p \otimes a) = F((p \otimes 1)(\epsilon \otimes a)) = F(p \otimes 1)F(\epsilon \otimes a) = f(p)a$. Thus F is uniquely determined. To establish existence we define $F : D(\mathcal{L}) \longrightarrow H$ by $F(p \otimes a) = f(p)a$ for $p \in \mathcal{L}^*$ and $a \in \mathcal{L}$. Since $f : \mathcal{L}^{*\ cop} \longrightarrow \mathcal{R}$ is a coalgebra map by part d) of Proposition 6.3.1 and multiplication $m : H \otimes H \longrightarrow H$ is a coalgebra map, it follows that $F = m(f \otimes 1_{\mathcal{L}})$ is a coalgebra map. Now f is also an algebra map. Therefore, to show that F is an algebra map, by (6.21) it suffices to show that

$$f(a_{(1)} \cdot p \cdot (s^{-1}(a_{(3)})))a_{(2)} \quad = \quad af(p)$$

holds for all $a \in \mathcal{L}$ and $p \in \mathcal{L}^*$. But this equation is the commutation relation of (6.12). Consequently F is an algebra map, and thus F is a Hopf algebra map.

To show that $F(\mathfrak{R}) = R$, we write $\mathfrak{R} = \sum_{i=1}^{r}(\epsilon \otimes a_i) \otimes (a^i \otimes 1)$, where $\{a_1, \ldots, a_r\}$ is a linear basis for $\mathcal{L}$ and $\{a^1, \ldots, a^r\}$ is the dual basis for $\mathcal{L}^*$. Now $R = \sum_{i=1}^{r} a_i \otimes b_i$ for some $b_1, \ldots, b_r \in \mathcal{R}$. By the way f is defined it follows that $f(a^i) = b_i$ for all $1 \leq i \leq r$. Therefore $R = \sum_{i=1}^{r} a_i \otimes f(a^i)$. Now observe that $(F \otimes F)(\mathfrak{R}) = \sum_{i=1}^{r} a_i \otimes f(a^i) = R$ so part a) follows. Since $H_R = \mathcal{R}\mathcal{L}$ by Theorem 6.3.1, part c) now follows also. ∎

Suppose that (H, R) is a quasitriangular Hopf algebra over the field k. By part c) of Theorem 6.3.1 we have that $\mathrm{Rank}R \mid \mathrm{Dim}H$. When (H, R) is minimal $\mathrm{Dim}H \mid (\mathrm{Rank}R)^2$ by the next corollary. We have observed that $(D(\mathcal{L}), \mathfrak{R})$ is minimal. Notice that

$$\mathrm{Dim}D(\mathcal{L}) = (\mathrm{Rank}\mathfrak{R})^2.$$

The following is [Radford, 1993b, Corollary 3].

Corollary 6.4.1 *Suppose that (H, R) is a finite-dimensional minimal quasitriangular Hopf algebra over the field k. Then:*

a) $\mathrm{Dim}H \mid (\mathrm{Rank}R)^2$.

b) *(H, R) is a quantum double if and only if* $\mathrm{Dim}H = (\mathrm{Rank}R)^2$.

Proof: Let $\mathcal{L} = R_{(\ell)}$. By Theorem 6.4.2 there is a morphism of quasitriangular Hopf algebras

$$(D(\mathcal{L}), \mathfrak{R}) \xrightarrow{F} (H, R)$$

such that $\mathrm{Rank}\mathfrak{R} = \mathrm{Rank}R$ and, since (H, R) is minimal, F is onto. The fact that F is onto means $F^* : H^* \longrightarrow D(\mathcal{L})^*$ is a one-one Hopf algebra map. By [Nichols and Zoeller, 1989, Theorem 8] any finite-dimensional Hopf algebra over k is a free module over any of its sub-Hopf algebras. We conclude that $\mathrm{Dim}H^* \mid \mathrm{Dim}D(\mathcal{L})^*$. Therefore $\mathrm{Dim}H \mid \mathrm{Dim}D(\mathcal{L})$. As $\mathrm{Dim}D(\mathcal{L}) = (\mathrm{Rank}\mathfrak{R})^2$ and $\mathrm{Rank}\mathfrak{R} = \mathrm{Rank}R$, we have established part a). To establish part b) we need only show that $\mathrm{Dim}H = (\mathrm{Rank}R)^2$ implies that (H, R) is a quantum double. Now $\mathrm{Dim}H = (\mathrm{Rank}R)^2$

implies $\mathrm{Dim}D(\mathcal{L}) = \mathrm{Dim}H$. Therefore in this case F must be an isomorphism since it is onto. ▮

It can be the case that $\mathrm{Dim}H = (\mathrm{Rank}R)^2$ but (H, R) is not a quantum double. For example, assume that the characteristic of k is not 2 and consider $\mathrm{T}_{2,-1}(k)$ of Section 6.5.2 with quasitriangular structure $(\mathrm{T}_{2,-1}(k), R_0)$. Then $\mathrm{Dim}\mathrm{T}_{2,-1}(k) = (\mathrm{Rank}R_0)^2$. We claim $\mathrm{T}_{2,-1}(k)$ is not the underlying Hopf algebra of a quantum double. For $\mathrm{T}_{2,-1}(k) \simeq D(\mathcal{L})$ means $\mathrm{Dim}\mathcal{L} = 2$, and therefore $\mathcal{L}$ is commutative and cocommutative. But $D(\mathcal{L}) = \mathcal{L}^{*\ cop} \otimes \mathcal{L}$ as a coalgebra means that $D(\mathcal{L})$ is cocommutative. Since $\mathrm{T}_{2,-1}(k)$ is not cocommutative we conclude that $\mathrm{T}_{2,-1}(k) \not\simeq D(\mathcal{L})$.

Exercise 6.4.1 Suppose that A is a finite-dimensional bialgebra over the field k, and let $\mathcal{C} = \sum_{i=1}^{r} p_i \otimes a_i \in A^* \otimes A$ be the canonical element of $A^* \otimes A$.

a) Show that $\sum_{i=1}^{r} p_i \otimes \Delta(a_i) = \sum_{i,j=1}^{r} p_i p_j \otimes a_i \otimes a_j$. [Hint: for $c \in A$ apply the operator $\iota(c) \otimes 1_A \otimes 1_A$ to both sides of expression and show that the results are equal. As usual $\iota : A \longrightarrow A^{**}$ is defined by $\iota(a)(a^*) = a^*(a)$.]

b) Show that $\sum_{i=1}^{r} \Delta(p_i) \otimes a_i = \sum_{i,j=1}^{r} p_i \otimes p_j \otimes a_i a_j$. [Hint: for $q \in A^*$ apply the operator $1_{A^*} \otimes 1_{A^*} \otimes q$ to both sides of expression and show that the results are equal.]

Exercise 6.4.2 Let $\mathcal{L}$ be a finite-dimensional Hopf algebra over the field k.

a) Show that the following are equivalent:

   i) $\mathcal{L}$ is commutative and cocommutative.

   ii) $D(\mathcal{L})$ is commutative.

   iii) $D(\mathcal{L})$ is cocommutative.

b) Show that if any one of these conditions hold then $D(\mathcal{L}) = \mathcal{L}^* \otimes \mathcal{L}$ as Hopf algebras.

Exercise 6.4.3 Let $H = k[G]$ be the group algebra of a finite group G over the field k. For $g \in G$ let $e_g \in H^*$ be defined by $e_g(h) = \delta_{g,h}$ for all $h \in G$. (Thus $\{e_g\}_{g \in G}$ is the dual basis for $k[G]^*$.)

a) Show that $e_g e_h = \delta_{g,h} e_g$ for all $g, h \in G$ and $\epsilon = \sum_{g \in G} e_g$.

b) Show that $\Delta(e_g) = \sum_{ab=g} e_a \otimes e_b$ and $\epsilon(e_g) = \delta_{e,g}$ for all $g \in G$, where e is the neutral element of G.

c) Show that the set of symbols $g{*}h = e_g \otimes h$, where $g, h \in G$, is a basis for $D(k[G])$ which satisfies

$$\Delta(g{*}h) = \sum_{ab=g} (b{*}h) \otimes (a{*}h), \qquad \epsilon(g{*}h) = \delta_{g,e}$$

and

$$(g{*}h)(g'{*}h') = \delta_{g,\, hg'h^{-1}}\, g{*}hh'.$$

d) Show that the set of symbols $g{*}h = g \otimes e_h$, where $g, h \in G$, is a basis for $D(k[G]^*)$ which satisfies

$$\Delta(g{*}h) = \sum_{ab=h} (g{*}a) \otimes (g{*}b), \qquad \epsilon(g{*}h) = \delta_{h,e}$$

and

$$(g'{*}h')(g{*}h) = \delta_{h',ghg^{-1}} g'g{*}h.$$

Exercise 6.4.4 Let $\mathcal{L}$ be a finite-dimensional Hopf algebra over the field k. Show that the left $D(\mathcal{L})$-modules are the left QYB $\mathcal{L}$-modules, where

$$(p \otimes a){\cdot}m = (a{\cdot}m)^{<1>} <p, (a{\cdot}m)^{(2)}>$$

for all $p \in \mathcal{L}^*$ and $a \in \mathcal{L}$.

The equivalent of this result was established in [Majid, 1991a].

Exercise 6.4.5 Let H be a finite-dimensional Hopf algebra over the field k. Show that $H \simeq A(R)$ for some solution $R : M \otimes M \longrightarrow M \otimes M$ to the QYBE, where M is a finite-dimensional vector space over k.

6.5 Some Fundamental Examples of Pointed Hopf Algebras

Some of the earlier examples of non-commutative and non-cocommutative Hopf algebras [Taft, 1971] involve the relation

$$xa = qax, \tag{6.24}$$

where x and a are certain key elements in the Hopf algebra and $q \in k$ is a primitive root of unity. Understanding the expansion of $(a + x)^n$ is important for the analysis of these examples.

Let A be an algebra over k with elements x and a which satisfy the relation $xa = qax$ where $q \in k\backslash 0$. Expanding $(a + x)^n$ and using (6.24) it is easy to see that

$$(a + x)^n = \sum_{m=0}^{n} \binom{n}{m}_q a^{n-m} x^m \tag{6.25}$$

for some scalars $\binom{n}{m}_q \in k$ which we will determine.

Definition 6.5.1 *The coefficients* $\binom{n}{m}_q$ *are* q-binomial coefficients.

A discussion of q-binomial coefficients and generalizations can be found in many references, particularly in [Andrews, 1989], and [Andrews and Baxter, 1987].

6.5.1 Q-Binomial Coefficients

We will consider the expansion $(a+x)^n$ in the free algebra $\mathcal{A}_q$ on symbols a and x over k modulo the single relation (6.24). The monomials of the form $a^\ell x^m$ for $\ell, m \geq 0$ form a linear basis for $\mathcal{A}_q$. See Exercise 6.5.1 at the end of this section for details.

The expansion of $(a+x)^n$ is a linear combination of monomials in a and x having n factors. Each of these monomials can be written as $a^{n-m}x^m$ for some $0 \leq m \leq n$. Let $\binom{n}{m}_q$ be the coefficient of $a^{n-m}x^m$ in the expansion. By convention $(a+x)^0 = 1$. Therefore

$$\binom{0}{0}_q = 1. \tag{6.26}$$

For integers n, m we set $\binom{n}{m}_q = 0$ if it is not the case that $0 \leq m \leq n$. The expansion of $(a+x)^{n+1} = (a+x)^n(a+x)$ yields

$$\binom{n+1}{m}_q = q^m \binom{n}{m}_q + \binom{n}{m-1}_q \tag{6.27}$$

for $0 \leq m \leq n+1$. Notice that the values $\binom{n}{m}_q$ are uniquely determined by (6.26) and (6.27). Since there is an algebra homomorphism $\mathcal{A}_q \longrightarrow A$, where A is as in the introduction of this section, determined by $a \longmapsto a$ and $x \longmapsto x$ the q-binomial coefficients for $\mathcal{A}_q$ satisfy (6.25) in A. By induction on n it follows that

$$\binom{n}{0}_q = 1 = \binom{n}{n}_q \quad \text{for} \quad n \geq 0. \tag{6.28}$$

Define $(0)_q = 0$ and $(n)_q = 1 + q + \cdots + q^{n-1}$ for $n \geq 1$. Set $(0)_q! = 1$ and $(n)_q! = (n)_q(n-1)_q \cdots (1)_q$ for $n \geq 1$. If q is not an nth root of unity for all $n \geq 1$ then $(n)_q \neq 0$ for all $n \geq 1$. If q is a primitive n^{th} root of unity, where $n > 1$, then $(1)_q, (2)_q, \ldots, (n-1)_q$ are not 0 (and hence $(n-1)_q! \neq 0$) but $(n)_q = 0$. Using (6.26) and (6.27) it is a straightforward matter to show by induction that if $n \geq 1$ and $(n-1)_q! \neq 0$ then

$$\binom{n}{m}_q = \frac{(n)_q!}{(m)_q!(n-m)_q!} \quad \text{for} \quad 0 < m < n. \tag{6.29}$$

One interesting consequence of (6.29) is that $\binom{n}{m}_q = 0$ for all $0 < m < n$ whenever q is a primitive n^{th} root of unity, $n > 1$. We have shown:

Lemma 6.5.1 *Suppose that A is an algebra over the field k and $a, x \in A$ satisfy $xa = qax$ where $q \in k$ is a primitive N^{th} root of unity, $N \geq 1$. Then $(a+x)^N = a^N + x^N$.*

∎

We have a complete description of the symbols $\binom{n}{m}_q$ in (6.28) and (6.29) when q is not a root of unity. Suppose that q is a primitive N^{th} root of unity, where $N > 1$. Generally the relation $xa = qax$ implies that $x^m a^n = q^{mn} a^n x^n$ for all $m, n \geq 0$. Since $q^N = 1$ it follows that a^N and x^N are central elements of A_q. Recall that any integer $n \geq 0$ has a unique decomposition $n = n_D N + n_R$, where n_D and n_R are integers and $0 \leq n_R < N$. By Lemma 6.5.1 we compute

$$\begin{aligned}
(a+x)^N &= \\
&= (a+x)^{n_R}(a+x)^{n_D N} \\
&= (\sum_{\ell=0}^{n_R} \binom{n_R}{\ell}_q a^{n_R-\ell} x^\ell)(\sum_{m=0}^{n_D} \binom{n_D}{m} a^{N(n_D-m)} x^{Nm}) \\
&= \sum_{\ell=0}^{n_R} \sum_{m=0}^{n_D} \binom{n_R}{\ell}_q \binom{n_D}{m} a^{n_R-\ell+N(n_D-m)} x^{\ell+Nm}.
\end{aligned}$$

Note that $n_R = n\text{mod}N$ and $n_D = n\text{div}N$. In general, for an integer n, let $n_R = n\text{mod}N$ and $n_D = n\text{div}N$. We have shown:

Proposition 6.5.1 *Suppose that $q \in k$ is neither 0 nor 1.*

a) *If q is not a root of unity then*

$$\binom{n}{m}_q = \frac{(n)_q!}{(m)_q!(n-m)_q!} \quad \textit{for} \quad 0 \leq m \leq n.$$

b) *Suppose that q is a primitive N^{th} root of unity. Then*

$$\binom{n}{m}_q = \binom{n_R}{m_R}_q \binom{n_D}{m_D} \quad \textit{for} \quad 0 \leq m \leq n.$$

∎

Exercise 6.5.1 Let $T(V)$ be the tensor algebra on the vector space V over k with basis $\{x, a\}$. Fix a non-zero $q \in k$ and let I be the ideal of $T(V)$ generated by $xa - qax$. Show that the cosets of $A_q = T(V)/I$ represented by the monomials $a^\ell x^m$, where $\ell, m \geq 0$, form a linear basis for $A_q = T(V)/I$ in two ways:

a) by using the Diamond Lemma [Bergman, 1978],

b) by finding an algebra B over k and an algebra map $\pi : T(V) \longrightarrow B$ such that the $\pi(a)^\ell \pi(x)^m$'s, where $\ell, m \geq 0$, form a linearly independent set in B.

[Hint: For the latter show that the vector space B with basis $\{v_{\ell,m}\}_{0 \leq \ell, m}$ is an associative algebra with multiplicative identity $v_{0,0}$ where we define $v_{\ell,m} v_{\ell',m'} = q^{m\ell'} v_{\ell+\ell',m+m'}$.]

6.5.2 Construction of the Examples

In this section we construct the examples found in [Taft, 1971], carefully laying out the details of the construction. In [Taft, 1982] the construction is redone using q-binomial coefficients as we do here. The methods we describe are typically used in the construction of many examples of finite-dimensional Hopf algebras.

Let $N > 1$ and suppose that $q \in k$ is a primitive N^{th} root of unity. Then there is an N^2-dimensional Hopf algebra $\mathrm{T}_{N,q}(k)$ over the field k described as follows. As a k-algebra $\mathrm{T}_{N,q}(k)$ is generated by a and x subject to the relations

$$a^N = 1, \quad x^N = 0, \quad \text{and} \quad xa = qax.$$

The coalgebra structure on $\mathrm{T}_{N,q}(k)$ is determined by

$$\Delta(a) = a \otimes a \quad \text{and} \quad \Delta(x) = x \otimes a + 1 \otimes x.$$

By virtue of the algebra relations and the fact that $\mathrm{Dim}\mathrm{T}_{N,q}(k) = N^2$ it follows that the monomials $a^\ell x^m$, where $0 \leq \ell, m < N$, form a linear basis for $\mathrm{T}_{N,q}(k)$ over k.

Now suppose that s is the antipode of $\mathrm{T}_{N,q}(k)$. Then $s(a) = a^{-1} = a^{N-1}$ and $s(x) = -xa^{-1} = -q^{-1}a^{N-1}x$. We have enough of a description of $\mathrm{T}_{N,q}(k)$ at this point to proceed with the formal construction.

Let C be the coalgebra over k with basis $\{e, a, x\}$ whose structure is determined by

$$\Delta(e) = e \otimes e, \quad \Delta(a) = a \otimes a, \quad \text{and} \quad \Delta(x) = x \otimes a + e \otimes x.$$

Now let $(\iota, T(C))$ be the tensor bialgebra of C. We will identify C with $\iota(C)$. Let I be the ideal of $T(C)$ generated by

$$a^N - 1, \quad x^N, \quad e - 1, \quad \text{and} \quad xa - qax.$$

We will show that I is a coideal of $T(C)$. This will mean that I is a bi-ideal of $T(C)$. Since ϵ is an algebra map and vanishes on generators it follows that $\epsilon(I) = (0)$. To show that $\Delta(I) \subseteq T(C) \otimes I + I \otimes T(C) = \mathcal{I}$ it suffices to show that $\Delta(z) \in \mathcal{I}$ for generators z since Δ is an algebra map and $\mathcal{I}$ is an ideal of $T(C) \otimes T(C)$. Now in any coalgebra C the difference of two grouplike elements spans a coideal as does the difference of two elements in $P_{g,h}(C)$ for $g, h \in G(C)$. Let $\pi : T(C) \longrightarrow T(C)/I$ be the projection. We are down to showing that $\Delta(x^N) \in \mathcal{I}$, or equivalently that $(\pi \otimes \pi)\Delta(x^N) = 0$ since $\mathrm{Ker}(\pi \otimes \pi) = \mathcal{I}$. Let $\mathcal{X} = \pi(1) \otimes \pi(x)$ and $\mathcal{A} = \pi(x) \otimes \pi(a)$. Then $\mathcal{X}\mathcal{A} = q\mathcal{A}\mathcal{X}$. Since Δ and π are algebra maps we calculate by Lemma 6.5.1

$$\begin{aligned}
(\pi \otimes \pi)(\Delta(x^N)) &= ((\pi \otimes \pi)(\Delta(x)))^N \\
&= (\mathcal{A} + \mathcal{X})^N \\
&= \mathcal{A}^N + \mathcal{X}^N \\
&= 0
\end{aligned}$$

since $\mathcal{A}^N = \mathcal{X}^N = 0$. Therefore I is a coideal of $T(C)$, and hence I is a bi-ideal of $T(C)$. Let $\mathrm{T}_{N,q}(k) = T(C)/I$.

To see that $\mathrm{T}_{N,q}(k)$ has an antipode, consider the linear map $\varsigma : C \longrightarrow T(C)^{op}$ determined by

$$\varsigma(e) = e, \quad \varsigma(a) = a^{N-1}, \quad \text{and} \quad \varsigma(x) = -q^{-1}a^{N-1}x.$$

By the universal mapping property of the tensor algebra there is an algebra map $S : T(C) \longrightarrow T(C)^{op}$ which uniquely extends ς. Let $\pi : T(C) \longrightarrow T(C)/I$ be the projection. Since $\pi S : T(C) \longrightarrow (T(C)/I)^{op}$ is an algebra map it is not hard to see that $\pi S(z) = 0$ for the generators z of I. Therefore $\pi(S(I)) = (0)$ which means that $S(I) \subseteq I$. Consequently there exists an algebra map $s : T(C)/I \longrightarrow T(C)^{op}/I$ such that $\pi S = s\pi$.

We claim that s is an antipode for $T(C)/I$. Generally if A is a bialgebra over k and $s : A \longrightarrow A^{op}$ is an algebra map then the solutions $a \in A$ to the pair of equations $s(a_{(1)})a_{(2)} = \epsilon(a)1 = a_{(1)}s(a_{(2)})$ is a subalgebra of A. In our case we need only show that these equations are satisfied for $\pi(a)$ and $\pi(x)$. We leave the reader with this easy exercise.

To complete the construction we need to establish that $\mathrm{Dim}\mathrm{T}_{N,q}(k) = N^2$. One solution is to use the Diamond Lemma [Bergman, 1978]. Another is to show that there exists an algebra B over k and an algebra map $f : T(C) \longrightarrow B$ such that $f(I) = (0)$ and the $f(a)^\ell f(x)^m$'s, where $0 \leq \ell, m < N$, form a linearly independent set. For the latter show that the vector space B over k with basis $\{v_{\ell,m}\}_{\ell \in Z_N, 0 \leq m < N}$ is an associative algebra, where

$$v_{\ell,m}v_{\ell',m'} = \begin{cases} 0, & \text{if } \ m + m' \geq N \\ q^{\ell' m} v_{\ell+\ell',m+m'}, & \text{otherwise.} \end{cases}$$

The desired map is determined by $f(x) = v_{0,1}$ and $f(a) = v_{1,0}$.

We will conclude this section by computing powers of the antipode s on a and x and computing the coproduct on basis elements. Since $s(a) = a^{-1}$ it follows that

$$s^{2n}(a) = a \quad \text{and} \quad s^{2n+1}(a) = a^{-1}$$

for $n \geq 0$. We have shown that $s(x) = -xa^{-1}$. Therefore

$$s^2(x) = s(-xa^{-1}) = -s(a^{-1})s(x) = -(a(-xa^{-1})) = q^{-1}x.$$

As a result

$$s^{2n}(x) = q^{-n}x \quad \text{and} \quad s^{2n+1}(x) = -q^{-(n+1)}a^{-1}x$$

for $n \geq 0$. We conclude from these calculations that the order of s as a linear endomorphism of $\mathrm{T}_{N,q}(k)$ is $2N$.

To calculate the coproduct we write $\Delta(x) = \mathcal{A}+\mathcal{X}$ where $\mathcal{A} = x\otimes a$ and $\mathcal{X} = 1\otimes x$. Observe that $\mathcal{X}\mathcal{A} = q\mathcal{A}\mathcal{X}$. Therefore

$$\begin{aligned}\Delta(x^n) &= \sum_{m=0}^{n} \binom{n}{m}_q \mathcal{A}^{n-m}\mathcal{X}^m \\ &= \sum_{m=0}^{n} \binom{n}{m}_q (x\otimes a)^{n-m}(1\otimes x)^m \\ &= \sum_{m=0}^{n} \binom{n}{m}_q x^{n-m}\otimes a^{n-m}x^m\end{aligned}$$

and hence

$$\Delta(a^\ell x^n) = \sum_{m=0}^{n} \binom{n}{m}_q a^\ell x^{n-m}\otimes a^{\ell+(n-m)}x^m$$

for $0 \leq \ell, n < N$, where the coefficients are defined by (6.28) and (6.29).

Exercise 6.5.2 Suppose that k is *any* field. Show that there is a 4-dimensional Hopf algebra H over k with antipode s of order 4 described as follows. As an algebra H is generated by a and x subject to the relations

$$a^2 = 1, \quad x^2 = x, \quad \text{and} \quad xa + ax = a - 1.$$

The coalgebra structure of H is determined by

$$\Delta(a) = a\otimes a \quad \text{and} \quad \Delta(x) = x\otimes a + 1\otimes x.$$

What is the connection between H and $\mathrm{T}_{2,-1}(k)$? See [Radford, 1971], [Taft, 1971].

Exercise 6.5.3 Let H be a Hopf algebra over the field k with antipode s. Then the order of s in the convolution algebra $\mathrm{End}(H)$ is the order of its convolution inverse 1_H.

a) For the Hopf algebra $\mathrm{T}_{N,q}(k)$ find the minimal polynomial of 1_H in the convolution algebra $\mathrm{End}(\mathrm{T}_{N,q}(k))$.

b) Determine the order of s in the convolution algebra $\mathrm{End}(\mathrm{T}_{N,q}(k))$ when k has characteristic 0.

Exercise 6.5.4 Let $H = \mathrm{T}_{N,q}(k)$. Show each of $H^{op}, H^{cop}, H^{op\ cop}, H^*, H^{*\ op}, H^{*\ cop}$ and $H^{*\ op\ cop}$ is isomorphic as a Hopf algebra to either $\mathrm{T}_{N,q}(k)$ or $\mathrm{T}_{N,q^{-1}}(k)$.

6.6 A Family of Quasitriangular Hopf Algebras and Their Associated Quantum Doubles

We will construct a rather extensive family of finite-dimensional quasitriangular Hopf algebras $(U_{(N,\,\nu,\,\omega)}, R)$ from related quantum doubles defined on a family of Hopf algebras $H_{(N,\,\nu\,\omega)}$. The quasitriangular Hopf algebras $(U_{(N,\,\nu\,\omega)}, R)$ seem to hold promise for producing invariants of knots and links through their distinguished elements R. This family includes the familiar $U_q(\mathrm{sl}(2,k))'$ when q is a root of unity.

The families of Hopf algebras described in this section and results about them are found in [Radford, 1994a, Section 5]. Also see [Gelaki, 1996].

6.6.1 Construction and Properties of $\mathrm{H}_{(\mathrm{N},\,\nu,\,\omega)}$

The Hopf algebras $H_{(N,\,\nu\,\omega)}$ are members of a 4-parameter family of pointed Hopf algebras over the field k. The Hopf algebras $U_{(N,\,\nu,\,\omega)}$ are quotients of quantum doubles of members of this 4-parameter family we are about to describe.

Let n, N, and ν be positive integers which satisfy the conditions n divides N and $1 \leq \nu < N$. We will assume that k has a primitive n^{th} root of unity q and we let r be the order of q^ν. We will construct a Hopf algebra $H = H_{n,\,q,\,N,\,\nu}$ over k which is determined by the following properties. As an algebra H is generated by a and x subject to the relations

$$a^N = 1, \quad x^r = 0, \quad \text{and} \quad xa = qax.$$

As a coalgebra H is determined by

$$\Delta(a) = a \otimes a \quad \text{and} \quad \Delta(x) = x \otimes a^\nu + 1 \otimes x.$$

The antipode of H is the algebra map $s : H \longrightarrow H^{op}$ determined by

$$s(a) = a^{-1} \quad \text{and} \quad s(x) = -xa^{-\nu} = -q^{-\nu}a^{-\nu}x.$$

As a result $s^2(a) = a$, $s^2(x) = q^{-\nu}x$ and therefore $s^2(h) = a^\nu h a^{-\nu}$ for all $h \in H$. Our calculations show that s^2 is a diagonalizable inner automorphism of H.

When $r = 1$ the Hopf algebra H is the group algebra of the cyclic group of order N.

To construct $H_{n,\,q,\,N,\,\nu}$ we start with the 4-dimensional coalgebra C over k with basis $\{A, B, D, X\}$ whose structure is determined by

$$\Delta(A) = A \otimes A, \quad \Delta(B) = B \otimes B, \quad \Delta(D) = D \otimes D,$$

and

$$\Delta(X) = X \otimes B + D \otimes X.$$

We let I be the ideal of the free bialgebra $T(C)$ on the coalgebra C generated by

$$A^N - 1, \quad B - A^\nu, \quad D - 1, \quad XA - qAX, \quad \text{and} \quad X^r.$$

Then I is a coideal of $T(C)$ as well and $H_{n,\,q,\,N,\,\nu} = T(C)/I$ is a Hopf algebra. The reader is referred to the construction of the family of Hopf algebras $\mathrm{T}_{N,q}(k)$ of Section

6.5.2 as a guide for filling in the details which establish that $T(C)/I$ is a Hopf algebra. The following is [Radford, 1994a, Proposition 7].

Proposition 6.6.1 *Suppose that $n > 1$ and the field k has a primitive n^{th} root of unity q. Let $H = H_{n,\,q,\,N,\,\nu}$ be the Hopf algebra described above. Then:*

a) *H is a pointed coalgebra and $G(H) = (a)$.*

b) *The set $\wp = \{a^\ell x^m \mid 0 \le \ell < N,\ 0 \le m < r\}$ is linear basis for H. Therefore $\mathrm{Dim}H = Nr$.*

c) *$\Delta(a^\ell x^m) = \sum_{i=0}^{m} \binom{m}{i}_{q^\nu} a^\ell x^{m-i} \otimes a^{\ell+(m-i)\nu} x^i$ for all $0 \le \ell < N$ and $0 \le m < r$.*

Proof: Part a) follows by Corollary 1.7.3. To establish part c), we write $\Delta(x) = \mathcal{A} + \mathcal{X}$, where $\mathcal{A} = x \otimes a^\nu$ and $\mathcal{X} = 1 \otimes x$, and apply (6.25) to $\Delta(x^m) = (\mathcal{A} + \mathcal{X})^m$.

To show part b) we use the Diamond Lemma [Bergman, 1978] on $T(C) = k\{A, B, D, X\}$, where $A < B < D < X$ and the substitution rules are:

$$A^N \longrightarrow 1,\ B \longrightarrow A^\nu,\ D \longrightarrow 1,\ XA \longrightarrow qAX, \text{ and } X^r \longrightarrow 0.$$

The only ambiguities to resolve are the following overlap ambiguities:

$$(A^\ell A^{N-\ell})A^\ell = A^\ell(A^{N-\ell}A^\ell), \quad (X^\ell X^{r-\ell})X^\ell = X^\ell(X^{r-\ell}X^\ell),$$

$$(XA)A^{N-1} = X(AA^{N-1}), \quad \text{and} \quad (X^{r-1}X)A = X^{r-1}(XA).$$

These are resolved in a straightforward manner. This completes our proof. ▮

The Hopf algebra $H = H_{n,\,q,\,N,\,\nu}$ is self dual in reasonable circumstances, according to [Radford, 1994a, Proposition 8], which is our next result.

Proposition 6.6.2 *Suppose that $n > 1$ and the field k has a primitive n^{th} root of unity q. Let $H = H_{n,\,q,\,N,\,\nu}$ and assume that $r > 1$. Then $H \simeq H^*$ as Hopf algebras if and only if k has a primitive N^{th} root of unity ω such that $q = \omega^\nu$.*

Proof: Let $\pi : H \longrightarrow H^*$ be a Hopf algebra isomorphism, $A = \pi(a)$, $\omega = A(a)$, and $X = \pi(x)$. We claim that ω is a primitive N^{th} root of unity. To see this, let $\ell \ge 0$. Since $a \in G(H)$ it follows that $A^\ell \in G(H^*) = \mathrm{Alg}(H, k)$. Since x is nilpotent $A^\ell(x) = 0$, and since a is a grouplike element $A^\ell(a) = \omega^\ell$. Now $A^\ell \neq \epsilon$ for $1 \le \ell < N$ and $A^N = \epsilon$. Therefore ω is a primitive N^{th} root of unity, and our claim is established.

We now show that $q = \omega^\nu$. We first apply both sides of $XA = qAX$ to $h \in G(H)$ and use the fact that $A(h) \neq 0$ to establish $X(h) = qX(h)$. Since $q \neq 1$ it follows that $X(h) = 0$ for all $h \in G(H)$. Suppose that $X(hx) = 0$ for all $h \in G(H)$. Then using part c) of Proposition 6.6.1 it follows by induction on j that $X^j(hx^i) = 0$ for all

$h \in G(H)$ whenever $1 \leq j < r$ and $0 \leq i \leq j$. As a consequence $X^{r-1} = 0$. This contradiction means $X(hx) \neq 0$ for some $h \in G(H)$. Now we apply both sides of $XA = qAX$ to hx to conclude that $A(ha^{\nu}) = qA(h)$. Thus $\omega^{\nu} = q$ follows.

Now suppose that $\omega \in k$ is a primitive N^{th} root of unity which satisfies $q = \omega^{\nu}$ and let $A : H \longrightarrow k$ be the algebra homomorphism determined by

$$A(a) = \omega \quad \text{and} \quad A(x) = 0.$$

Define a functional $X : H \longrightarrow k$ by

$$X(a^{\ell}x^{m}) = \delta_{1,m} \quad \text{for all} \quad 0 \leq \ell < N, \; 0 \leq m < r.$$

Now $\Delta(A) = A \otimes A$ since A is an algebra homomorphism. Establishing $X(uv) = X(u)A^{\nu}(v) + \epsilon(u)X(v)$ for $u, v \in \wp$ described in part b) of Proposition 6.6.1 will show that $\Delta(X) = X \otimes A^{\nu} + \epsilon \otimes X$. There are basically two cases to work out: $u = a^{i}, \; v = a^{j}x$ and $u = a^{i}x, \; v = a^{j}$. Using part c) of Proposition 6.6.1 we derive the formulas

$$A^{\ell}X^{m}(a^{u}x^{v}) = \delta_{m,v}(m)_{q^{\nu}}!\omega^{\ell u} \tag{6.30}$$

and $XA(a^{u}x^{v}) = \delta_{1,v}\omega^{u+\nu}$ for all $\ell, u \in Z$ and $0 \leq m \leq r$ and $0 \leq v < r$, the former by induction on m. The relations $A^{N} = \epsilon, X^{r} = 0$, and $XA = qAX$ follow by the formulas.

To complete our proof we need only show that

$$\{A^{\ell}X^{m} \mid 0 \leq \ell < N, \; 0 \leq m < r\}$$

is linearly independent. Suppose that $\sum_{\ell=0}^{N-1}\sum_{m=0}^{r-1} \alpha_{\ell,m}A^{\ell}X^{m} = 0$ is a dependence relation, where $\alpha_{\ell,m} \in k$. Evaluating both sides of this equation on $a^{u}x^{v}$ we derive that $\sum_{\ell=0}^{N-1} \alpha_{\ell,v}\omega^{\ell u} = 0$ by (6.30). Thus for all $0 \leq v < r$ it follows that

$$(\alpha_{0,v}, \ldots, \alpha_{N-1,v})\mathcal{A} = 0,$$

where $\mathcal{A} = (a_{ij})_{0 \leq i,j < N}$ is the $N \times N$ matrix with entries given by $a_{ij} = \omega^{ij}$. Now $\mathcal{A}$ is invertible since ω is a primitive N^{th} root of unity. Consequently $\alpha_{\ell,v} = 0$ for all $0 \leq \ell < N$ and $0 \leq v < r$. This concludes our proof. ∎

Set $H_{(N,\,\nu,\,\omega)} = H_{n,\,\omega^{\nu},\,N,\,\nu}$, where n is the order of $q = \omega^{\nu}$. When $H_{(N,\,\nu,\,\omega)}$ admits a quasitriangular structure is given in the following corollary, [Radford, 1994a, Corollary 3].

Corollary 6.6.1 *Suppose that $N > 1$ and the field k has a primitive N^{th} root of unity ω and let $H = H_{(N,\,\nu,\,\omega)}$. Then*

a) *H admits a quasitriangular structure if and only if $N = 2\nu$ and ν is odd.*

Suppose that $N = 2\nu$ and ν is odd.

b) *The quasitriangular structures which H admits are described by $(H, R_{s,\beta})$ where*

$$R_{s,\beta} = \frac{1}{N}(\sum_{i,\ell=0}^{N-1} \omega^{-i\ell} a^i \otimes a^{s\ell}) + \frac{\beta}{N}(\sum_{i,\ell=0}^{N-1} \omega^{-i\ell} a^i x \otimes a^{s\ell+\nu} x),$$

$1 \leq s < N$ is odd and $\beta \in k$.

Proof: We will sketch an outline for the argument. Let $R \in H \otimes H$. Then (H, R) is a quasitriangular Hopf algebra if and only if $F_R : H^* \longrightarrow H^{cop}$ is a bialgebra map which satisfies

$$(p_{(1)} \rightharpoonup h) F_R(p_{(2)}) = F_R(p_{(1)})(h \leftharpoonup p_{(2)}) \tag{6.31}$$

for all $p \in H^*$ and $h \in H$.

Suppose that $F_R : H^* \longrightarrow H^{cop}$ is a bialgebra map and set $G = (a)$. By Proposition 6.6.2 there is an isomorphism of Hopf algebras $H \simeq H^*$ which sends a to A and x to X as defined above. Thus F is a composite of bialgebra maps

$$H^* \simeq H \xrightarrow{f} H^{cop}.$$

The reader can check that (6.31) holds for F_R generally if and only if it holds for subcoalgebras C of H and P of H^* which generate H and H^* respectively as algebras. Note that (6.31) holds for $p = \epsilon$ or $h = 1$. Thus (6.31) holds for $H = H_{(N,\nu,\omega)}$ if and only if it holds when $p = A, A^\nu$ or X and $h = a, a^\nu$ or x; that is

$$(A \rightharpoonup h) f(a) = f(a)(h \leftharpoonup A),$$

$$(A^\nu \rightharpoonup h) f(a^\nu) = f(a^\nu)(h \leftharpoonup A^\nu)$$

and

$$(X \rightharpoonup h) f(a^\nu) + h f(x) = f(x)(h \leftharpoonup A^\nu) + (h \leftharpoonup X)$$

hold for $h = a, a^\nu$ and x. These conditions are equivalent to

$$qxf(a) = f(a)x, \quad qf(x)a = af(x), \quad f(a^\nu) = a^\nu, \quad \text{and} \quad xf(x) = f(x)x.$$

Observe that $f(a) = a^s$ for some integer s. Thus the first and third equations imply $q^{s+1} = 1$ and $q^s = q$. Therefore $q^2 = 1$, so s is odd and N divides 2ν. This means that ν is odd and $N = 2\nu$. Computing, using Proposition 6.6.1, we derive $f(x) = \beta a^{-\nu} x$ for some $\beta \in k$.

Now suppose that $N = 2\nu$, where ν is odd, and $1 \leq s < N$ is an odd integer. Then the order of $q = \omega^\nu$ is $r = 2$. Let $\beta \in k$. With the aid of Proposition 6.6.1 the reader can show there is a unique bialgebra map $f : H \longrightarrow H^{cop}$ determined by $f(a) = a^s$ and $f(x) = \beta a^{-\nu} x = \beta a^\nu x$. Let $F : H^* \longrightarrow H^{cop}$ be the composite of f with the isomorphism $H^* \simeq H$ described above.

Now we will determine the R such that $F = F_R$. Let $G = (a)$. By Exercise 6.3.2 the elements $e_\ell = \sum_{i=0}^{N-1} \frac{1}{N}\omega^{-i\ell}a^i \in k[G]$ form an orthogonal set of idempotents for the group algebra $k[G]$. By (6.30) it follows that

$$A^\ell X^m (\frac{1}{(j)_{q^\nu}!} e_i x^j) = \delta_{\ell,\, i}\delta_{m,\, j}$$

for all $0 \leq \ell, i < N$ and $0 \leq m, j < r$. We conclude for any linear map $T : H^* \longrightarrow H$ that $T = F_R$, where

$$R = \sum_{\ell=0}^{N-1} \sum_{m=0}^{r-1} \frac{1}{(m)_{q^\nu}!} e_\ell x^m \otimes T(A^\ell X^m).$$

Therefore for $R_{s,\, \beta}$ satisfying $F = F_{R_{s,\, \beta}}$, we have

$$R_{s,\, \beta} = \sum_{\ell=0}^{N-1} \sum_{m=0}^{r-1} \frac{1}{(m)_{q^\nu}!} e_\ell x^m \otimes a^{s\ell}(\beta a^{-\nu} x)^m.$$

This concludes our proof. ∎

For $\mathrm{T}_{2,-1}(k) = H_{(2,\, 1,\, \omega)}$ observe that $R_{1,\, \beta}$ is the R_β of Section 6.5.

6.6.2 Construction and Properties of $\mathrm{U}_{(\mathrm{N},\, \nu,\, \omega)}$

Let N and ν be positive integers such that $1 \leq \nu < N$ and N does not divide ν^2. Suppose that $\omega \in k$ is a primitive N^{th} of unity. Let $q = \omega^\nu$ and r be the order of $q^\nu = \omega^{\nu^2}$. We define a Hopf algebra $U_{(N,\, \nu,\, \omega)}$ as follows. As an algebra $U = U_{(N,\, \nu,\, \omega)}$ is generated by a, x, and y subject to the relations

$$a^N = 1, \quad x^r = 0, \quad y^r = 0,$$

$$xa = qax, \quad ya = q^{-1}ay,$$

and

$$yx - q^{-\nu}xy = a^{2\nu} - 1.$$

The coalgebra structure of $U_{(N,\, \nu,\, \omega)}$ is determined by

$$\Delta(a) = a \otimes a, \quad \Delta(x) = x \otimes a^\nu + 1 \otimes x, \quad \text{and} \quad \Delta(y) = y \otimes a^\nu + 1 \otimes y.$$

The antipode of U is the algebra map $s : U \longrightarrow U^{op}$ determined by

$$s(a) = a^{-1}, \quad s(x) = -q^{-\nu}a^{-\nu}x, \quad \text{and} \quad s(y) = -q^\nu a^{-\nu}y.$$

Thus $s^2(a) = a$, $s^2(x) = q^{-\nu}x$, $s^2(y) = q^\nu y$, and $s^2(h) = a^\nu h a^{-\nu}$ for all $h \in U$; consequently s^2 is a diagonalizable inner automorphism.

The details of the construction of $U_{(N,\,\nu,\,\omega)}$ follow those of $H_{(N,\,\nu,\,\omega)}$. We start with the 5-dimensional coalgebra C over k with basis $\{A, B, D, X, Y\}$ whose structure is determined by

$$\Delta(A) = A \otimes A, \quad \Delta(B) = B \otimes B, \quad \Delta(D) = D \otimes D,$$

$$\Delta(X) = X \otimes B + D \otimes X, \quad \text{and} \quad \Delta(Y) = Y \otimes B + D \otimes Y,$$

and we let I be the ideal of the free bialgebra $T(C)$ on the coalgebra C generated by

$$A^N - 1, \quad B - A^\nu, \quad D - 1, \quad X^r, \quad Y^r,$$

$$XA - qAX, \quad YA - q^{-1}AY, \quad \text{and} \quad YX - q^{-\nu}XY - A^{2\nu} + 1.$$

The details of the construction are left to the reader.

As one might suspect $\wp = \{a^\ell x^m y^p \mid 0 \le \ell < N,\ 0 \le m, p < r\}$ is a linear basis for $U_{(N,\,\nu,\,\omega)}$. To see this apply the Diamond Lemma to $T(C) = k\{A, B, D, X, Y\}$ with $A < B < D < X < Y$ and the substitution rules

$$A^N \longrightarrow 1, \quad B \longrightarrow A^\nu, \quad D \longrightarrow 1, \quad X^r \longrightarrow 0, \quad Y^r \longrightarrow 0,$$

$$XA \longrightarrow qAX, \quad YA \longrightarrow q^{-1}AY, \quad \text{and}$$
$$YX \longrightarrow q^{-\nu}XY + A^{2\nu} - 1.$$

There are ten ambiguities to resolve, all of them overlap:

$$(A^\ell A^{N-\ell})A^\ell = A^\ell(A^{N-\ell}A^\ell),$$

$$(X^\ell X^{r-\ell})X^\ell = X^\ell(X^{r-\ell}X^\ell), \quad (Y^\ell Y^{r-\ell})Y^\ell = Y^\ell(Y^{r-\ell}Y^\ell),$$
$$(XA)A^{N-1} = X(AA^{N-1}), \quad (YA)A^{N-1} = Y(AA^{N-1}),$$
$$(X^{r-1}X)A = X^{r-1}(XA), \quad (Y^{r-1}Y)A = Y^{r-1}(YA),$$
$$(YX)A = Y(XA),$$
$$(YX)X^{r-1} = Y(XX^{r-1}), \quad \text{and} \quad (Y^{r-1}Y)X = Y^{r-1}(YX).$$

All but the last two have immediate resolution. By induction on ℓ the monomial YX^ℓ reduces to

$$q^{-\nu\ell}X^\ell Y + (1 + q^\nu + \ldots + q^{\nu(\ell-1)})A^{2\nu}X^{\ell-1} - (1 + q^{-\nu} + \ldots + q^{-\nu(\ell-1)})X^{\ell-1}$$

and the monomial $Y^\ell X$ reduces to

$$q^{-\nu\ell}XY^\ell + q^{-\nu(\ell-1)}(1 + q^{-\nu} + \ldots + q^{-\nu(\ell-1)})A^{2\nu}Y^{\ell-1} -$$
$$(1 + q^{-\nu} + \ldots + q^{-\nu(\ell-1)})Y^{\ell-1}$$

for $1 \leq \ell < r$. From these reductions resolution of the last two ambiguities follows. Properties of $U_{(N,\,\nu,\,\omega)}$ we need are recorded in our next proposition [Radford, 1994a, Proposition 10], whose proof is left to the reader.

Proposition 6.6.3 *Suppose that $N > 1$ and the field k has a primitive N^{th} root of unity ω. Let $U = U_{(N,\,\nu,\,\omega)}$. Then:*

a) *U is a pointed coalgebra and $G(U) = (a)$.*

b) *The set $\wp = \{a^\ell x^m y^p \mid 0 \leq \ell < N,\ 0 \leq m, p < r\}$ is linear basis for U. Thus $\mathrm{Dim} U_{(N,\,\nu,\,\omega)} = \frac{N^3}{(N,\nu^2)^2}$.*

c)

$$\Delta(a^\ell x^m y^p) =$$

$$\sum_{i=0}^{m}\sum_{j=0}^{p}\binom{m}{i}_{q^\nu}\binom{p}{j}_{q^{-\nu}} q^{\nu i(p-j)} a^\ell x^{m-i} y^{p-j} \otimes a^{\ell+\nu(m-i)+\nu(p-j)} x^i y^j$$

for all $0 \leq \ell < N$ and $0 \leq m, p < r$.

d) *$H_{(N,\,\nu,\,\omega)}$ is the sub-Hopf algebra of $U_{(N,\,\nu,\,\omega)}$ generated by a and x.*

∎

Observe that $G(U^*) = \{\epsilon\}$ if and only if N is odd and ν and N are relatively prime.

There is a very nice relationship between $D(H_{(N,\,\nu,\,\omega)})$ and $U_{(N,\,\nu,\,\omega)}$. Let $H = H_{(N,\,\nu,\,\omega)}$ and recall that $D(H) = H^{*\,cop} \otimes H$ as a coalgebra, and $\jmath_H : H \longrightarrow D(H)$ and $\jmath_{H^*} : H^{*\,cop} \longrightarrow D(H)$ defined by $\jmath_H(h) = \epsilon \otimes h$ and $\jmath_{H^*}(p) = p \otimes 1$ for $h \in H$ and $p \in H^*$ are one-one Hopf algebra maps. Let $s, S = s^*$ and s be the antipodes of H, H^* and $D(H)$ respectively. Let $Y = S^{-1}(X) = -A^{-\nu}X$ and set

$$\mathbf{a} = \jmath_H(a), \quad \mathbf{x} = \jmath_H(x), \quad \mathbf{A} = \jmath_{H^*}(A), \quad \text{and} \quad \mathbf{y} = \jmath_{H^*}(Y).$$

The relations

$$\mathbf{a}^N = 1, \quad \mathbf{x}^r = 0, \quad \mathbf{xa} = q\mathbf{ax}$$

$$\mathbf{A}^N = 1, \quad \mathbf{y}^r = 0, \quad \text{and} \quad \mathbf{A}^{-1}\mathbf{y} = q\mathbf{y}\mathbf{A}^{-1}$$

hold in $D(H)$ since they hold in H or H^*. Using (6.19), (6.20), and (6.30) we have

$$\mathbf{Aa} = \mathbf{aA} \quad \text{and} \quad \mathbf{yx} = q^{-\nu}\mathbf{xy} + \mathbf{A}^{-\nu}\mathbf{a}^\nu - 1,$$

$$\mathbf{xA} = q^{-1}\mathbf{Ax}, \quad \text{and} \quad \mathbf{ay} = q\mathbf{ya}.$$

It is clear that these ten relations provide a complete description of the algebra structure of $D(H)$. Hence there is an onto algebra map

$$D(H_{(N,\,\nu,\,\omega)}) \xrightarrow{\pi} U_{(N,\,\nu,\,\omega)}$$

determined by

$$\pi(\mathbf{a}) = a, \quad \pi(\mathbf{x}) = x, \quad \pi(\mathbf{A}) = a^{-1}, \quad \text{and} \quad \pi(\mathbf{y}) = y.$$

It is easy to see that π is a coalgebra map; hence π is a Hopf algebra map. Computing dimensions gives $\mathrm{Ker}\pi = (\mathbf{A}^{-1} - \mathbf{a})$. We have shown that

$$D(H_{(N,\,\nu,\,\omega)})/(\mathbf{A}^{-1} - \mathbf{a}) \simeq U_{(N,\,\nu,\,\omega)}$$

as Hopf algebras. Since π is onto, $(U_{(N,\,\nu,\,\omega)}, R)$ is quasitriangular, where $R = (\pi \otimes \pi)(\mathfrak{R})$.

Let $\{h_1, \ldots, h_n\}$ be a linear basis for $H = H_{(N,\,\nu,\,\omega)}$ and $\{\overline{h_1}, \ldots, \overline{h_n}\}$ be the dual basis for H^*. Then we may write $\mathfrak{R} = \sum_{i=1}^{n} (\epsilon \otimes h_i) \otimes (\overline{h_i} \otimes 1)$. Let $\wp = \{a^\ell x^m \mid 0 \leq \ell < N, 0 \leq m < r\}, G = (a)$, and $\mathcal{G} = (A)$. For $\ell \in Z$ set

$$E_\ell = \sum_{i=0}^{N-1} \frac{\omega^{-i\ell}}{N} A^i \quad \text{and} \quad e_\ell = \sum_{i=0}^{N-1} \frac{\omega^{-i\ell}}{N} a^i.$$

Then $\{E_0, \ldots, E_{N-1}\}$ and $\{e_0, \ldots, e_{N-1}\}$ are orthogonal sets of idempotents for the group algebras $k[\mathcal{G}]$ and $k[G]$ respectively. Note that

$$ye_\ell = e_{\ell+\nu} y$$

for all $\ell \in Z$. The elements in the dual basis of $\wp$ for $H^*_{(N,\,\nu,\,\omega)}$ are described by

$$\overline{a^\ell x^m} = \frac{1}{(m)_{q^\nu}!} E_\ell X^m \quad \text{for all} \quad 0 \leq \ell < N,\ 0 \leq m < r$$

by (6.30). Therefore

$$\mathfrak{R} = \sum_{\ell=0}^{N-1} \sum_{m=0}^{r-1} \frac{1}{(m)_{q^\nu}!} (\epsilon \otimes a^\ell x^m) \otimes (E_\ell X^m \otimes 1).$$

Since $\pi : D(H_{(N,\,\nu,\,\omega)}) \longrightarrow U_{(N,\,\nu,\,\omega)}$ a Hopf algebra map we have:

Theorem 6.6.1 *Suppose that $N > 1$ and the field k has a primitive N^{th} root of unity ω. Let $(U_{(N,\,\nu,\,\omega)}, R)$ be defined as above. Then*

$$R = \sum_{m=0}^{r-1} \sum_{\ell=0}^{N-1} \frac{1}{(m)_{q^\nu}!} a^\ell x^m \otimes s^{-1}(y^m e_\ell).$$

∎

The previous theorem is [Radford, 1994a, Theorem 5]. Lastly, we note that $(U_Q(\mathrm{sl}(2,k))', R) = (U_{(n,\,2,\,\omega)}, R)$, where $Q = \omega^{-2} = q^{-1}$. Set $e = ya^{-1}/(q^2-1)$, $f = xa^{-1}$, and $k = a^{-1}$. Then $t = \omega^{-1}$ is a primitive n^{th} root of unity, $Q = t^2$, r is the order of $t^4 = Q^2$, the algebra structure of $U_{(n,\,2,\,\omega)}$ is defined by the relations

$$k^n = 1, \quad ke = Qek, \quad kf = Q^{-1}fk, \quad e^r = 0, \quad f^r = 0$$

and

$$ef - fe = \frac{k^2 - k^{-2}}{Q - Q^{-1}}.$$

The coalgebra structure of $U_{(n,\,2,\,\omega)}$ is determined by

$$\Delta(k) = k \otimes k, \quad \Delta(e) = e \otimes k^{-1} + k \otimes e, \quad \text{and} \quad \Delta(f) = f \otimes k^{-1} + k \otimes f.$$

It is shown in [Radford, 1998, Section 7] that this quasitriangular structure is the usual one.

7 COQUASITRIANGULAR STRUCTURES

Coquasitriangular bialgebras, objects dual to quasitriangular bialgebras, provide in a natural way solutions to the QYBE as do quasitriangular bialgebras. The FRT construction is the prime example of a coquasitriangular bialgebra. See [Faddeev et al., 1990], [Faddeev et al., 1989], and [Faddeev et al., 1988]. Typical references for coquasitriangular bialgebras include [Larson and Towber, 1991] and [Schauenburg, 1992a]. Also see [Majid, 1990b, Section 3.2.3].

7.1 Further Properties of $\mathrm{A}(\mathrm{R})$

Let M be a finite-dimensional vector space with basis $B = \{m_1, \ldots, m_n\}$ over the field k and suppose $R \in \mathrm{End}(M \otimes M)$ is a constant QYBE solution. As motivation for the abstract axioms to follow in this chapter, we study, through a sequence of exercises, a notion related to the bialgebra map defined in Exercise 4.3.6. All of this will be generalized to the one-parameter case in the last section of this chapter.

Let $T(C)$ be the free bialgebra on the coalgebra $C = \mathrm{C}_n(k)$ with standard basis $\{t_i^j\}$ from Section 2.7.1. We are interested in certain bilinear forms

$$T(\mathrm{C}) \times T(\mathrm{C}) \xrightarrow{\beta} k.$$

Given any choice of scalars $b_{i,k}^{j,l}$ there exists such a bilinear form β determined by

$$\beta(t_i^j, t_k^l) = b_{i,k}^{j,l},$$

$$\beta(1, a) = 1 = \beta(a, 1),$$

and

$$\beta(ab, c) = \beta(a, c_{(1)})\beta(b, c_{(2)}), \quad \beta(a, cd) = \beta(a_{(1)}, d)\beta(a_{(2)}, c)$$

for all $a, b, c, d \in T(C)$. Now let $\beta = \beta_R$ satisfy

$$\beta_R(t_i^j, t_k^l) = R_{i,k}^{j,l}. \tag{7.1}$$

Observe that

$$\beta_R(t_k^l t_i^j, t_u^v) = R_{k,s}^{l,v} R_{i,u}^{j,s}.$$

Recall the FRT-relations in the constant case are given by (2.10)

$$R_{s_1,s_2}^{a,b} t_i^{s_1} t_j^{s_2} - t_{s_2}^b t_{s_1}^a R_{i,j}^{s_1,s_2}.$$

Straightforward computations show that

$$\beta_R(R_{s_1,s_2}^{a,b} t_i^{s_1} t_j^{s_2} - t_{s_2}^b t_{s_1}^a R_{i,j}^{s_1,s_2}, t_u^v) = 0$$

and

$$\beta_R(t_u^v, R_{s_1,s_2}^{a,b} t_i^{s_1} t_j^{s_2} - t_{s_2}^b t_{s_1}^a R_{i,j}^{s_1,s_2}) = 0.$$

Exercise 7.1.1 Show that the bilinear form β_R passes to a form on the quotient $A(R) = T/I$, which we also denote by β_R, where I is the ideal of $T(C)$ generated by the FRT relations, and which satisfies

$$\beta_R(1, a) = 1 = \beta_R(a, 1),$$

$$\beta_R(ab, c) = \beta_R(a, c_{(1)})\beta_R(b, c_{(2)}),$$

$$\beta_R(a, cd) = \beta_R(a_{(1)}, d)\beta_R(a_{(2)}, c),$$

and

$$\beta_R(a_{(1)}, b_{(1)})a_{(2)}b_{(2)} = b_{(1)}a_{(1)}\beta_R(a_{(2)}, b_{(2)}).$$

In the next section we will study bialgebras A which possess bilinear forms $\beta : A \times A \longrightarrow k$ satisfying the properties of β_R we have just discussed. Such bialgebras will be called *coquasitriangular*.

Now recall the $A(R)$-module structure on M is determined by $t_i^j{\cdot}m_k = R_{i,k}^{j,s}m_s$. Using β_R, we can express this action in terms of the comodule structure given by

$$\rho(m_i) = m_s \otimes t_i^s,$$

viz.

$$t{\cdot}m = m^{<1>}\beta_R(t, m^{(2)})$$

for $t \in A(R)$. As we will see, the last observation will generalize to an arbitrary coquasitriangular bialgebra A, i.e. any right A-comodule M over A gives rise to a left A-module structure and that these structures satisfy the compatibility condition (2.17). Thus, right A-comodules over such bialgebras A will be seen to give rise to QYB A-modules [Radford and Towber, 1993], [Schauenburg, 1992a].

7.2 Coquasitriangular Coalgebras

To motivate the notion of coquasitriangular coalgebra we look to quasitriangular algebras discussed in Section 6.1.

Let (A, R) be an quasitriangular algebra over the field k and write $R = \sum_{i=1}^r a_i \otimes b_i \in A \otimes A$. Recall from Section 6.1 that for all left A-modules M the linear map $R_M : M \otimes M \longrightarrow M \otimes M$ defined by

$$R_M(m \otimes n) = \sum_{i=1}^{r} a_i{\cdot}m \otimes b_i{\cdot}n$$

for all $m, n \in M$ is a solution to the QYBE if and only if R_A is a solution, which is to say (6.2)

$$\sum_{i,j,k=1}^{r} a_ia_j \otimes b_ia_k \otimes b_jb_k = \sum_{i,j,k=1}^{r} a_ja_i \otimes a_kb_i \otimes b_kb_j.$$

Let $C = A^o$ be the dual coalgebra and define a bilinear form $\beta_R : C \times C \longrightarrow k$ by $\beta_R(c, d) = (c \otimes d)(R)$. For $c, d, e \in C$ we apply $c \otimes d \otimes e$ to both sides of (6.2) and deduce

$$\beta(c_{(1)}, d_{(1)})\beta(c_{(2)}, e_{(1)})\beta(d_{(2)}, e_{(2)}) = \beta(c_{(2)}, d_{(2)})\beta(c_{(1)}, e_{(2)})\beta(d_{(1)}, e_{(1)}) \quad (7.2)$$

where $\beta = \beta_R$. Let M be a finite-dimensional left A-module. By Exercise 1.4.9 recall that (M, μ) has a right A^o-comodule (M, ρ_μ) structure such that

$$a{\cdot}m = m^{<1>}{<}m^{(2)}, a{>}$$

for $a \in A$ and $m \in M$. Thus

$$\begin{aligned} R_M(m \otimes n) &= \sum_{i=1}^{r} a_i{\cdot}m \otimes b_i{\cdot}n \\ &= m^{<1>} \otimes n^{<1>} <m^{(2)}, a_i><n^{(2)}, b_i>, \end{aligned}$$

and hence

$$R_M(m \otimes n) = m^{<1>} \otimes n^{<1>} \beta(m^{(2)}, n^{(2)}) \tag{7.3}$$

for all $m, n \in M$.

Definition 7.2.1 *A* coquasitriangular coalgebra *over k is a pair (C, β) where C is a coalgebra over k and $\beta : C \times C \longrightarrow k$ is a bilinear form satisfying (7.2).*

Definition 7.2.2 *Suppose that (C, β) and (C', β') are coquasitriangular coalgebras over k. Then $f : (C, \beta) \longrightarrow (C', \beta')$ is a* map of coquasitriangular coalgebras *if $f : C \longrightarrow C'$ is a coalgebra map such that $\beta(c, d) = \beta'(f(c), f(d))$ for all $c, d \in C$.*

If U is a vector space over k and $R \in U \otimes U$ we let $\beta_R : U^* \times U^* \longrightarrow k$ denote the bilinear form defined by $\beta_R(u^*, v^*) = (u^* \otimes v^*)(R)$ for $u^*, v^* \in U^*$. We will let the reader complete the proof of the following.

Lemma 7.2.1 *Suppose that A is an algebra over the field k and $R \in A \otimes A$.*

a) *If (A, R) is a quasitriangular algebra over k then (A^o, β_R) is a coquasitriangular coalgebra over k.*

b) *Suppose that A is finite-dimensional. If (A^*, β_R) is a coquasitriangular coalgebra over k then (A, R) is a quasitriangular algebra over k.*

∎

If C is a coalgebra over k, we say that a bilinear form $\beta : C \times C \longrightarrow k$ is invertible if there exists a bilinear form $\eta : C \times C \longrightarrow k$ such that

$$\beta(c_{(1)}, d_{(1)})\eta(c_{(2)}, d_{(2)}) = \epsilon(c)\epsilon(d) = \eta(c_{(1)}, d_{(1)})\beta(c_{(2)}, d_{(2)})$$

for all $c, d \in C$. If these equations are satisfied we say that η is an inverse of β. Observe that if $\widehat{\beta}, \widehat{\eta} : C \otimes C \longrightarrow k$ are the linear forms associated to β and η respectively, to say that η is an inverse of β is to say that $\widehat{\eta}$ is an inverse of $\widehat{\beta}$ in the dual algebra $(C \otimes C)^*$. In particular β has at most one inverse which we denote by β^{-1} when it exists.

Proposition 7.2.1 *Suppose that (C, β) is a coquasitriangular coalgebra over the field k and let M be a right C-comodule. Let R_M be the linear map defined by (7.3). Then:*

a) *R_M is a solution to the QYBE.*

b) *If β is invertible then R_M is a linear automorphism of $M \otimes M$.*

Proof: To show part a) we let $\ell, m, n \in M$ and calculate

$$
\begin{aligned}
&R_{(1,2)}R_{(1,3)}R_{(2,3)}(\ell \otimes m \otimes n) \\
&\quad = R_{(1,2)}R_{(1,3)}(\ell \otimes m^{<1>} \otimes n^{<1>}\beta(m^{(2)}, n^{(2)})) \\
&\quad = R_{(1,2)}(\ell^{<1>} \otimes m^{<1>} \otimes n^{<1><1>}\beta(\ell^{(2)}, n^{<1>(2)})\beta(m^{(2)}, n^{(2)})) \\
&\quad = \ell^{<1><1>} \otimes m^{<1><1>} \otimes n^{<1><1>} \\
&\qquad\qquad * \beta(\ell^{<1>(2)}, m^{<1>(2)})\beta(\ell^{(2)}, n^{<1>(2)})\beta(m^{(2)}, n^{(2)}) \\
&\quad = \ell^{<1>} \otimes m^{<1>} \otimes n^{<1>}\beta({\ell^{(2)}}_{(1)}, {m^{(2)}}_{(1)})\beta({\ell^{(2)}}_{(2)}, {n^{(2)}}_{(1)})\beta({m^{(2)}}_{(2)}, {n^{(2)}}_{(2)})
\end{aligned}
$$

and

$$
\begin{aligned}
&R_{(2,3)}R_{(1,3)}R_{(1,2)}(\ell \otimes m \otimes n) \\
&\quad = R_{(2,3)}R_{(1,3)}(\ell^{<1>} \otimes m^{<1>} \otimes n\beta(\ell^{(2)}, m^{(2)})) \\
&\quad = R_{(2,3)}(\ell^{<1><1>} \otimes m^{<1>} \otimes n^{<1>}\beta(\ell^{(2)}, m^{(2)})\beta(\ell^{<1>(2)}, n^{(2)})) \\
&\quad = \ell^{<1><1>} \otimes m^{<1><1>} \otimes n^{<1><1>} \\
&\qquad\qquad * \beta(\ell^{(2)}, m^{(2)})\beta(\ell^{<1>(2)}, n^{(2)})\beta(m^{<1>(2)}, n^{<1>(2)}) \\
&\quad = \ell^{<1>} \otimes m^{<1>} \otimes n^{<1>}\beta({\ell^{(2)}}_{(2)}, {m^{(2)}}_{(2)})\beta({\ell^{(2)}}_{(1)}, {n^{(2)}}_{(2)})\beta({m^{(2)}}_{(1)}, {n^{(2)}}_{(1)}).
\end{aligned}
$$

Therefore $R_{(2,3)}R_{(1,3)}R_{(1,2)} = R_{(1,2)}R_{(1,3)}R_{(2,3)}$.

To show part b) we let $R = R_M$ and define $S : M \otimes M \longrightarrow M \otimes M$ by $S(m \otimes n) = m^{<1>} \otimes n^{<1>}\eta(m^{(2)}, n^{(2)})$, where η is the inverse of β. We calculate

$$
\begin{aligned}
RS(m \otimes n) &= R(m^{<1>} \otimes n^{<1>}\eta(m^{(2)}, n^{(2)})) \\
&= m^{<1><1>} \otimes n^{<1><1>}\beta(m^{<1>(2)}, n^{<1>(2)})\eta(m^{(2)}, n^{(2)}) \\
&= m^{<1>} \otimes n^{<1>}\beta({m^{(2)}}_{(1)}, {n^{(2)}}_{(1)})\eta({m^{(2)}}_{(2)}, {n^{(2)}}_{(2)}) \\
&= m^{<1>} \otimes n^{<1>}\epsilon(m^{(2)})\epsilon(n^{(2)}) \\
&= m \otimes n
\end{aligned}
$$

for all $m, n \in M$. Therefore $RS = 1_{M\otimes M}$. Likewise $SR = 1_{M\otimes M}$ and part b) follows. ∎

Exercise 7.2.1 Suppose that U and V are finite-dimensional vector spaces over the field k.

a) Show that the map $\iota : U\otimes V \longrightarrow (U^*\otimes V^*)^*$ defined by $\iota(u\otimes v)(u^*\otimes v^*) = u^*(u)v^*(v)$ is a linear isomorphism.

b) For $\nu \in U\otimes V$ define a bilinear form $\beta_\nu : U^* \times V^* \longrightarrow k$ by $\beta_\nu(u^*, v^*) = (u^*\otimes v^*)(\nu)$. Show that the correspondence $\nu \longmapsto \beta_\nu$ determines a linear isomorphism $U\otimes V \simeq \mathrm{B}(U^* \times V^*, k)$ of $U\otimes V$ and the space of bilinear forms $U^* \times V^* \longrightarrow k$.

c) Suppose that $U = V = A$ is a finite-dimensional algebra over the field k and $R \in A\otimes A$. Show that the bilinear form β_R has an inverse if and only R has an inverse in the algebra $A\otimes A$, in which case $\beta_R^{-1} = \beta_{R^{-1}}$.

Exercise 7.2.2 Let (C, β) be a coquasitriangular coalgebra over the field k.

a) Show that (C^{cop}, β) is a coquasitriangular coalgebra over k.

b) Show that $(C, \tilde{\beta})$ is a coquasitriangular coalgebra over k, where $\tilde{\beta}(c, d) = \beta(d, c)$ for all $c, d \in C$.

c) If β has an inverse η show that (C, η) is a coquasitriangular coalgebra over k.

Exercise 7.2.3 Let $f : (A, R) \longrightarrow (A', R')$ be a map of quasitriangular algebras over k. Show that the coalgebra map $f^o : A'^o \longrightarrow A^o$ is a map of coquasitriangular coalgebras $f^o : (A'^o, \beta_{R'}) \longrightarrow (A^o, \beta_R)$.

Exercise 7.2.4 Suppose that $f : (C, \beta) \longrightarrow (C', \beta')$ is a map of *finite-dimensional* coquasitriangular coalgebras over k. Show that the algebra map $f^* : C'^* \longrightarrow C^*$ is a map of quasitriangular algebras $f^* : (C'^*, R') \longrightarrow (C^*, R)$, where $\beta = \beta_R$ and $\beta' = \beta_{R'}$ and we identify $C \simeq C^{**}$ and $C' \simeq C'^{**}$ in the usual way.

Exercise 7.2.5 Let (C, β) and (C', β') be coquasitriangular coalgebras over the field k. Show that $(C\otimes C', \beta'')$ is a coquasitriangular coalgebra over k, where

$$\beta''(c\otimes c', d\otimes d') = \beta(c, d)\beta'(c', d')$$

for all $c, d \in C$ and $c', d' \in C'$ and $C\otimes D$ has the tensor product coalgebra structure.

Exercise 7.2.6 Show that the category of coquasitriangular coalgebras over k has direct sums.

Exercise 7.2.7 Suppose that (C, β) is a quasitriangular coalgebra over the field k and $f : C' \longrightarrow C$ is a coalgebra map.

a) Show that C' has a coquasitriangular coalgebra structure (C', β') such that

$$(C', \beta') \xrightarrow{f} (C, \beta)$$

is a map of coquasitriangular coalgebras.

b) What is the counterpart of part a) for quasitriangular algebras?

7.3 Coquasitriangular Bialgebras and Hopf Algebras

Definition 7.3.1 *Let A be a bialgebra over the field k and suppose that $\beta : A \times A \longrightarrow k$ is a bilinear form. Then the pair (A, β) is said to be a* coquasitriangular bialgebra *if the following axioms are satisfied:*

(CoQT.1) $\beta(ab, c) = \beta(a, c_{(1)})\beta(b, c_{(2)})$,

(CoQT.2) $\beta(1, b) = \epsilon(b)$,

(CoQT.3) $\beta(a, bc) = \beta(a_{(2)}, b)\beta(a_{(1)}, c)$

(CoQT.4) $\beta(a, 1) = \epsilon(a)$, *and*

(CoQT.5) $\beta(a_{(1)}, b_{(1)})a_{(2)}b_{(2)} = b_{(1)}a_{(1)}\beta(a_{(2)}, b_{(2)})$

for all $a, b, c \in A$.

Observe that if (A, β) is a coquasitriangular bialgebra then $(A^{op\ cop}, \beta)$, $(A^{op}, \tilde{\beta})$, and hence $(A^{cop}, \tilde{\beta})$, are also, where $\tilde{\beta}(a, b) = \beta(b, a)$ for $a, b \in A$.

We are using the definition of coquasitriangular bialgebra found in [Schauenburg, 1992a]. In [Larson and Towber, 1991] a different definition of coquasitriangular bialgebra is given. The two are basically the same in the sense that (A, β) satisfies one definition if and only if $(A, \tilde{\beta})$ satisfies the other. In addition, the theory extends non-trivially to the case of $A(R)$ for one-parameter QYBE solution. We will present these results in the last section of this chapter.

There is a fundamental connection between finite-dimensional quasitriangular bialgebras and finite-dimensional coquasitriangular bialgebras. By virtue of Exercise 7.3.7 it follows that (A, R) is a finite-dimensional quasitriangular bialgebra if and only if (A^*, β_R) is a finite-dimensional coquasitriangular bialgebra.

For a bilinear form $\beta : V \times V \longrightarrow k$ on a vector space V over the field k define linear maps $\beta_\ell, \beta_r : V \longrightarrow V^*$ by

$$\beta_\ell(u)(v) = \beta(u, v) = \beta_r(v)(u)$$

for $u, v \in V$. Let $\iota : V \longrightarrow V^{**}$ be the linear map defined by $\iota(v)(v^*) = v^*(v)$ for $v^* \in V^*$ and $v \in V$. Observe that

$$\beta_\ell = \beta_r^* \iota \qquad \text{and} \qquad \beta_r = \beta_\ell^* \iota. \tag{7.4}$$

Axioms (CoQT.1)–(CoQT.4) can be formulated in terms of algebraic properties of β_ℓ and β_r.

Proposition 7.3.1 *Suppose that A is a bialgebra over the field k and $\beta : A \times A \longrightarrow k$ is a bilinear form. Then:*

a) *The following are equivalent:*

i) (CoQT.1) – (CoQT.2) *are satisfied for all* $a, b, c \in A$.

ii) $\beta_\ell : A \longrightarrow A^*$ *is an algebra map.*

iii) Im $\beta_r \subseteq A^o$ *and* $\beta_r : A \longrightarrow A^o$ *is a coalgebra map.*

b) *The following are equivalent:*

i) (CoQT.3) – (CoQT.4) *are satisfied for all* $a, b, c \in A$.

ii) $\beta_r : A \longrightarrow A^{*\ op}$ *is an algebra map.*

iii) Im $\beta_\ell \subseteq A^o$ *and* $\beta_\ell : A \longrightarrow A^{o\ cop}$ *is a coalgebra map.*

c) *Suppose that* β *has inverse* η. *If* (CoQT.1) – (CoQT.4) *are satisfied for* β *then they are satisfied for* $\tilde{\eta}$.

Proof: It is easy to see that (CoQT.1) and (CoQT.2) hold if and only if $\beta_\ell : A \longrightarrow A^*$ is an algebra map and it is easy to see that (CoQT.3) and (CoQT.4) hold if and only if $\beta_r : A \longrightarrow A^{*\ op}$ is an algebra map.

Assume (CoQT.1) and let $c \in C$. Then $\Delta_{\beta_r(c)}$ exists and is $\beta_r(c_{(1)}) \otimes \beta_r(c_{(2)})$. In particular $\beta_r(c) \in A^o$. If (CoQT.2) holds then $\epsilon_{A^o}(\beta_r(c)) = \beta_r(c)(1) = \beta(1, c) = \epsilon(c)$. Thus if (CoQT.1) and (CoQT.2) hold then $\beta_r : A \longrightarrow A^o$ is a coalgebra map. The converse follows easily from definitions. We have completed the proof of part a). That (CoQT.3) and (CoQT.4) hold if and only if Im$\beta_\ell \in A^o$ and $\beta_\ell : A \longrightarrow A^{o\ cop}$ is a coalgebra map follows by a similar argument. This completes the proof of part b).

To show part c) we set $\eta = \beta^{-1}$. An easy calculation shows that the pairs η_ℓ, β_ℓ and η_r, β_r are inverses in the convolution algebra $\mathrm{Hom}(A, A^*)$. By Exercise 1.6.1 it follows that $\eta_\ell : A \longrightarrow A^{*\ op}$ is an algebra map since $\beta_\ell : A \longrightarrow A^*$ is an algebra map and $\eta_r : A \longrightarrow A^*$ is an algebra map since $\beta_r : A \longrightarrow A^{*\ op}$ is an algebra map. Since $\tilde{\eta}_\ell = \eta_r$ and $\tilde{\eta}_r = \eta_\ell$, part c) follows by parts a) and b). This completes our proof. ∎

We can use Proposition 7.3.1 to reformulate the axioms for coquasitriangular bialgebras.

Corollary 7.3.1 *Suppose that* A *is a bialgebra over the field* k. *Then the following are equivalent:*

a) A *is a coquasitriangular bialgebra.*

b) *There is a bialgebra map* $A \xrightarrow{r} A^{o\ op}$ *such that*

$$r(b_{(1)})(a_{(1)})a_{(2)}b_{(2)} = b_{(1)}a_{(1)}r(b_{(2)})(a_{(2)})$$

for all $a, b \in A$.

c) *There is a bialgebra map* $A \xrightarrow{r} A^{o\ cop}$ *such that*

$$r(a_{(1)})(b_{(1)})a_{(2)}b_{(2)} = b_{(1)}a_{(1)}r(a_{(2)})(b_{(2)})$$

for all $a, b \in A$.

∎

We can also use Proposition 7.3.1 to show that (A^{cop}, β^{-1}) is a coquasitriangular bialgebra when (A, β) is and β is invertible.

Proposition 7.3.2 *Suppose that* (A, β) *is a coquasitriangular bialgebra over the field* k *and the bilinear form* β *has an inverse. Then* (A^{cop}, β^{-1}) *is a coquasitriangular bialgebra.*

Proof: Let $\eta = \beta^{-1}$. We will show that $(A, \tilde{\eta})$ is a coquasitriangular bialgebra. For then $(A^{cop}, \eta) = (A^{cop}, \tilde{\tilde{\eta}})$ is a coquasitriangular bialgebra.

By part c) of Proposition 7.3.1 to show that $(A, \tilde{\eta})$ is a coquasitriangular bialgebra it suffices to show that

$$\tilde{\eta}(a_{(1)}, b_{(1)})a_{(2)}b_{(2)} = b_{(1)}a_{(1)}\tilde{\eta}(a_{(2)}, b_{(2)})$$

or

$$\eta(b_{(1)}, a_{(1)})a_{(2)}b_{(2)} = b_{(1)}a_{(1)}\eta(b_{(2)}, a_{(2)})$$

for all $a, b \in A$. But

$$\begin{aligned}
\eta(b_{(1)}, a_{(1)})a_{(2)}b_{(2)} &= \eta(b_{(1)}, a_{(1)})a_{(2)}b_{(2)}\epsilon(b_{(3)})\epsilon(a_{(3)}) \\
&= \eta(b_{(1)}, a_{(1)})a_{(2)}b_{(2)}\beta(b_{(3)}, a_{(3)})\eta(b_{(4)}, a_{(4)}) \\
&= \eta(b_{(1)}, a_{(1)})\beta(b_{(2)}, a_{(2)})b_{(3)}a_{(3)}\eta(b_{(4)}, a_{(4)}) \\
&= \epsilon(b_{(1)})\epsilon(a_{(1)})b_{(2)}a_{(2)}\eta(b_{(3)}, a_{(3)}) \\
&= b_{(1)}a_{(1)}\eta(b_{(2)}, a_{(2)}).
\end{aligned}$$

Therefore $(A, \tilde{\eta})$ is a coquasitriangular bialgebra. ∎

Theorem 7.3.1 *Suppose that* (A, β) *is a coquasitriangular bialgebra over the field* k *and that* A *has an antipode* s. *Then:*

a) *The bilinear form* β *has an inverse which is given by* $\beta^{-1}(a, b) = \beta(s(a), b)$ *for* $a, b \in A$.

b) $\beta(a, b) = \beta(s(a), s(b))$ *for all* $a, b \in A$.

Proof: We will give a direct argument. A proof of part a) based on the convolution algebra will be outlined in the exercises.

Part a) follows from the calculations

$$\begin{aligned}\beta(a_{(1)}, b_{(1)})\beta(s(a_{(2)}), b_{(2)}) &= \beta(a_{(1)}s(a_{(2)}), b)\\ &= \beta(\epsilon(a)1, b)\\ &= \epsilon(a)\beta(1, b)\\ &= \epsilon(a)\epsilon(b)\end{aligned}$$

and

$$\begin{aligned}\beta(s(a_{(1)}), b_{(1)})\beta(a_{(2)}, b_{(2)}) &= \beta(s(a_{(1)})a_{(2)}, b)\\ &= \beta(\epsilon(a)1, b)\\ &= \epsilon(a)\beta(1, b)\\ &= \epsilon(a)\epsilon(b)\end{aligned}$$

for all $a, b \in A$.

The calculation

$$\begin{aligned}\beta(s(a_{(1)}), b_{(1)})\beta(s(a_{(2)}), s(b_{(2)})) &= \beta(s(a)_{(2)}, b_{(1)})\beta(s(a)_{(1)}, s(b_{(2)}))\\ &= \beta(s(a), b_{(1)}s(b_{(2)}))\\ &= \beta(s(a), \epsilon(b)1)\\ &= \beta(s(a), 1)\epsilon(b)\\ &= \epsilon(s(a))\epsilon(b)\\ &= \epsilon(a)\epsilon(b)\end{aligned}$$

for all $a, b \in A$ shows that $\xi : A \times A \longrightarrow k$ defined by $\xi(a, b) = \beta(s(a), s(b))$ is a right inverse for β^{-1}. Therefore $\xi = \beta$ and part b) follows. ∎

Suppose that (A, β) is a coquasitriangular bialgebra over the field k. For $c \in A$, application of $\beta(\ , c)$ to both sides of (CoQT.5) gives the relation

$$\beta(a_{(1)}, b_{(1)})\beta(a_{(2)}, c_{(1)})\beta(b_{(2)}, c_{(2)}) = \beta(a_{(2)}, b_{(2)})\beta(a_{(1)}, c_{(2)})\beta(b_{(1)}, c_{(1)})$$

by virtue of (CoQT.1). Thus (A_c, β) is a coquasitriangular coalgebra, where A_c is the underlying coalgebra structure of A. In particular (7.3) defines a solution to the QYBE for right A-comodules M. We formally note:

Lemma 7.3.1 *A coquasitriangular bialgebra over the field k is a coquasitriangular coalgebra.*

We leave the straightforward proof of the following proposition to the reader. Part a) is a variation of [Radford and Towber, 1993, Proposition 8] when k is a field. If (A, β) is a coquasitriangular bialgebra over k then $\beta_\ell : A \longrightarrow A^*$ is an algebra map by part a) of Proposition 7.3.1.

Proposition 7.3.3 *Let (A, β) be a coquasitriangular bialgebra over the field k and let (M, ρ) be a right A-comodule. Let (M, μ) be the left A-module structure on M derived from the associated left rational A^*-module structure (M, μ_ρ) by pullback along $A \xrightarrow{\beta_\ell} A^*$. Then:*

a) *$a{\cdot}m = m^{<1>}\beta(a, m^{(2)})$ for all $a \in A$ and $m \in M$, and the triple (M, ρ, μ) is a left QYB A-module.*

b) *The associated QYBE solution R is given by*

$$R(m \otimes n) = m^{<1>} \otimes n^{<1>}\beta(m^{(2)}, n^{(2)})$$

for all m, $n \in M$.

∎

In fact, it is easy to see that the construction in Proposition 7.3.3 is functorial so that we have a functor

$$\mathcal{M}^A \longrightarrow {}_A\mathcal{QYB}.$$

Note that since the module structure corresponding to a given comodule is locally finite, these functors are generally far from being onto.

Let (A, β) be a coquasitriangular bialgebra over the field k. We have noted that (CoQT.1) and (CoQT.5) imply that (A_c, β) is a coquasitriangular coalgebra, where A_c is the underlying coalgebra structure of A. Thus the map R_M defined by (7.3) which is defined for right A-comodules M is a solution to the QYBE. Give $M \otimes M$ the tensor product A^{op}-module structure. The reader should note that (CoQT.5) implies

$$M \otimes M \xrightarrow{\sigma_{M,M}} M \otimes M \tag{7.5}$$

defined by

$$\sigma_M(m \otimes n) = R_M\tau_{M,M}(m \otimes n) = n^{<1>} \otimes m^{<1>}\beta(n^{(2)}, m^{(2)})$$

for $m, n \in M$ is a map of right A-comodules See Exercise 7.3.6 for details.

Exercise 7.3.1 Prove Proposition 7.3.3

Exercise 7.3.2 Suppose that A and B are bialgebras over the field k and $\beta, \eta : A \times B \longrightarrow k$ are bilinear forms. We will say that β and η are inverses if

$$\beta(a_{(1)}, b_{(1)})\eta(a_{(2)}, b_{(2)}) = \epsilon(a)\epsilon(b) = \eta(a_{(1)}, b_{(1)})\beta(a_{(2)}, b_{(2)})$$

for all $a \in A$ and $b \in B$.

a) Show that β has at most one inverse (which we denote by β^{-1} if it exists).

b) Show that the following are equivalent:

i) The bilinear forms β and η are inverses.

ii) The bilinear forms $\tilde{\beta}$ and $\tilde{\eta}$ are inverses.

iii) β_ℓ and η_ℓ are inverses in the convolution algebra $\mathrm{Hom}(A, B^*)$.

iv) β_r and η_r are inverses in the convolution algebra $\mathrm{Hom}(B, A^*)$.

[Hint: Noting that $\beta_r = \tilde{\beta}_\ell$ and $\beta_\ell = \tilde{\beta}_r$ will save a little work.]

Exercise 7.3.3 Use results on the convolution algebra to prove part a) of Theorem 7.3.1 by noting $\beta_\ell : A \longrightarrow A^*$ is an algebra map, so β_ℓ has an inverse in the convolution algebra $\mathrm{Hom}(A, A^*)$ which is $(\beta^{-1})_\ell = (\beta_\ell)^{-1} = \beta_\ell s$.

Exercise 7.3.4 Let A and B be bialgebras over the field k and suppose that $\beta : A \times B \longrightarrow k$ is a bilinear form.

a) Show that the following are equivalent:

i) $\beta_\ell : A \longrightarrow B^*$ is an algebra map.

ii) $\beta(aa', b) = \beta(a, b_{(1)})\beta(a', b_{(2)})$ and $\beta(1, b) = \epsilon(b)$ for all $a, a' \in A$ and $b \in B$.

iii) $\mathrm{Im}\ \beta_r \subseteq A^o$ and $\beta_r : B \longrightarrow A^o$ is a coalgebra map.

b) Deduce from part a) that $\beta_r : B \longrightarrow A^{*\ op}$ is an algebra map if and only if $\mathrm{Im}\ \beta_\ell \in B^o$ and $\beta_\ell : A \longrightarrow B^{o\ cop}$ is a coalgebra map.

c) Use parts a) and b) to give a proof of Proposition 7.3.1.

The pairing of a coalgebra and an algebra is discussed in [Radford, 1973]. Pairings of bialgebras are discussed in [Majid, 1990b, p. 9].

Exercise 7.3.5 Let (A, β) be a coquasitriangular bialgebra. Show that the fact that (A_c, β) is a coquasitriangular coalgebra is derivable from (CoQT.3) and (CoQT.5), where A_c is the underlying coalgebra structure on A.

Exercise 7.3.6 Suppose that (A, β) is a coquasitriangular bialgebra and that M, N are right A-comodules. Define $\sigma_{M,N} : M \otimes N \longrightarrow N \otimes M$ by

$$\sigma_{M,N}(m \otimes n) = n^{<1>} \otimes m^{<1>} \beta(n^{(2)}, m^{(2)})$$

for $m, n \in M$. Show that $\sigma_{M,N}$ is a map of right A-comodules, where the tensor product $M \otimes N$ of right A-comodules is given the tensor product A^{op}-comodule structure (thus $\rho(m \otimes n) = (m^{<1>} \otimes n^{<1>}) \otimes n^{(2)} m^{(2)}$ for $m \in M$ and $n \in N$). Note that $\sigma_{M,M} = R_M \tau_{M,M}$. See [Larson and Towber, 1991], [Schauenburg, 1992a].

Exercise 7.3.7 Suppose that A is a finite-dimensional bialgebra (respectively Hopf algebra) over the field k and suppose that $R \in A \otimes A$. Show that (A, R) is quasitriangular bialgebra (respectively Hopf algebra) if and only if (A^*, β_R) is a coquasitriangular bialgebra (respectively Hopf algebra).

7.4 The Free Coquasitriangular Bialgebra on a Coquasitriangular Coalgebra

We will construct the free coquasitriangular bialgebra $(\jmath, \mathrm{T}_\beta(C), \mathfrak{R})$ on a coquasitriangular coalgebra (C, β) over k in this section. We will show that β has an inverse if and only if $\mathfrak{R}$ has an inverse.

There are several formulas which will facilitate our discussion which are relegated to our first lemma. Suppose that A is a bialgebra over k and $\beta : A \times A \longrightarrow k$ is a bilinear form which satisfies (CoQT.1) – (CoQT.4). Then $\beta_\ell : A \longrightarrow A^*$ and $\beta_r : A \longrightarrow A^{*\,op}$ are algebra maps. Therefore A has left A-module structures $(A, \succ_r)$ and $(A, \succ_\ell)$ defined by

$$a \succ_\ell b = \beta_\ell(a) \rightharpoonup b = b_{(1)}\beta(a, b_{(2)})$$

and

$$a \succ_r b = b \leftharpoonup \beta_r(a) = \beta(b_{(1)}, a)b_{(2)}$$

for all $a, b \in A$, and A has two right A-module structures $(A, {}_r\!\prec)$ and $(A, {}_\ell\!\prec)$ defined by

$$b \,{}_\ell\!\prec a = b \leftharpoonup \beta_\ell(a) = \beta(a, b_{(1)})b_{(2)}$$

and

$$b \,{}_r\!\prec a = \beta_r(a) \rightharpoonup b = b_{(1)}\beta(b_{(2)}, a)$$

for all $a, b \in A$.

Lemma 7.4.1 *Suppose that A is a bialgebra over the field k and $\beta : A \times A \longrightarrow k$ is a bilinear form which satisfies* (CoQT.1) – (CoQT.4). *Define $d_\beta : A \times A \longrightarrow A$ by*

$$d_\beta(a, b) = \beta(a_{(1)}, b_{(1)})a_{(2)}b_{(2)} - b_{(1)}a_{(1)}\beta(a_{(2)}, b_{(2)})$$

for $a, b \in A$. Then:

a) $\epsilon(d_\beta(a, b)) = 0$,

b) $\Delta(d_\beta(a, b)) = d_\beta(a_{(1)}, b_{(1)}) \otimes a_{(2)}b_{(2)} + b_{(1)}a_{(1)} \otimes d_\beta(a_{(2)}, b_{(2)})$,

c) $d_\beta(1, b) = 0 = d_\beta(a, 1)$,

d) $d_\beta(ab, c) = (c_{(1)} \succ_r a)d_\beta(b, c_{(2)}) + d_\beta(a, c_{(1)})(b \,{}_r\!\prec c_{(2)})$, *and*

e) $d_\beta(a, bc) = d_\beta(a_{(2)}, b)(c \,{}_\ell\!\prec a_{(1)}) + (a_{(2)} \succ_\ell b)d_\beta(a_{(1)}, c)$

for all $a, b, c \in A$.

Proof: Parts a) and c) are straightforward. We will leave parts d) and e) as exercises for the reader. To show part b) we calculate

$$\begin{aligned}
&\Delta(d_\beta(a,b))\\
&\quad= \beta(a_{(1)},b_{(1)})a_{(2)(1)}b_{(2)(1)}\otimes a_{(2)(2)}b_{(2)(2)}\\
&\qquad -b_{(1)(1)}a_{(1)(1)}\otimes b_{(1)(2)}a_{(1)(2)}\beta(a_{(2)},b_{(2)})\\
&\quad= \beta(a_{(1)(1)},b_{(1)(1)})a_{(1)(2)}b_{(1)(2)}\otimes a_{(2)}b_{(2)}\\
&\qquad -b_{(1)}a_{(1)}\otimes b_{(2)(1)}a_{(2)(1)}\beta(a_{(2)(2)},b_{(2)(2)})\\
&\quad= (\beta(a_{(1)(1)},b_{(1)(1)})a_{(1)(2)}b_{(1)(2)} - b_{(1)(1)}a_{(1)(1)}\beta(a_{(1)(2)},b_{(1)(2)}))\otimes a_{(2)}b_{(2)}\\
&\qquad +b_{(1)}a_{(1)}\otimes(\beta(a_{(2)(1)},b_{(2)(1)})a_{(2)(2)}b_{(2)(2)} - b_{(2)(1)}a_{(2)(1)}\beta(a_{(2)(2)},b_{(2)(2)}))\\
&\quad= d_\beta(a_{(1)},b_{(1)})\otimes a_{(2)}b_{(2)} + b_{(1)}a_{(1)}\otimes d_\beta(a_{(2)},b_{(2)}).
\end{aligned}$$

This concludes our proof. ∎

Now we proceed with the construction of $\mathrm{T}_\beta(C)$. Let (C,β) be a coquasitriangular coalgebra over k, and let $(\jmath, T(C))$ be the tensor bialgebra on the coalgebra C. Ignoring the coalgebra structure of $T(C)$, the pair $(\jmath, T(C))$ is the tensor algebra on the vector space C. Therefore there is a unique algebra map $f : T(C) \longrightarrow C^{*\ op}$ such that $f\jmath = \beta_r$. Let $\iota : C \longrightarrow C^{*\ o}$ be the coalgebra map defined by $\iota(c)(c^*) = c^*(c)$ and consider the composite of coalgebra maps

$$C \xrightarrow{\ \iota\ } C^{*\ o} \xrightarrow{\ f^o\ } T(C)^{o\ cop}.$$

By the universal mapping property of the tensor bialgebra $(\jmath, T(C))$ there is a unique bialgebra map $F : T(C) \longrightarrow T(C)^{o\ cop}$ such that $F\jmath = f^o\iota$. The linear map $F : T(C) \longrightarrow T(C)^*$ defines a unique bilinear form $\mathfrak{R}' : T(C)\times T(C) \longrightarrow k$ such that $\mathfrak{R}'_\ell = F$. By Proposition 7.3.1 the bilinear form $\mathfrak{R}'$ satisfies (CoQT.1)–(CoQT.4).

We next show that

$$\mathfrak{R}'(\jmath(c),\jmath(d)) = \beta(c,d) \tag{7.6}$$

for $c, d \in C$ by computing

$$\begin{aligned}
\mathfrak{R}'(\jmath(c),\jmath(d)) &= (\mathfrak{R}'_\ell(\jmath(c)))(\jmath(d))\\
&= F(\jmath(c))(\jmath(d))\\
&= (f^o(\iota(c)))(\jmath(d))\\
&= \iota(c)(f(\jmath(d)))\\
&= \iota(c)(\beta_r(d))\\
&= \beta_r(d)(c)\\
&= \beta(c,d).
\end{aligned}$$

Now let $\mathcal{I}$ be the span of the $d_{\mathfrak{R}'}(\jmath(c),\jmath(d))$'s where $c, d \in C$. Since $\jmath(C)$ is a subcoalgebra of $T(C)$ it follows that $\mathcal{I}$ is a coideal of $T(C)$ by parts a) and b) of

Lemma 7.4.1. Let I be the ideal of $T(C)$ generated by $\mathcal{I}$. Then I is a bi-ideal of $T(C)$ since $\mathcal{I}$ is a coideal of $T(C)$.

Set $\mathrm{T}_\beta(C) = T(C)/I$. We will show that $\mathfrak{R}'$ lifts to a (unique) bilinear form $\mathfrak{R} : \mathrm{T}_\beta(C) \times \mathrm{T}_\beta(C) \longrightarrow k$ such that

$$\mathfrak{R}(\pi(a), \pi(b)) = \mathfrak{R}'(a, b) \tag{7.7}$$

for $a, b \in T(C)$, where $\pi : T(C) \longrightarrow T(C)/I$ is the projection. This can be done if and only if $\mathfrak{R}'(I, T(C)) = (0) = \mathfrak{R}'(T(C), I)$. Since $\jmath : C \longrightarrow T(C)$ is a coalgebra map and (C, β) is a coquasitriangular coalgebra, we can use (7.6) to deduce $\mathfrak{R}'(\jmath(c), d_{\mathfrak{R}'}(\jmath(d), \jmath(e))) = 0 = \mathfrak{R}'(d_{\mathfrak{R}'}(\jmath(c), \jmath(d)), \jmath(e))$ for all $c, d, e \in C$. Therefore $\mathfrak{R}'(\jmath(C), \mathcal{I}) = (0) = \mathfrak{R}'(\mathcal{I}, \jmath(C))$. Since $\mathcal{I}$ is a coideal of $T(C)$ it follows from (CoQT.1) and (CoQT.3) that $\mathfrak{R}'(\jmath(C)), I) = (0) = \mathfrak{R}'(I, \jmath(C))$. Since I is a coideal of $T(C)$ it follows that the set of all $x \in T(C)$ such that $\mathfrak{R}'(x, I) = (0)$ is a subalgebra of $T(C)$. Since $\jmath(C)$ generates $T(C)$ as an algebra we now have $\mathfrak{R}'(T(C), I) = (0)$. One can use a similar argument to show that $\mathfrak{R}'(I, T(C)) = (0)$.

Let $\mathfrak{R} : \mathrm{T}_\beta(C) \times \mathrm{T}_\beta(C) \longrightarrow k$ be the unique bilinear form which satisfies (7.7). Since π is an onto bialgebra map and $\mathfrak{R}'$ satisfies (CoQT.1)–(CoQT.4) necessarily $\mathfrak{R}$ does as well. Let $j = \pi\jmath$. Then j is a coalgebra map. Set $\mathcal{C} = j(C)$. For $c, d \in C$ observe that $d_{\mathfrak{R}}(\pi(\jmath(c)), \pi(\jmath(d))) = d_{\mathfrak{R}'}(j(c), j(d)) = 0$. Therefore $d_{\mathfrak{R}'}(\mathcal{C}, \mathcal{C}) = (0)$. Since $\mathcal{C}$ is a subcoalgebra of $\mathrm{T}_\beta(C)$ it follows by parts c) and d) of Lemma 7.4.1 that the set of all $x \in \mathrm{T}_\beta(C)$ such that $d_{\mathfrak{R}'}(x, \mathcal{C}) = (0)$ is a subalgebra of $A = \mathrm{T}_\beta(C)$. Since $\mathcal{C}$ generates A as an algebra we conclude that $d_{\mathfrak{R}'}(A, \mathcal{C}) = (0)$. By a similar argument $d_{\mathfrak{R}'}(A, A) = (0)$. We have shown that $(\mathrm{T}_\beta(C), \mathfrak{R})$ is a coquasitriangular bialgebra. By (7.6) and (7.7) the coalgebra map $j = \pi\jmath$ is a map of coquasitriangular coalgebras. Part a) of the following has been established.

Theorem 7.4.1 *Suppose that (C, β) is a coquasitriangular coalgebra over the field k. Then the triple $(j, \mathrm{T}_\beta(C), \mathfrak{R})$ constructed above satisfies the following:*

a) $j : (C, \beta) \longrightarrow (\mathrm{T}_\beta(C), \mathfrak{R})$ *is a map of coquasitriangular coalgebras, and*

b) *If (A, β') is a coquasitriangular bialgebra over k and $f : (C, \beta) \longrightarrow (A, \beta')$ is a map of coquasitriangular coalgebras then there exists a map of coquasitriangular bialgebras $F : (\mathrm{T}_\beta(C), \mathfrak{R}) \longrightarrow (A, \beta')$ uniquely determined by $Fj = f$.*

Proof: It remains to show part b). By virtue of the universal mapping property of the free bialgebra $(\jmath, T(C))$ on the coalgebra C there exists a unique bialgebra map $F' : T(C) \longrightarrow A$ such that $F'\jmath = f$. For $c, d \in C$ we compute

$$\begin{aligned} \beta'(F'(\jmath(c)), F'(\jmath(d))) &= \beta'(f(c), f(d)) \\ &= \beta(c, d) \\ &= \mathfrak{R}'(\jmath(c), \jmath(d)). \end{aligned}$$

Since F' is a bialgebra map, by (CoQT.1) and (CoQT.2) the set of all $a \in T(C)$ such that $\beta'(F'(a), F'(\jmath(d))) = \Re'(a, \jmath(d))$ for all $d \in C$ is a subalgebra of $T(C)$. Since $\jmath(C)$ generates $T(C)$ as an algebra it now follows that $\beta'(F'(a), F'(\jmath(d))) = \Re'(a, \jmath(d))$ for all $a \in T(C)$. A similar argument shows that $\beta'(F'(a), F'(b)) = \Re'(a, b)$ for all $a, b \in T(C)$. Therefore

$$\beta'(F'(a), F'(b)) = \Re'(a, b) = \Re(\pi(a), \pi(b)) \tag{7.8}$$

for all $a, b \in T(C)$.

To conclude the proof we need only show that $F'(I) = (0)$. For if this is the case there exists a unique bialgebra map $F : T(C)/I \longrightarrow A$ such that $F\pi = F'$. Thus $\Re(a, b) = \beta'(F(a), F(b))$ for all $a, b \in \mathrm{T}_\beta(C)$ by (7.8). Since F' is a bialgebra map $F'(d_{\Re'}(a, b)) = d_{\beta'}(F'(a), F'(b)) = 0$ for all $a, b \in T(C)$ by (7.8). Therefore $F'(I) = (0)$, and the proof of part b) is complete. ∎

Definition 7.4.1 *Let (C, β) be a coquasitriangular coalgebra over the field k. The triple $(\jmath, \mathrm{T}_\beta(C), \Re)$ of the theorem s called the* free coquasitriangular bialgebra on (C, β).

Corollary 7.4.1 *Suppose that (C, β) is a coquasitriangular coalgebra over the field k and $(\jmath, \mathrm{T}_\beta(C), \Re)$ is the free coquasitriangular bialgebra on (C, β). Then β is invertible if and only if $\Re$ is invertible.*

Proof: If $\Re$ is invertible then β is invertible since $\jmath : (C, \beta) \longrightarrow (\mathrm{T}_\beta(C), \Re)$ is a map of coquasitriangular coalgebras.

Suppose that β is invertible. We will use the notation of the proof of Theorem 7.4.1. Using Exercise 7.4.2 we see that $\Re'$ has an inverse $\wp'$. Since β' satisfies (CoQT.1)–(CoQT.4) it follows that $\tilde{\wp}'$ does also by part c) of Exercise 7.2.2. Observe that (CoQT.1)–(CoQT.4) hold for $\tilde{\wp}'$ by part c) of Proposition 7.3.1.

We need only show that there exists a bilinear form $\wp : \mathrm{T}_\beta(C) \times \mathrm{T}_\beta(C) \longrightarrow k$ such that $\wp'(a, b) = \wp(\pi(a), \pi(b))$ for $a, b \in T(C)$, where $\pi : T(C) \longrightarrow T(C)/I$ is the projection. To this end it will suffice to show that

$$\wp'(d_{\Re'}(a, b), c) = 0 = \wp'(a, d_{\Re'}(b, c))$$

for $a, b, c \in T(C)$. See the discussion following (7.7) for details. We will verify the first equation and leave the reader with the exercise of verifying the second. Observe that

$$\begin{aligned}
&\wp'(\Re'(a_{(1)}, b_{(1)})a_{(2)}b_{(2)}, c)\\
&= \Re'(a_{(1)}, b_{(1)})\wp'(a_{(2)}, c_{(2)})\wp'(b_{(2)}, c_{(1)})\\
&= \Re'(a_{(1)}, b_{(1)})\wp'(a_{(2)}, c_{(2)})\wp'(b_{(2)}, c_{(1)})\wp'(a_{(3)}, b_{(3)})\Re'(a_{(4)}, b_{(4)})\\
&= \Re'(a_{(1)}, b_{(1)})\wp'(a_{(3)}, b_{(3)})\wp'(a_{(2)}, c_{(2)})\wp'(b_{(2)}, c_{(1)})\Re'(a_{(4)}, b_{(4)})\\
&= \Re'(a_{(1)}, b_{(1)})\wp'(a_{(2)}, b_{(2)})\wp'(a_{(3)}, c_{(1)})\wp'(b_{(3)}, c_{(2)})\Re'(a_{(4)}, b_{(4)}).
\end{aligned}$$

The last equation follows since $(T(C), \wp')$ is a coquasitriangular coalgebra. On the other hand

$$\begin{aligned}
&\wp'(b_{(1)}a_{(1)}\Re'(a_{(2)}, b_{(2)}), c)\\
&= \wp'(b_{(1)}, c_{(2)})\wp'(a_{(1)}, c_{(1)})\Re'(a_{(2)}, b_{(2)})\\
&= \Re'(a_{(1)}, b_{(1)})\wp'(a_{(2)}, b_{(2)})\wp'(b_{(3)}, c_{(2)})\wp'(a_{(3)}, c_{(1)})\Re'(a_{(4)}, b_{(4)}).
\end{aligned}$$

Thus $\wp'(d_{\Re'}(a, b), c) = 0$. This concludes our proof. ∎

Exercise 7.4.1 Suppose that A is a bialgebra over the field k and $\beta : A \times A \longrightarrow k$ is a bilinear form satisfying (CoQT.1)–(CoQT.4). If C is a subcoalgebra of A which generates A as an algebra, show that (A_c, β) is a coquasitriangular coalgebra if and only if $(C, \beta|_{C\times C})$ is a coquasitriangular coalgebra.

Exercise 7.4.2 Let C and D be coalgebra over the field k. Suppose that $f : C \longrightarrow D^*$ is linear and has an inverse in the convolution algebra $\mathrm{Hom}(C, D^*)$. Let $(\jmath_C, T(C))$ be the tensor algebra on C and $F : T(C) \longrightarrow D^*$ be the unique algebra map satisfying $F\jmath_C = f$. Now let $\iota : D \longrightarrow D^{*\,o}$ be the coalgebra map defined by $\iota(d)(d^*) = d^*(d)$ and let $\mathcal{F} : T(D) \longrightarrow T(C)^o$ be the unique bialgebra map determined by $\mathcal{F}\jmath_D = F^o\iota$. Show that $\mathcal{F}$ has an inverse in the convolution algebra $\mathrm{Hom}(T(D), T(C)^*)$.

Exercise 7.4.3 Verify the formulas of parts d) and e) of Lemma 7.4.1.

Exercise 7.4.4 Suppose that (C, β) is a coquasitriangular coalgebra over the field k and that $\varsigma : (C, \beta) \longrightarrow (C^{cop}, \beta)$ is a map of coquasitriangular coalgebras. Assume further that $\beta_\ell : C \longrightarrow C^*$ is invertible in the convolution algebra $\mathrm{Hom}(C, C^*)$ and $\beta_\ell^{-1} = \beta_\ell\varsigma$. Show that there exists a triple $(\iota, \mathrm{H}_{\beta,\,\varsigma}(C), \Re)$ which satisfies the following property:

a) $(\mathrm{H}_{\beta,\,\varsigma}(C), \Re)$ is a coquasitriangular Hopf algebra and $\iota : (C, \beta) \longrightarrow (\mathrm{H}_{\beta,\,\varsigma}(C), \Re)$ is a map of coquasitriangular coalgebras, and

b) if (A, β') is a coquasitriangular Hopf algebra and $f : (C, \beta) \longrightarrow (A, \beta')$ is a map of coquasitriangular coalgebras then there exists a unique map of coquasitriangular Hopf algebras $F : (\mathrm{H}_{\beta,\,\varsigma}(C), \Re) \longrightarrow (A, \beta')$ such that $F\iota = f$.

7.5 One-Parameter Quantum Yang–Baxter Equation, Coquasitriangularity, and Tensor Product

This section builds on the material from [Cotta-Ramusino et al., 1993] which gives the FRT construction a coquasitriangular structure in the one-parameter case and presents an elegant setting for understanding the tensor product of one-parameter QYBE solutions given in [Lambe, 1994] (see (3.16)).

If R is a constant QYBE solution, Section 7.1 established the existence of a coquasitriangular structure

$$A(R) \times A(R) \xrightarrow{\beta_R} k$$

which is the multiplicative extension of

$$\beta_R(t_i^j, t_k^l) = R_{k,i}^{l,j}. \tag{7.9}$$

As we have noted in Section 7.3, there is a twisted version of coquasitriangularity. In this case, it is realized on $A(R)$ by extending

$$\beta'_R(t_i^j, t_k^l) = R_{k,i}^{j,l}$$

to all of $A(R) \times A(R)$ in such a manner that the axioms of Exercise 7.1.1 are satisfied for $A(R)^{cop}$. Both of these were presented in [Cotta-Ramusino et al., 1993] for the FRT construction associated to a one-parameter QYBE solution. but were not named as such.

7.5.1 R-commutative Spectral Parameter

Definition 7.5.1 *The one-parameter QYBE solution*

$$X \xrightarrow{R} \mathrm{End}(\otimes^2 M)$$

has an R-commutative spectral parameter *if*

$$R(xy) = R(yx)$$

where, as usual, $ab = \varphi(a,b)$ *and* $\varphi : Z \longrightarrow X$ *and* $Z \subseteq X \times X$. *Assume that* φ *has an identity element* 1.

Consider again the free tensor bialgebra $T(C)$ from Section 2.7.1 and let R be a one-parameter R-commutative QYBE solution.

Exercise 7.5.1 Following Section 2.7.1 (or otherwise), prove that there is a unique linear map

$$(A(R) \otimes A(R))_Z \xrightarrow{\beta_R} k$$

given on generators by

$$\beta_R(t_i^j(x) \otimes t_k^l(z)) = R_{i,k}^{j,l}(xz)$$

satisfying (CoQT.1)-(CoQT.4). Here $(A(R) \otimes A(R))_Z$ is the sub-bialgebra of $A(R) \otimes A(R)$ generated by elements of the form $1 \otimes t_i^k(x), t_j^l(y) \otimes 1$, and $t_a^b(z) \otimes t_c^d(w)$ for $(z,w) \in Z$ and $(1,x),(y,1) \in Z$.

Does β_R provide a "coquasitriangular structure on $A(R)$", i.e. is

$$\beta_R(a_{(1)}, b_{(1)})a_{(2)}b_{(2)} = b_{(1)}a_{(1)}\beta_R(a_{(2)}, b_{(2)})$$

valid (whenever both sides are defined)?

7.5.2 Constructions when X is a Group

Assume that $R : X \longrightarrow \mathrm{End}(\otimes^2 M)$ and we have $\varphi : Z \longrightarrow X$ where $Z \subseteq X \times X$. Write $xy = \varphi(x, y)$ as usual.

We say that φ is R-associative if $R(x(yz)) = R((xy)z)$ whenever both sides are defined. Assume that φ has an identity element 1, i.e. for all $x \in X, (x, 1), (1, x) \in Z$ and $R(\varphi(x, 1)) = R(x)$ and $R(\varphi(1, x)) = R(x)$. Finally, assume that for all $x \in X$ there is an $x' \in X$ such that $(x, x') \in X$ and $R(xx') = R(1)$. By abuse of notation we write $x' = x^{-1}$ for any such element. We *do not* assume that $R(xy) = R(yx)$. Of course, if φ is a group law, all of these properties hold. In general, when the conditions just given hold for a one-parameter QYBE solution, we will simply say that "X is a group".

Exercise 7.5.2 a) Following your calculations (and the notation) from the previous exercise, prove that there is a unique linear map

$$(A(R) \otimes A(R))_Z \xrightarrow{\beta_R} k$$

given on generators by

$$\beta_R(t_i^j(x) \otimes t_k^l(z)) = R_{i,k}^{j,l}(xz^{-1}),$$

assuming that X is a group.

b) Prove that

$$\beta_R(a_{(1)} \otimes b_{(1)})a_{(2)}b_{(2)} = b_{(1)}a_{(1)}\beta_R(a_{(2)} \otimes b_{(2)})$$

whenever both sides make sense so that $A(R)$ inherits a "coquasitriangular structure" [Cotta-Ramusino et al., 1993].

We will use the bilinear form notation for β_R from now on.

A crucial fact used in our computations below is given by the

Lemma 7.5.1 *Suppose that R is a one-parameter QYBE solution and that X is a group, not necessarily assumed to be commutative. Consider the FRT construction $A(R)$ with standard multiplicative generators $t_i^j(x)$. Then*

$$\beta_R(t_{i_1}^{j_1}(x) \dots t_{i_r}^{j_r}(x), t_{k_1}^{l_1}(y) \dots t_{k_s}^{l_s}(y))$$

is a product of coefficients of the form $R_{a,b}^{c,d}(xy^{-1})$ whenever these quantities are defined.

∎

We leave the easy proof by induction to the reader.

We now present a method by which we can transform a given one-parameter QYBE solution R into another using a given family of coalgebras.

Definition 7.5.2 *Let M be a vector space over k. Let*

$$\mathrm{Comod}(M, A) \subseteq \mathrm{Hom}(M, M \otimes A)$$

be the space of all comodule structures on M over the bialgebra A. A map

$$X \xrightarrow{\rho} \mathrm{Comod}(M, A(R))$$

is a parameterized family of coalgebra structures *on M over R if* $\mathrm{Im}\rho(x) \subseteq M \otimes A(R)_x$ *where $A(R)_x$ is the subalgebra of $A(R)$ generated by $t_i^j(x)$.*

We will use the notation $\rho_x = \rho(x)$.

From the multiplicativity of $\beta_R(\ ,\ )$ in the first slot (to ensure associativity of the action), the following is easily seen.

Proposition 7.5.1 *Let $X \xrightarrow{\rho} \mathrm{Comod}(M, A(R))$ be a parameterized family of coalgebra structures on M over R and assume that X is a group. Then*

$$a{\cdot}m = m^{<1>_1}\beta_R(a, m^{(2)_1}), \tag{7.10}$$

where we have written,

$$\rho_1(m) = m^{<1>_1} \otimes m^{(2)_1},$$

defines a left $A(R)$-module structure on M. ∎

Note that the module action is defined everywhere for $A(R)$ since we are only using the constant part ρ_1 of the comodule structure in the formula above.

We have an immediate corollary of Lemma 7.5.1.

Corollary 7.5.1 *Let $X \xrightarrow{\rho} \mathrm{Comod}(M, A(R))$ be a parameterized family of coalgebra structures on a finite-dimensional vector space M over R and assume that X is a group. Assume further that for all $(x, z) \in Z \subseteq X \times X$, we have that $(xz, z^{-1}) \in Z$. Let $\{m_i\}$ be a basis for N over k. We have unique elements $\alpha_p^q(x)$ of the subalgebra of $A(R)$ generated by $t_i^j(x)$ given by*

$$\rho_x(m_i) = m_s \otimes \alpha_i^s(x).$$

The elements $\alpha_p^q(x)$ satisfy

$$\beta_R(\alpha_i^j(xz), \alpha_k^l(z)) = \beta_R(\alpha_i^j(x), \alpha_k^l(1)).$$

for all $x, z \in X$. ∎

For the rest of this section, we will assume the hypotheses of this corollary.

Definition 7.5.3 *Let* $X \xrightarrow{\rho} \mathrm{Comod}(M, A(R))$ *be a parameterized family of coalgebra structures on* M *over* R *and let* (M, μ) *be the associated* $A(R)$*-module structure from Proposition 7.5.1. The operator*

$$X \xrightarrow{R_\rho} \mathrm{End}(\otimes^2 M) \tag{7.11}$$

given by

$$R_\rho(x)(m \otimes n) = m^{<1>_x} \otimes m^{(2)_x} \cdot n,$$

a ρ-perturbation *of* R.

We have come to the main result of this section.

Theorem 7.5.1 *Let* $\rho : X \longrightarrow \mathrm{Comod}(M, A(R))$ *be a parameterized family of coalgebra structures on* M *over* R *and let* R_ρ *be the* ρ*-perturbation of* R*. Then* R_ρ *is a one-parameter QYBE solution.*

Proof: We will use coordinates and let the reader work out an invariant version of the proof. Let $R_{i,j}^{k,l}(x)$ be the coordinates of $R(x)$ with respect to the basis $\{m_i\}$ for M. and similarly assume that

$$\rho_x(m_i) = m_s \otimes \alpha_i^s(x).$$

Notice that

$$R_\rho(x)(m_i \otimes m_j) = m_k \otimes m_l \beta_R(\alpha_i^k(x), \alpha_j^l(1))$$

by definition. But by Corollary 7.5.1 we then have that

$$R_{s_1,s_2}^{a,b}(x) = \beta_R(\alpha_{s_1}^a(x), \alpha_{s_2}^b(1)) = \beta_R(\alpha_{s_1}^a(xz), \alpha_{s_2}^b(z)),$$

$$R_{i,s_3}^{s_1,c}(xz) = \beta_R(\alpha_i^{s_1}(xz), \alpha_{s_3}^c(1)),$$

and

$$R_{r_2,r_3}^{b,c}(z) = \beta_R(\alpha_j^{s_2}(z), \alpha_k^{s_3}(1)).$$

Similarly, we have

$$\begin{aligned} \beta_R(t_i^{r_1}(xz), t_j^{r_2}(z)) &= R_{i,j}^{r_1,r_2}(x), \\ \beta_R(t_{r_1}^a(xz), t_k^{r_3}(1)) &= R_{r_1,k}^{a,r_3}(xz), \\ \beta_R(t_{r_2}^b(z), t_{r_3}^c(1)) &= R_{r_2,r_3}^{b,c}(z) \end{aligned}$$

for any $x, z \in X$. Thus analogous to the proof of Proposition 7.2.1, the one-parameter QYBE holds by the coquasitriangularity of $A(R)$. ∎

7.5.3 *Tensor Product of One-Parameter Quantum Yang–Baxter Solutions*

Let $R : X \longrightarrow \mathrm{End}(\otimes^2 M)$ be a one-parameter QYBE solution. We have already seen a natural example of a parameterized comodule structures for M over R, viz.

$$\rho_x(m_i) = m_s \otimes t_i^s(x)$$

where $\{m_i\}$ is a k-basis for M and $t_i^j(x)$ is the usual collection of multiplicative generators of $A(R)$.

If

$$X \xrightarrow{\rho,\rho'} \mathrm{Comod}(M_i, A(R))$$

are two families of parameterized comodules over R, we define their tensor product by

$$(\rho \otimes \rho')_x(m) = (m^{<1>_x} \otimes m^{<1>_x{}'}) \otimes (m^{(1)_x{}'} m^{(1)_x}).$$

Note that this allows us to define the tensor product of two ρ-perturbations as the ρ-perturbation of the tensor product family of comodules. In fact, we can take tensor powers $\otimes^k \rho$ of this comodule structure. By Theorem 7.5.1, we thus get a sequence of tensor power one-parameter QYBE solutions

$$\otimes^k R = R_{\otimes^k \rho}.$$

These are the tensor powers described in [Cotta-Ramusino et al., 1993] and [Lambe, 1994, p. 46] and Section 3.9.

Exercise 7.5.3 Let A be a bialgebra over the field k. Let M_i for $i = 1, 2$ be two left-A-modules

$$A \otimes M_i \xrightarrow{\mu_i} M_i$$

and let

$$M_i \xrightarrow{\rho_i(x)} M_i \otimes A$$

be two parameterized comodule structures over A in the sense of Section 2.10. Suppose that the one-parameter compatibility condition (2.21) holds for $i = 1, 2$. Does Proposition 3.8.1 generalize to this setting?

8 SOME CLASSES OF SOLUTIONS

Various techniques are discussed for solving the QYBE. The two-dimensional case is treated in great detail and a theoretical determination of all upper triangular solutions is made. The results of Section 2.11.2 are placed in a theoretical context. We apply some of the techniques developed in earlier chapters in Section 8.6.1 to find some one-parameter QYBE solutions.

Let M be a finite-dimensional vector space over k. We illustrate how to find invertible QYB operators $R : M \otimes M \longrightarrow M \otimes M$ by several means in this chapter, principally through an analysis of $\widetilde{A(R)}$. When R is congruent to an upper triangular solution, $\widetilde{A(R)}$ is a pointed bialgebra by Theorem 4.4.2. Since there are good techniques for studying the structure of pointed bialgebras, there is the possibility of producing interesting solutions to the QYBE through an analysis of $\widetilde{A(R)}$. We do this when $\mathrm{Dim}M = 2$ in Sections 8.2 and 8.3.

For the case $\mathrm{Dim}M = 2$ partial determination of $\widetilde{A(R)}$ naturally groups the upper triangular solutions into six congruence classes. Solutions belonging to different classes are not congruent. Recall that the reduced FRT construction $\widetilde{A(R)}$ is an congruence invariant. We use $\widetilde{A(R)}$ in establishing noncongruence.

All solutions to the QYBE equation when $\mathrm{Dim}M = 2$ were found in [Hietarinta, 1993b] by symmetry reductions and computer methods (also see [Hietarinta, 1993a]). For higher dimensions, computer techniques to find all solutions seem infeasible, although progress was made in [Hietarinta, 1993c] for dimension three. A combination of bialgebra and computer methods could lead to the discovery of large classes of upper triangular solutions.

There are higher dimensional upper triangular examples which arise from finite-dimensional Hopf algebras, namely from the quantum double in certain cases and from the examples of Section 6.5.2. See Exercise 6.4.5. The reader is encouraged investigate them after reading this chapter.

Suppose that A is a bialgebra over k and M is a finite-dimensional vector space over k. Consider a triple (M, μ, ρ), where (M, μ) is a left A-module and $\rho : M \longrightarrow M \otimes A$ is a *linear* map. Recall that $R = R_{(\mu,\rho)}$ is defined by $R(m \otimes n) = m^{<1>} \otimes m^{(2)}{\cdot}n$ for $m, n \in M$. In Section 2.11.2 computer algebra was used to find such solutions R in the case $A = \mathbb{C}[\mathbb{Z}_2]$ is the group algebra of the cyclic group of order 2 over the field of complex numbers and $\mathrm{Dim}M = 3$. An interesting set of solutions R to the QYBE emerged where ρ is *not* a comodule structure map. This combination of computer and algebra techniques suggests a fruitful approach to finding solutions to the QYBE. We place these particular calculations in the context of $H = k[\mathbb{Z}_n]$ and develop a theory for such triples in Section 8.5.

Sections 8.1–8.3 are basically reformulations of material found in [Radford, 1994b].

8.1 Some Consequences of $\mathbb{M}$-Reduction

The results in this section are very elementary. We isolate them because of their importance for computation. The following lemma is [Radford, 1994b, Lemma 8].

Lemma 8.1.1 *Suppose that A is a bialgebra over the field k. Let $\pi : A \longrightarrow \mathrm{End}(M)$ be the representation of A afforded by a left A-module structure (M, μ). Suppose that A is M-reduced.*

a) *Let $g \in G(A)$ and suppose that $\pi(g) \in \mathrm{End}(M)$ is invertible. For $x \in A$, either $xg = 0$ or $gx = 0$ implies that $x = 0$.*

b) *Suppose that $g, g' \in G(A)$ and $x \in P_{g,g'}(A)$. Then $\pi(x) = 0$ implies $x = 0$.*

c) *The map $\iota : G(A) \longrightarrow \mathrm{End}(M)$ defined by $\iota(g) = \pi(g)$ is a one-one map of semigroups.*

d) *Suppose that C is a coalgebra over k and that $f, g : C \longrightarrow A$ are coalgebra maps. Then $\pi f = \pi g$ implies that $f = g$.*

Proof: To show part a) we first note that $xg = 0$ implies $0 = \pi(xg) = \pi(x)\pi(g)$. Thus $xg = 0$ implies $\pi(x) = 0$ since $\pi(g)$ is invertible. Likewise $gx = 0$ implies $\pi(x) = 0$. Now left or right multiplication in A by a grouplike element of A is a coalgebra map. Since the kernel of coalgebra map is a coideal, part a) follows.

To establish part b) we need only note that $x \in P_{g,g'}(A)$ spans a coideal of A. For $g, g' \in G(A)$ observe that $g - g' \in P_{g,g'}(A)$. If $\pi(g) = \pi(g')$ then $\pi(g - g') = 0$. Thus part c) follows from part b).

Part d) follows with the observation that if $f, g : C \longrightarrow D$ are coalgebra maps then $\text{Im}(f - g)$ is a coideal of D. This concludes the proof of the lemma. ∎

Apropos of the lemma, the hypothesis of part a) is satisfied in a very natural way. Suppose that (M, μ, ρ) is a finite-dimensional left QYB A-module and the associated QYBE solution R is invertible. If $m \in M$ generates a one-dimensional sub-comodule of M then $\rho(m) = m \otimes g$ for some $g \in G(A)$ and $\pi(g)$ is an invertible operator. For the calculation $m \otimes (\pi(g)(n)) = m \otimes (g{\cdot}n) = R(m \otimes n)$ for $n \in M$ shows that $\pi(g)$ is one-one, hence is an invertible operator.

More generally, suppose that L and N are sub-comodules of M. Assume that $L \subseteq N$ and let $\varphi : N \longrightarrow N/L$ be the projection. Let $(N/L, \rho_\varphi)$ be the right A-comodule structure of the quotient N/L. Since $R(L \otimes M) \subseteq L \otimes M$ and $R(N \otimes M) \subseteq N \otimes M$ there is a unique linear map $\mathcal{R} : (N/L) \otimes M \longrightarrow (N/L) \otimes M$ such that $(\varphi \otimes 1_M)R = \mathcal{R}(\varphi \otimes 1_M)$. Since R is one-one necessarily $\mathcal{R}$ is one-one. Thus if $\varphi(m) \in N/L$ is not zero and $\rho_\varphi(\varphi(m)) = \varphi(m) \otimes g$ for some $g \in G(A)$, it follows that $\pi(g)$ is invertible.

For a coalgebra C over k and $g \in G(C)$ we set $M_g = \{m \in M \mid \rho(m) = m \otimes g\}$.

Corollary 8.1.1 *Let A be a bialgebra over the field k. Suppose that (M, μ, ρ) is a left QYB A-module. Suppose that A is M-reduced and $\pi : A \longrightarrow \text{End}(M)$ is the representation of A afforded by (M, μ). Let $g, h \in G(A)$ satisfy $M_g \neq (0)$ and $\pi(h)$ is invertible. Then $h{\cdot}M_g \subseteq M_\ell$ for some $\ell \in G(A)$ uniquely determined by $hg = \ell h$.*

Proof: Let $m \in M_g$ and assume that $m \neq 0$. Now $h{\cdot}m = \pi(h)(m) \neq 0$ since $\pi(h)$ is one-one. Write $0 \neq \rho(h{\cdot}m) = \sum_{i=1}^r m_i \otimes a_i \in M \otimes A$ where $r = \text{Rank}\rho(h{\cdot}m)$. By Lemma A.4.1 the m_i's and the a_i's form linearly independent sets. Observe that $h{\cdot}m = (1_M \otimes \epsilon)(\rho(h{\cdot}m)) = \sum_{i=1}^r \epsilon(a_i)m_i$.

For h and m the compatibility condition $h_{(1)}{\cdot}m^{<1>} \otimes h_{(2)}m^{(2)} = (h_{(2)}{\cdot}m)^{<1>} \otimes (h_{(2)}{\cdot}m)^{(2)}h_{(1)}$ comes down to $h{\cdot}m \otimes hg = (h{\cdot}m)^{<1>} \otimes (h{\cdot}m)^{(2)}h$. We rewrite this last equation

$$\sum_{i=1}^r m_i \otimes \epsilon(a_i)hg = \sum_{i=1}^r m_i \otimes a_i h.$$

Since $\{m_1, \ldots, m_r\}$ is linearly independent it follows that

$$\epsilon(a_i)hg = a_i h$$

for all $1 \leq i \leq r$. Since $\{a_1, \ldots, a_r\}$ is linearly independent $\{a_1h, \ldots, a_rh\}$ is also by part a) of Lemma 8.1.1. Therefore $r = 1$, and consequently $\rho(h{\cdot}m) = h{\cdot}m \otimes \ell$ for some $\ell \in G(A)$. Rewriting the compatibility condition for h and m we obtain $h{\cdot}m \otimes hg = h{\cdot}m \otimes \ell h$. Since $h{\cdot}m \neq 0$ it follows that $hg = \ell h$. Since $hg = xh$ has at most one solution $x \in A$ by part a) of Lemma 8.1.1, the corollary now follows. ∎

8.2 When the Reduced FRT Construction is Generated by Grouplike Elements

In this section we will find necessary and sufficient conditions for an invertible solution $R : M \otimes M \longrightarrow M \otimes M$ to the QYBE in the finite-dimensional case to have the property that $\widetilde{A(R)}$ is generated by grouplike elements.

Let $R : M \otimes M \longrightarrow M \otimes M$ be such an operator. Suppose that A is a bialgebra over k and (M, μ, ρ) is a left QYB A-module such that R on M such that R is the associated QYBE solution.

Suppose that A is generated as an algebra by grouplike elements. Then A is spanned by $G(A)$. Since $G(A)$ is linearly independent, it follows that $M = \oplus_{g \in G(A)} M_g$. Let $\{g_1, \ldots, g_s\}$ be the subset of $G(A)$ such that $M_{g_i} \neq (0)$. Then

$$M = \oplus_{i=1}^{s} M_{g_i}.$$

The operators $\pi(g_1), \ldots, \pi(g_s) \in \mathrm{End}(M)$ are invertible by the discussion following the proof of Lemma 8.1.1.

Let $R_{g_i} \in \mathrm{End}(M)$ be defined by $R_{g_i}(m) = g_i{\cdot}m$ for $m \in M$. Then for $m \in M_{g_i}$ and $n \in M$ we compute

$$R(m \otimes n) = m^{<1>} \otimes m^{(2)}{\cdot}n = m \otimes g_i{\cdot}n = m \otimes R_{g_i}(n).$$

For $1 \leq i, j \leq s$ there exists $1 \leq \ell \leq s$ such that $R_{g_i}(M_{g_j}) \subseteq M_{g_\ell}$ and that $R_{g_i}R_{g_j} = R_{g_\ell}R_{g_i}$ by Corollary 8.1.1. The following is [Radford, 1994b, Theorem 4] for the operator R^τ.

Theorem 8.2.1 *Suppose that M is a finite-dimensional vector space over the field k and that $R : M \otimes M \longrightarrow M \otimes M$ is an invertible solution to the QYBE. Then the following are equivalent:*

a) *$\widetilde{A(R)}$ is spanned by its grouplike elements.*

b) *There is a direct sum decomposition $M = \oplus_{i=1}^{s} M_i$ of M and a family of operators $\{R_1, \ldots, R_s\} \subseteq \mathrm{End}(M)$ such that*

   i) *for $1 \leq i, j \leq s$ there is an $1 \leq \ell \leq s$ such that $R_i(M_j) \subseteq M_\ell$ and $R_iR_j = R_\ell R_i$, and*

ii) $R = E_1 \otimes R_1 + \cdots + E_s \otimes R_s$, *where* E_i *is the projection of* $M = \oplus_{i=1}^s M_i$ *onto* M_i.

Proof: We have shown that part a) implies part b). Suppose that the hypothesis of part b) holds. Let C be the coalgebra over k with basis $\{g_1, \ldots, g_s\}$ of grouplike elements. Define a linear map $F : C \longrightarrow \mathrm{End}(M)$ by $f(g_i) = R_i$. Let $(\jmath, T(C))$ be the tensor bialgebra of C. Since $\jmath$ is one-one, we may think of $\jmath$ as the inclusion and thus regard C as a subcoalgebra of $T(C)$.

Since $T(C)$ is also the free algebra on the vector space C, there is a unique algebra map $F : T(C) \longrightarrow \mathrm{End}(M)$ which extends f. Let (M, μ) be the left A-module structure on M afforded by the representation F. We endow M with a right $T(C)$-comodule structure (M, ρ) where $\rho(m) = m \otimes g_i$ for $m \in M_i$.

Now let I be the ideal of $T(C)$ generated by the differences $g_i g_j - g_\ell g_i$ where ℓ is chosen so that condition i) of part b) is satisfied. Since these differences span a coideal of $T(C)$ it follows that I is a bi-ideal of $T(C)$. Let $A = T(C)/I$ and $\pi : T(C) \longrightarrow A$ be the projection. Since $F(g_i g_j - g_\ell g_i) = R_i R_j - R_\ell R_i = 0$ the differences $g_i g_j - g_\ell g_i \in \mathrm{Ker} F = \mathrm{ann}_{T(C)}(M)$. We appeal to Exercise 4.1.3 at this point for the existence of a triple (M, μ_π, ρ_π) which is a left QYB A-module such that $R = R_{(\mu_\pi, \rho_\pi)}$. By Theorem 4.2.1 it follows that $\widetilde{A(R)}$ is a quotient of a sub-bialgebra of A. Since subcoalgebras and quotients of coalgebras spanned by their grouplike elements are themselves spanned by their grouplike elements, it follows that $\widetilde{A(R)}$ is spanned by its grouplike elements. This completes the proof of the theorem. ∎

We end this section with a discussion of the $\mathrm{Dim} M = 2$ case when $\widetilde{A(R)}$ is generated by grouplike elements. This will be the first step of the analysis of the $\mathrm{Dim} M = 2$ case when $\widetilde{A(R)}$ is pointed. In the next section we complete this analysis, with a few technical exceptions when the field k has characteristic 2.

Suppose that $\mathrm{Dim} M = 2$ and $R : M \otimes M \longrightarrow M \otimes M$ is an invertible solution to the QYBE. Assume that A is a bialgebra over k, generated by grouplike elements, and that (M, μ, ρ) is a left QYB A-module structure on M such that R is the associated QYBE solution. Then $M = \oplus_{g \in G(A)} M_g$. Therefore $M = M_g$ for some $g \in G(A)$ or $M = M_g \oplus M_h$ where $g, h \in G(A)$ are distinct and $\mathrm{Dim} M_g = 1 = \mathrm{Dim} M_h$. Thus in any event we may assume that A is generated as an algebra by $\mathbf{a}$ and $\mathbf{d}$, where

$$\Delta(\mathbf{a}) = \mathbf{a} \otimes \mathbf{a} \quad \text{and} \quad \Delta(\mathbf{d}) = \mathbf{d} \otimes \mathbf{d},$$

and there is a basis $B = \{m, n\}$ for M such that

$$\rho(m) = m \otimes \mathbf{a} \quad \text{and} \quad \rho(n) = n \otimes \mathbf{d}.$$

Let $\pi : A \longrightarrow \mathrm{End}(M)$ be the representation of A afforded by (M, μ). We may assume that A is M-reduced as well. Observe that $\widetilde{A(R)}$ is the quotient of the subalgebra of

A generated by $A(\rho)$ by Theorem 4.2.2. Since R is invertible, by the discussion following the proof of Lemma 8.1.1 we have

$$\pi(\mathbf{a}), \pi(\mathbf{d}) \in \mathrm{End}(M) \quad \text{are invertible.}$$

Write

$$\pi(\mathbf{a})_B = \begin{pmatrix} a & b \\ x & c \end{pmatrix} \quad \text{and} \quad \pi(\mathbf{d})_B = \begin{pmatrix} d & e \\ y & f \end{pmatrix}.$$

Consider the basis $\mathcal{B} = \{m \otimes m, m \otimes n, n \otimes m, n \otimes n\}$ for $M \otimes M$. Then

$$R_{\mathcal{B}} = \begin{pmatrix} \pi(\mathbf{a})_B & \mathcal{O} \\ \mathcal{O} & \pi(\mathbf{d})_B \end{pmatrix}$$

has block form, where $\mathcal{O} \in \mathrm{M}_2(k)$ is the zero matrix.

Case 1: $\mathbf{a} = \mathbf{d}$.

The associated solution R to the QYBE equation is determined by

$$\textbf{C1)} \quad [R]_{\mathcal{B}} = \begin{pmatrix} \pi(\mathbf{a})_B & \mathcal{O} \\ \mathcal{O} & \pi(\mathbf{a})_B \end{pmatrix} = \left(\begin{array}{cc|cc} a & b & 0 & 0 \\ x & c & 0 & 0 \\ \hline 0 & 0 & a & b \\ 0 & 0 & x & c \end{array}\right)$$

where $\pi(\mathbf{a})$ is invertible.

We are not asserting that A is M-reduced. Observe that $\widetilde{A(R)}$ is the quotient of the subalgebra of A generated by $A(\rho)$ by Theorem 4.2.2. It could very well be the case that the algebra A does have other relations. For example, if $\pi(\mathbf{a})^n = 1$ then $\mathbf{a}^n = 1$ necessarily by part c) of Lemma 8.1.1 in order for A to be M-reduced.

Now suppose that $\mathbf{a} \neq \mathbf{d}$. Then $\mathbf{ad} = \mathbf{da}$ by Corollary 8.1.1. Since $\mathbf{a} \neq \mathbf{d}$ it follows that $\pi(\mathbf{a}) \neq \pi(\mathbf{d})$ by part c) of Lemma 8.1.1.

Case 2: $\mathbf{a} \neq \mathbf{d}$. Then:

$$\mathbf{ad} = \mathbf{da}.$$

The associated solution R to the QYBE is determined by

$$\textbf{C2)} \quad [R]_{\mathcal{B}} = \begin{pmatrix} \pi(\mathbf{a})_B & \mathcal{O} \\ \mathcal{O} & \pi(\mathbf{d})_B \end{pmatrix} = \left(\begin{array}{cc|cc} a & b & 0 & 0 \\ x & c & 0 & 0 \\ \hline 0 & 0 & d & e \\ 0 & 0 & y & f \end{array}\right)$$

where $\pi(\mathbf{a}), \pi(\mathbf{d})$ commute, are distinct, and are invertible.

Exercise 8.2.1 Show that Case 1 and Case 2 are in fact realized as described by constructing a bialgebra A over k with the indicated properties and a left QYB A-module structure on M so that R is the associated QYBE solution.

Exercise 8.2.2 Show that solutions R, R' to the QYBE which come from Case 1 and Case 2 respectively are not congruent.

Exercise 8.2.3 Suppose that A and (M, μ, ρ) satisfy the description of Case 1 or Case 2. Let $\pi(\mathbf{a})_B = \mathcal{A}$ and $\pi(\mathbf{d})_B = \mathcal{D}$. Show that A is a Hopf algebra if and only if $\mathcal{A}^m\mathcal{D}^n = I$ for some positive integers m and n.

Exercise 8.2.4 For each of Case 1 and Case 2 find a complete set of representatives for the congruence classes of solutions to the QYBE which arise in that case.

For the remainder of these exercises M is a finite-dimensional vector space over k, $T : M \longrightarrow M$ is linear, and $R : M \otimes M \longrightarrow M \otimes M$ is defined by $R = 1_M \otimes T$. (Thus R arises from Case 1.)

Exercise 8.2.5 Show, working from definitions, that R is a solution to the QYBE and $R^\tau = T \otimes 1_M$.

Exercise 8.2.6 Show that if the minimal polynomial of T does not divide $p(X) = X^m - X^\ell$ for all $0 \leq \ell < m$ then $\widetilde{A(R)} = k[a]$ and is infinite-dimensional, where a is grouplike. (In this case $k[a]$ is the polynomial algebra in indeterminant a over k.)

Exercise 8.2.7 Suppose that the minimal polynomial of T does divide $p(X) = X^m - X^\ell$, where $0 \leq \ell < m$ and m is the smallest such integer. Show that $\widetilde{A(R)} = k[a]$ and is m-dimensional, where a is grouplike and $a^m = a^\ell$.

Exercise 8.2.8 Find a necessary and sufficient condition in terms of T for $\widetilde{A(R)}$ to be a Hopf algebra.

Exercise 8.2.9 Show that $\widetilde{A(R^\tau)}$ is spanned by its grouplike elements if and only if T is diagonalizable. See Exercise 4.3.4.

Exercise 8.2.10 When $\mathrm{Dim} M \geq 2$ find an example of invertible solutions $R', R'' : M \otimes M \longrightarrow M \otimes M$ to the QYBE which satisfy

a) $\widetilde{A(R')} \simeq \widetilde{A(R'')}$ as bialgebras,

b) R' and R'' are similar operators of $M \otimes M$, and

c) there is no invertible operator $U \in \mathrm{End}(M)$ such that $R'' = (U \otimes U)R'(U \otimes U)^{-1}$.

Exercise 8.2.11 When $\mathrm{Dim} M \geq 3$ find an example of invertible solutions $R', R'' : M \otimes M \longrightarrow M \otimes M$ to the QYBE which satisfies

a) $\widetilde{A(R')} \simeq \widetilde{A(R'')}$ as bialgebras and are finite-dimensional, and

b) R' and R'' are not similar operators of $M \otimes M$.

8.3 Solutions when Dim M is Two and the Reduced FRT Construction is Pointed but not Generated by Grouplike Elements

Throughout this section M is a 2-dimensional vector space over the field k. We will complete our determination of all invertible solutions $R : M \otimes M \longrightarrow M \otimes M$ to the QYBE such that $\widetilde{A(R)}$ is pointed, with a few exceptions when the characteristic of k is 2. By Theorem 4.4.1 requiring $\widetilde{A(R)}$ to be pointed is equivalent to the existence of a non-zero $m \in M$ such that $R(m \otimes M) \subseteq m \otimes M$.

Suppose that $R : M \otimes M \longrightarrow M \otimes M$ is an invertible solution to the QYBE. Let A be a bialgebra over k and suppose that (M, μ, ρ) is a left QYB A-module such that R is the associated QYBE. Assume that A is M-reduced. Observe that $\widetilde{A(R)}$ is the subalgebra of A generated by $A(\rho)$ by Theorem 4.2.2.

Let $\pi : A \longrightarrow \mathrm{End}(M)$ be the representation of A afforded by (M, μ). Suppose $0 \neq m \in M$ satisfies $R(m \otimes M) \subseteq m \otimes M$. By Exercise 4.4.3 we have

$$\rho(m) = m \otimes \mathbf{a} \tag{8.1}$$

for some $\mathbf{a} \in G(A)$. Extend $\{m\}$ to a basis $B = \{m, n\}$ for M. Writing

$$\rho(n) = m \otimes \mathbf{x} + n \otimes \mathbf{d}$$

it follows by the comodule axioms that

$$\Delta(\mathbf{a}) = \mathbf{a} \otimes \mathbf{a}, \quad \Delta(\mathbf{d}) = \mathbf{d} \otimes \mathbf{d}, \quad \text{and} \quad \Delta(\mathbf{x}) = \mathbf{a} \otimes \mathbf{x} + \mathbf{x} \otimes \mathbf{d}. \tag{8.2}$$

Let C be the subcoalgebra of A spanned by $\mathbf{a}, \mathbf{d}$ and $\mathbf{x}$. We may assume that A is generated as an algebra by C.

The case when $\widetilde{A(R)}$ is spanned by grouplike elements was covered in the previous section. Therefore we will assume $\widetilde{A(R)}$ is *not* generated by grouplike elements in this section. In particular we will assume that

$$C \quad \text{is } \textit{not} \text{ spanned by grouplike elements.} \tag{8.3}$$

Since R is invertible, from the discussion following Lemma 8.1.1 we infer that

$$\pi(\mathbf{a}) \quad \text{and} \quad \pi(\mathbf{d}) \quad \text{are invertible.} \tag{8.4}$$

Now $\mathbf{x} \neq 0$ by (8.3). Therefore

$$\pi(\mathbf{x}) \neq 0 \tag{8.5}$$

by part b) of Lemma 8.1.1. By definition $m \in M_g$, where $g = \mathbf{a}$, by (8.1). If $M = \sum_{g \in G(A)} M_g$ then $A(\rho) = C$ is spanned by grouplike elements. Therefore $M_g \subseteq km$ for all $g \in G(A)$. By Corollary 8.1.1 it follows that

$$h{\cdot}m \in km \quad \text{for} \quad h = \mathbf{a}, \mathbf{d}, \quad \text{and} \quad \mathbf{ad} = \mathbf{da}. \tag{8.6}$$

We now consider the compatibility condition

$$a_{(1)}\cdot m^{<1>} \otimes a_{(2)}m^{(2)} = (a_{(2)}\cdot m)^{<1>} \otimes (a_{(2)}\cdot m)^{(2)}a_{(1)}$$

which we write

$$\Delta(a)\rho(u) = \rho(a_{(2)}\cdot u)(1 \otimes a_{(1)}) \tag{8.7}$$

for $a \in A$ and $u \in M$ where multiplication is interpreted in the obvious way. There are six cases to consider: $a = \mathbf{a}, \mathbf{d}$ or $\mathbf{x}$ and $u = m$ or n. Since $\mathbf{a}\cdot m, \mathbf{d}\cdot m \in km$ we may write the matrix representations of $\mathbf{a}, \mathbf{d}$ and $\mathbf{x}$ as

$$\pi(\mathbf{a})_B = \begin{pmatrix} a & b \\ 0 & c \end{pmatrix}, \ \pi(\mathbf{d})_B = \begin{pmatrix} d & e \\ 0 & f \end{pmatrix}, \text{ and } \pi(\mathbf{x})_B = \begin{pmatrix} p & r \\ q & s \end{pmatrix}.$$

Again, $\pi(\mathbf{a})_B$ and $\pi(\mathbf{d})_B$ are invertible, commute, and $\pi(\mathbf{x})_B \neq 0$.

Let $a = \mathbf{a}$ or $\mathbf{d}$. When $u = m$ then (8.7) holds by virtue of (8.6). When $u = n$ then (8.7) holds if and only if

$$a\mathbf{ax} + b\mathbf{ad} = b\mathbf{a}^2 + c\mathbf{xa} \tag{8.8}$$

and

$$d\mathbf{dx} + e\mathbf{d}^2 = e\mathbf{ad} + f\mathbf{xd}. \tag{8.9}$$

Let $a = \mathbf{x}$. When $u = m$ then (8.7) holds if and only if

$$(a - q)\mathbf{xa} + p\mathbf{da} = p\mathbf{a}^2 + d\mathbf{ax}. \tag{8.10}$$

When $u = n$ then (8.7) holds if and only if

$$(f - q)\mathbf{dx} + s\mathbf{da} = s\mathbf{d}^2 + c\mathbf{xd} \tag{8.11}$$

and

$$a\mathbf{x}^2 + b\mathbf{xd} + p\mathbf{dx} + r\mathbf{d}^2 = r\mathbf{a}^2 + s\mathbf{xa} + e\mathbf{ax} + f\mathbf{x}^2. \tag{8.12}$$

Lemma 8.3.1 *With the assumption* $\mathbf{ad} = \mathbf{da}$ *the conditions (8.8) – (8.12) are collectively equivalent to the list of conditions:*

a) $a\mathbf{ax} + b\mathbf{ad} = b\mathbf{a}^2 + c\mathbf{xa}$,

b) $d\mathbf{dx} + e\mathbf{d}^2 = e\mathbf{ad} + f\mathbf{xd}$,

c) $a\mathbf{x}^2 + b\mathbf{xd} + p\mathbf{dx} + r\mathbf{d}^2 = r\mathbf{a}^2 + s\mathbf{xa} + e\mathbf{ax} + f\mathbf{x}^2$,

d) $(\frac{p}{d} + \frac{b}{a})(\mathbf{d} - \mathbf{a}) = 0 = (\frac{s}{c} + \frac{e}{f})(\mathbf{d} - \mathbf{a})$, *and*

e) $f = a$ *or* $f = -\frac{cd}{a}$, *and* $q = a - \frac{cd}{a}$.

Proof: Notice that (8.8) is equivalent to

$$\mathbf{ax} = (\frac{c}{a})\mathbf{xa} + (\frac{b}{a})(\mathbf{a}^2 - \mathbf{ad})$$

and (8.10) is equivalent to

$$\mathbf{ax} = (\frac{a-q}{d})\mathbf{xa} + (\frac{p}{d})(\mathbf{da} - \mathbf{a}^2).$$

Subtracting the first equation from the second we obtain

$$(\frac{a-q}{d} - \frac{c}{a})\mathbf{xa} + (\frac{p}{d} + \frac{b}{a})(\mathbf{da} - \mathbf{a}^2) = 0.$$

Since A is M-reduced and $\pi(\mathbf{a})$ is invertible, by part a) of Lemma 8.1.1 an $\mathbf{a}$ can be canceled from each term of the last equation leaving

$$(\frac{a-q}{d} - \frac{c}{a})\mathbf{x} + (\frac{p}{d} + \frac{b}{a})(\mathbf{d} - \mathbf{a}) = 0.$$

Now $\mathbf{x}$ is not in the span of $\mathbf{a}$ and $\mathbf{d}$ by (8.3). Therefore

$$\frac{a-q}{d} - \frac{c}{a} = 0 \quad \text{and} \quad (\frac{p}{d} + \frac{b}{a})(\mathbf{d} - \mathbf{a}) = 0. \tag{8.13}$$

Therefore (8.10) may be replaced by (8.13). Likewise (8.11) may be replaced by

$$\frac{f-q}{c} - \frac{d}{f} = 0 \quad \text{and} \quad (\frac{s}{c} + \frac{e}{f})(\mathbf{d} - \mathbf{a}) = 0.$$

A little further calculation shows that these two conditions are equivalent to parts d) and e) of the lemma. This concludes our proof. ∎

Consider the basis $\mathcal{B} = \{m \otimes m, m \otimes n, n \otimes m, n \otimes n\}$ for $M \otimes M$. Then $[R]_\mathcal{B}$ has the block form

$$[R]_\mathcal{B} = \begin{pmatrix} \pi(\mathbf{a})_\mathcal{B} & \pi(\mathbf{x})_\mathcal{B} \\ \mathcal{O} & \pi(\mathbf{d})_\mathcal{B} \end{pmatrix}$$

where $\mathcal{O} \in \mathrm{M}_2(k)$ is the zero matrix. We observe that A is a Hopf algebra if and only if the multiplicative semigroup of $\mathrm{M}_2(k)$ generated by $\pi(\mathbf{a})_\mathcal{B}$ and $\pi(\mathbf{d})_\mathcal{B}$ is a group.

We will treat the case $\mathbf{a} = \mathbf{d}$ first and the case $\mathbf{a} \neq \mathbf{d}$ in sub-cases. The essential arguments will be sketched and the details will be left to the reader.

First of all suppose that $\mathbf{a} = \mathbf{d}$. Then $q = a - c$ and either $c = a$ or $c = -c$ by part e) of Lemma 8.3.1. Assume that the characteristic of k is not 2. Then $c = a$ since $c \neq 0$. Consequently $q = 0$. By part b) of Lemma 8.3.1 we conclude $\mathbf{ax} = \mathbf{xa}$. Therefore the equation of part c) of the same lemma reduces to $(p - s)\mathbf{xa} = 0$. Since $\pi(\mathbf{a})$ is invertible and $\pi(\mathbf{x}) \neq 0$, necessarily $p - s = 0$.

Case 3: $\mathbf{a} = \mathbf{d}$ and characteristic $k \neq 2$. Then:

$$\mathbf{ad} = \mathbf{da} \quad \text{and} \quad \mathbf{xa} = \mathbf{ax}.$$

The associated solution R to the QYBE has the matrix representation

$$\text{C3)} \quad [R]_B = \begin{pmatrix} \pi(\mathbf{a})_B & \pi(\mathbf{x})_B \\ \mathcal{O} & \pi(\mathbf{d})_B \end{pmatrix} = \left(\begin{array}{cc|cc} a & b & p & r \\ 0 & a & 0 & p \\ \hline 0 & 0 & a & b \\ 0 & 0 & 0 & a \end{array}\right)$$

where $\pi(\mathbf{a})$ is invertible and $\pi(\mathbf{x}) \neq 0$.

Observe that any bialgebra realizing Case 3 can not be spanned by grouplike elements. For if this were the case $C = k\mathbf{a} + k\mathbf{x}$ would be also. But then $\mathbf{a} + \alpha\mathbf{x}$ would have to be grouplike for some $\alpha \in k$. It is not hard to see that $\mathbf{x} = 0$ in this case. As $\pi(\mathbf{x}) \neq 0$, it can not be the case that C is spanned by grouplike elements.

Now suppose that $\mathbf{a} \neq \mathbf{d}$. Since A is M-reduced, it follows that $\pi(\mathbf{a}) \neq \pi(\mathbf{d})$ by part c) Lemma 8.1.1. There are three basic cases: $a \neq c$, and $a = c$ with sub-cases $b \neq 0$ and $b = 0$.

Assume $a \neq c$. Then $\pi(\mathbf{a})$ is diagonalizable. This is the only point in our analysis we make a possible adjustment to our initial choice for n. Choose n to be an eigenvector of $\pi(\mathbf{a})$. With this choice $b = 0$. Since $b = 0$, and it follows that $p = 0$ by part d) of Lemma 8.3.1. Since $\mathbf{ad} = \mathbf{da}$ it follows that $\pi(\mathbf{ad}) = \pi(\mathbf{da})$. Consequently $e = 0$. Since $e = 0$ notice that $s = 0$ by part d) of Lemma 8.3.1 again. Observe that part a) of Lemma 8.3.1 reduces to $a\mathbf{ax} = c\mathbf{xa}$. When we apply π to both sides of this equation we derive $ra^2 = rc^2$. Likewise part b) of Lemma 8.3.1 becomes $d\mathbf{dx} = f\mathbf{xd}$. Consequently $rd^2 = rf^2$.

Case 4: $\mathbf{a} \neq \mathbf{d}, a \neq c$. Then:

$$\mathbf{ad} = \mathbf{da}, \quad a\mathbf{ax} = c\mathbf{xa}, \quad d\mathbf{dx} = f\mathbf{xd},$$

$$(a - f)\mathbf{x}^2 = r(\mathbf{a}^2 - \mathbf{d}^2),$$

and

$$ra^2 = rc^2, rd^2 = rf^2, f = a \quad \text{or} \quad f = -\frac{cd}{a}.$$

The associated solution R to the QYBE has the matrix representation

$$\text{C4)} \quad [R]_B = \begin{pmatrix} \pi(\mathbf{a})_B & \pi(\mathbf{x})_B \\ \mathcal{O} & \pi(\mathbf{d})_B \end{pmatrix} = \left(\begin{array}{cc|cc} a & 0 & 0 & r \\ 0 & c & a - \frac{cd}{a} & 0 \\ \hline 0 & 0 & d & 0 \\ 0 & 0 & 0 & f \end{array}\right)$$

where $\pi(\mathbf{a}), \pi(\mathbf{d})$ are distinct, invertible and $\pi(\mathbf{x}) \neq 0$.

A bialgebra which satisfies the conditions of Case 4 can not be cocommutative, and hence can not be generated by grouplike elements. To see this, first note that $\mathbf{x}$ is not a non-zero scalar multiple of a grouplike element. Since $\mathbf{a} \neq \mathbf{d}$ it follows that $\{\mathbf{a}, \mathbf{d}\}$ is a linearly independent set of grouplike elements. Thus $\Delta^{cop}(\mathbf{x}) = \Delta(\mathbf{x})$ would mean that $\pi(\mathbf{x}) = \alpha(\pi(\mathbf{a}) - \pi(\mathbf{d}))$ for some $\alpha \in k$. But there is no $\alpha \in k$ such that $\pi(\mathbf{x}) = \alpha(\pi(\mathbf{a}) - \pi(\mathbf{d}))$. Since $\pi(\mathbf{ax}) \neq \pi(\mathbf{xa})$, such a bialgebra can not be commutative either.

Assume that $a = c$. Then $q = a - d$, and $f = a$ or $f = -d$ by part e) of Lemma 8.3.1. Assume further that $b \neq 0$. Then applying π to the equation $\mathbf{ad} = \mathbf{da}$ yields $f = d$. Suppose that the characteristic of k is not 2. Then $f \neq d$ since $f = -d$ and $d \neq 0$. Therefore $d = f = a$. Since $\mathbf{a} \neq \mathbf{d}$ we use part d) of Lemma 8.3.1 to conclude that $p = -b$ since $a = d$ and also conclude that $s = -e$ since $c = a = f$. Since $\pi(\mathbf{a}) \neq \pi(\mathbf{d})$ it follows that $e \neq b$. Observe that part c) of Lemma 8.3.1 can be simplified using parts a) and b) of the same. Since the characteristic of k is not 2 and $b \neq e$ it follows that $\pi(\mathbf{a}^2) \neq \pi(\mathbf{d}^2)$. Therefore $\mathbf{a}^2 \neq \mathbf{d}^2$ and we conclude that $\frac{be}{a} + r = 0$.

Case 5: $\mathbf{a} \neq \mathbf{d}, a = c, b \neq 0$ and characteristic $k \neq 2$. Then:

$$\mathbf{ax} - \mathbf{xa} = \frac{b}{a}(\mathbf{a}^2 - \mathbf{ad}), \quad \mathbf{ad} = \mathbf{da}, \quad \text{and} \quad \mathbf{dx} - \mathbf{xd} = \frac{e}{a}(\mathbf{ad} - \mathbf{d}^2),$$

and

$$e \neq b.$$

The associated solution R to the QYBE has the matrix representation

$$\textbf{C5)} \quad [R]_B = \begin{pmatrix} \pi(\mathbf{a})_B & \pi(\mathbf{x})_B \\ \mathcal{O} & \pi(\mathbf{d})_B \end{pmatrix} = \left(\begin{array}{cc|cc} a & b & -b & -\frac{eb}{a} \\ 0 & a & 0 & -e \\ \hline 0 & 0 & a & e \\ 0 & 0 & 0 & a \end{array}\right)$$

where $\pi(\mathbf{a}), \pi(\mathbf{d})$ are distinct, invertible, and $\pi(\mathbf{x}) \neq 0$.

Since $\pi(\mathbf{x}) = \alpha(\pi(\mathbf{a}) - \pi(\mathbf{d}))$ has no solution for $\alpha \in k$, it follows that any bialgebra which satisfies Case 5 is not cocommutative, and hence is not generated by grouplike elements. Also note that $\pi(\mathbf{ad}) \neq \pi(\mathbf{a}^2)$. Therefore such a bialgebra is not commutative.

Now suppose that $b = 0$. Then $\mathbf{a}$ is central by part b) of Lemma 8.3.1, $p = 0$ and $fs = -ea$ by part e) of the same. Observe that the equation of part c) of the lemma reduces to $(a - f)\mathbf{x}^2 = (s + e)\mathbf{ax} + r(\mathbf{a}^2 - \mathbf{d}^2)$ since $\mathbf{a}$ and $\mathbf{x}$ commute. Applying π to both sides of this equation and comparing entries we see that $ra^2 = rd^2$ and $re(a + d) = 0$ when $f = a$.

Let $\alpha \in k$. Since A is M-reduced, $\mathbf{x} - \alpha(\mathbf{a} - \mathbf{d}) = 0$ if and only if $\pi(\mathbf{x} - \alpha(\mathbf{a} - \mathbf{d})) = 0$ by part b) of Lemma 8.1.1. Thus $\pi(\mathbf{x}) = \alpha(\pi(\mathbf{a}) - \pi(\mathbf{d}))$ is not possible since A is not generated by grouplike elements. The last equation has a solution α if and only if $a = d, f = -d$ and $2ar = -e^2$.

Case 6: $\mathbf{a} \neq \mathbf{d}, a = c, b = 0$. Then:

$$\mathbf{ax} = \mathbf{xa}, \quad \mathbf{ad} = \mathbf{da}, \quad d\mathbf{dx} + e\mathbf{d}^2 = e\mathbf{ad} + f\mathbf{xd},$$

and

$$(a - f)\mathbf{x}^2 = r(\mathbf{a}^2 - \mathbf{d}^2) + (e - \frac{ae}{f})\mathbf{ax}.$$

$$f = a \quad \text{or} \quad f = -d.$$

$$(\text{If} \quad f = a, \quad \text{then} \quad ra^2 = rd^2 \quad \text{and} \quad re(a + d) = 0;$$

$$\text{if} \quad f = -d, \quad \text{then} \quad a \neq d \quad \text{or} \quad 2ar \neq -e^2.)$$

The associated solution R to the QYBE has a matrix representation

$$\textbf{C6)} \quad [R]_{\mathcal{B}} = \begin{pmatrix} \pi(\mathbf{a})_{\mathcal{B}} & \pi(\mathbf{x})_{\mathcal{B}} \\ \mathcal{O} & \pi(\mathbf{d})_{\mathcal{B}} \end{pmatrix} = \left(\begin{array}{cc|cc} a & 0 & 0 & r \\ 0 & a & a-d & -\frac{ae}{f} \\ \hline 0 & 0 & d & e \\ 0 & 0 & 0 & f \end{array}\right)$$

where $\pi(\mathbf{a}), \pi(\mathbf{d})$ are invertible, distinct, and $\pi(\mathbf{x}) \neq 0$.

Any bialgebra satisfying Case 6 is not cocommutative, and hence is not generated by grouplike elements. Suppose that $\mathbf{xd} = \mathbf{dx}$. Since A is M-reduced it follows that $(d - f)\mathbf{x} - e(\mathbf{a} - \mathbf{d}) = 0$. Thus $d = f$ and $e = 0$. But $f \neq a$, so $f = -d$. This means that the characteristic of k is 2. We conclude that when the characteristic of k is not 2 any such bialgebra is neither commutative nor cocommutative.

Exercise 8.3.1 Show that Cases 3 – 6 are in fact realized as described by constructing a bialgebra A over k with the indicated properties and showing there exists a left QYB A-module structure on M so that R is the associated QYBE.

Exercise 8.3.2 Show that solutions R, R' to the QYBE which come from different cases among Cases 1 – 6 are not congruent, when the characteristic of k is not 2.

Exercise 8.3.3 Suppose that A and (M, μ, ρ) satisfy the description in one of Cases 1 – 6. Let $\pi(\mathbf{a})_{\mathcal{B}} = \mathcal{A}$ and $\pi(\mathbf{d})_{\mathcal{B}} = \mathcal{D}$. Show that A is a Hopf algebra if and only if $\mathcal{A}^m\mathcal{D}^n = I$ for some positive integers m and n.

Exercise 8.3.4 For each of Cases 3 – 6, find a complete set of representatives for the congruence classes of solutions to the QYBE which arise from that particular case.

Exercise 8.3.5 Suppose that $\text{Dim}M = 2$ and that $R : M \otimes M \longrightarrow M \otimes M$ is a solution to the QYBE.

a) Suppose that $R(M \otimes m) \subseteq M \otimes m$ for some non-zero $m \in M$. Explain how the results of this section can be used to identify R (up to congruence).

b) Suppose that $B = \{m, n\}$ is a basis for M and consider the basis $\mathcal{B} = \{m \otimes m, m \otimes n, n \otimes m, n \otimes n\}$ for $M \otimes M$. If $[R]_{\mathcal{B}}$ is upper triangular, show that R arises from one of Cases 1 - 6.

Exercise 8.3.6 Suppose that $\mathrm{Dim} M = 2$ and that $R : M \otimes M \longrightarrow M \otimes M$ is a solution to the QYBE. Suppose that $m \in M$ and $R(m \otimes M) \subseteq m \otimes M$ or $R(M \otimes m) \subseteq M \otimes m$ implies $m = 0$.

a) If A is a bialgebra over k and (M, μ, ρ) is a left QYB A-module structure on M such that R is the associated QYBE solution, show that M is a simple right A-comodule and a simple left A-module.

b) Show that neither $\widetilde{A(R)}$ nor $\widetilde{A(R^{\tau})}$ is pointed.

Exercise 8.3.7 Suppose that M is a finite-dimensional vector space over the field k and $\mathrm{Dim} M \geq 2$. Let $R : M \otimes M \longrightarrow M \otimes M$ be the solution to the QYBE given by $R = \tau_{M,M}$. Suppose that (M, μ, ρ) is a left QYB $\widetilde{A(R)}$-module such that R is the associated solution.

a) Show that (M, μ) is a simple left $\widetilde{A(R)}$-module and (M, ρ) is a simple right $\widetilde{A(R)}$-comodule.

b) Show that neither $\widetilde{A(R)}$ nor $\widetilde{A(R^{\tau})}$ is pointed.

8.4 Solutions Obtained by Patching and Solutions in Higher Dimension

In this section we will describe a method for obtaining solutions to the QYBE for vector spaces of dimension larger that two by "patching" solutions to the QYBE in the two-dimensional case in certain situations [Lambe and Radford, 1993, Section 10]. The QYB operator

$$\begin{aligned} R &= qb(\sum_{i=1}^{r} e^i_i \otimes e^i_i) + b(\sum_{i<j} e^i_i \otimes e^j_j) + \\ &\quad c(\sum_{i>j} e^i_i \otimes e^j_j) + (qb - q^{-1}c)(\sum_{i<j} e^i_j \otimes e^j_i) \end{aligned}$$

arises in this manner.

Let $q, b, c \in k \backslash 0$. We will base our discussion on the solutions $R_{q,b,c}$ and $R'_{q,b,c}$ in the $\mathrm{Dim} M = 2$ case. Let $B = \{m, n\}$ be a basis for M. The operator $R = R_{q,b,c}$ is defined by

$$\begin{aligned} R(m \otimes m) &= qb(m \otimes m), \\ R(m \otimes n) &= b(m \otimes n), \\ R(n \otimes m) &= c(n \otimes m) + (qb - q^{-1}c)(m \otimes n), \\ R(n \otimes n) &= qb(n \otimes n). \end{aligned}$$

The operator $R' = R'_{q,b,c}$ is defined by

$$\begin{aligned} R'(m\otimes m) &= qb(m\otimes m),\\ R'(m\otimes n) &= b(m\otimes n),\\ R'(n\otimes m) &= c(n\otimes m)+(qb-q^{-1}c)(m\otimes n),\\ R'(n\otimes n) &= -q^{-1}c(n\otimes n). \end{aligned}$$

The meaning of the term patching will emerge during the course of the proof of the following result which is [Lambe and Radford, 1993, Proposition 10.8] for the operator R^τ.

Proposition 8.4.1 *Suppose that M is a finite-dimensional vector space of dimension $r \geq 2$ over a field k. Let $\{m_1, \ldots, m_r\}$ be a basis for M and define $e^i_j \in \mathrm{End}(M)$ by $e^i_j(m_\ell) = m_i\delta_{j,\ell}$ for $1 \leq i, j, \ell \leq n$. Then for non-zero scalars $q, b, c \in k$*

$$\begin{aligned} R &= qb(\sum_{i=1}^{r-1} e^i_i\otimes e^i_i)+\alpha(e^r_r\otimes e^r_r)+b(\sum_{i<j} e^i_i\otimes e^j_j)\\ &\quad +c(\sum_{i>j} e^i_i\otimes e^j_j)+(qb-q^{-1}c)(\sum_{i<j} e^i_j\otimes e^j_i) \end{aligned}$$

is a solution to the QYBE, where $\alpha = qb$ or $\alpha = -q^{-1}c$.

Proof: We show directly that the equation

$$R_{(1,2)}R_{(1,3)}R_{(2,3)}(m_i\otimes m_j\otimes m_\ell) = R_{(2,3)}R_{(1,3)}R_{(1,2)}(m_i\otimes m_j\otimes m_\ell)$$

holds for all $1 \leq i, j, \ell \leq r$. Suppose that two or more of the subscripts i, j, ℓ are the same. Then m_i, m_j and m_ℓ belong to a two-dimensional span M' of basis elements. It is easy to check that $R(M'\otimes M') \subseteq M'\otimes M'$ and that $R|_{M'\otimes M'} = R_{q,b,c}$ or $R'_{q,b,c}$. So this case is settled.

Now suppose that i, j and ℓ are distinct. There are six cases to work out, based on the order relations among subscripts. They are straightforward to check. ∎

Recall the tensor algebra of a left QYB A-module has a structure of a left QYB A-module. There are very interesting solutions to the QYBE associated to the tensor algebra and its quotients. We direct the reader to [Lambe and Radford, 1993, Section 10] for examples.

8.5 A Class of Weak Quantum Yang–Baxter Modules

Suppose that A is a bialgebra and M is a finite-dimensional vector space over the field k. In this section we consider triples (M, μ, ρ), where (M, μ) is a left A-module and $\rho : M \longrightarrow M\otimes A$ is a *linear* map. We will write as usual $\rho(m) = m^{<1>}\otimes m^{(2)} \in$

$M \otimes A$ and $a^* \rightharpoonup m = m^{<1>}<a^*, m^{(2)}>$ for $m \in M$ and $a^* \in A^*$. In this section we do a mathematical analysis on solutions $R = R_{(\mu,\rho)}$ to the QYBE defined by (2.12)

$$R(m \otimes n) = m^{<1>} \otimes m^{(2)} \cdot n$$

for all $m, n \in M$, where (M, ρ) may *not* be a right A-comodule and $A = k[G]$ is the group algebra of the cyclic group G of order n. We will assume that k contains a primitive n^{th} root of unity. The elementary case $n = 2$ and $\mathrm{Dim} M = 3$ is interesting enough in its own right. The reader is referred to Section 2.11.2 where an account of the discovery and an analysis of these examples by computer is described.

We will still want the module structure map μ and the linear map ρ to be related by the compatibility condition (2.17)

$$a_{(1)} \cdot m^{<1>} \otimes a_{(2)} m^{(2)} = (a_{(2)} \cdot m)^{<1>} \otimes (a_{(2)} \cdot m)^{(2)} a_{(1)}$$

for all $a \in A$ and $m \in M$. The reader can show that (2.17) is equivalent to (2.18)

$$\rho(a \cdot m) = a \cdot m^{<1>} \otimes m^{(2)}$$

for all $a \in A$ and $m \in M$ when $A = H$ is a commutative cocommutative Hopf algebra, as is the case when $H = k[G]$ is the group algebra of a cyclic group. More generally:

Lemma 8.5.1 *Suppose that H is a commutative cocommutative Hopf algebra over the field k. Let M be a vector space over k and $\mu : H \otimes M \longrightarrow M$ and $\rho : M \longrightarrow M \otimes H$ be linear maps. Write $\mu(a \otimes m) = a \cdot m$ and $\rho(m) = m^{<1>} \otimes m^{(2)}$ for $a \in H$ and $m \in M$. Then (2.17) is equivalent to (2.18).* ∎

As an aid to computation:

Lemma 8.5.2 *Suppose that A is a bialgebra over the field k, (M, μ) is a left A-module, and $\rho : M \longrightarrow M \otimes A$ is a linear map. Then the set of all $a \in A$ such that (2.17) holds for all $m \in M$ is a subalgebra of A.* ∎

For the remainder of this section $H = k[G]$ is the group algebra of the cyclic group G of order n over k, (M, μ) is a finite-dimensional left H-module, and $\rho : M \longrightarrow M \otimes H$ is a linear map. We will assume that k contains a primitive n^{th} root of unity.

Suppose $G = (t)$ and $\pi : H \longrightarrow \mathrm{End}(M)$ is the representation afforded by the left H-module (M, μ). Let $T = \pi(t)$. Since $t^n = 1$ it follows that $T^n = 1_M$. Therefore $p(T) = 0$ where $p(X) = X^n - 1 \in k[X]$. Since k contains a primitive n^{th} root of unity it follows that $p(X)$ splits into a product of distinct linear factors over k. Therefore T is diagonalizable and the eigenvalues of T are n^{th} roots of unity. Let $B = \{m_1, \ldots, m_r\}$ be a basis for M of eigenvectors for T. Then there are $\alpha_1, \ldots, \alpha_r \in \mathrm{Alg}(H, k) = G(H^*)$ such that

$$a \cdot m_i = \alpha_i(a) m_i \tag{8.14}$$

for all $a \in H$ and $1 \leq i \leq r$. Define $e^i_j \in H$ by

$$\rho(m_j) = m_i \otimes e^i_j. \tag{8.15}$$

For $m = m_j$ we compute

$$t{\cdot}m^{<1>} \otimes m^{(2)} = t{\cdot}m_i \otimes e^i_j = m_i \otimes \alpha_i(t)e^i_j$$

and

$$(t{\cdot}m)^{<1>} \otimes (t{\cdot}m)^{(2)} = \alpha_j(t)m_j^{<1>} \otimes m_j^{(2)} = m_i \otimes \alpha_j(t)e^i_j.$$

Since H is commutative and cocommutative we conclude by Lemma 8.5.2 that the compatibility condition (2.17) holds if and only if $(\alpha_i(t) - \alpha_j(t))e^i_j = 0$ for all $1 \leq i, j \leq r$. Since t generates H as an algebra $\alpha_i = \alpha_j$ if and only if $\alpha_i(t) = \alpha_j(t)$. Thus (2.17) holds if and only if

$$e^i_j \neq 0 \quad \text{implies} \quad \alpha_i = \alpha_j.$$

Let $R : M \otimes M \longrightarrow M \otimes M$ be defined by (2.12). Then a straightforward calculation show that $R_{(1,2)}R_{(1,3)}R_{(2,3)} = R_{(2,3)}R_{(1,3)}R_{(1,2)}$ if and only if

$$\alpha_v(e^u_x)\alpha_k(e^x_i)\alpha_k(e^v_j) = \alpha_k(e^u_x)\alpha_j(e^x_i)\alpha_k(e^v_j) \tag{8.16}$$

for all $1 \leq i, j, k, u, v \leq r$.

Set

$$E = \begin{pmatrix} e^1_1 & \cdots & e^1_r \\ \vdots & & \vdots \\ e^r_1 & \cdots & e^r_r \end{pmatrix} \in \mathrm{M}_r(H)$$

and for $\alpha \in H^*$ set

$$E(\alpha) = \begin{pmatrix} \alpha(e^1_1) & \cdots & \alpha(e^1_r) \\ \vdots & & \vdots \\ \alpha(e^r_1) & \cdots & \alpha(e^r_r) \end{pmatrix} \in \mathrm{M}_r(k).$$

By reformulating (8.16) we obtain: R is a solution to the QYBE if and only if

$$E(\alpha_k)E(\alpha_j)\alpha_k(e^v_j) = E(\alpha_v)E(\alpha_k)\alpha_k(e^v_j)$$

for all $1 \leq k, j, v \leq r$.

Define operators $R_1, \ldots, R_r \in \mathrm{End}(M)$ by

$$R_j(m) = \alpha_j {\rightharpoonup} m = m^{<1>}{<}\alpha_j, m^{(2)}{>}$$

for $m \in M$. Then

$$R(m \otimes m_j) = R_j(m) \otimes m_j \tag{8.17}$$

for all $m \in M$ and $1 \leq j \leq r$. Observe that

$$[R_j]_B = E(\alpha_j)$$

for all $1 \leq j \leq r$. Order the basis $\mathcal{B} = \{m_i \otimes m_j\}_{1 \leq i,j \leq r}$ for $M \otimes M$ lexicographically, reading *right* to *left*. Then

$$[R]_{\mathcal{B}} = \begin{pmatrix} E(\alpha_1) & & \\ & \ddots & \\ & & E(\alpha_r) \end{pmatrix} \tag{8.18}$$

is a diagonal array. In particular R is invertible if and only if $E(\alpha_i)$ is invertible for all $1 \leq i \leq r$.

We next observe that (M, ρ) is a right H-comodule if and only if $\{e^i_j\}_{1 \leq i,j \leq r}$ satisfies the comatrix identities. Since $G(H^*)$ spans H^* this is the case if and only if $\epsilon(e^i_j) = \delta^i_j$ and $(\alpha \otimes \alpha')(\Delta(e^i_j)) = \alpha(e^i_\ell)\alpha'(e^\ell_j)$ for all $\alpha, \alpha' \in G(H^*)$. Thus (M, ρ) is a right H-comodule if and only if

$$E(\epsilon) = I \quad \text{and} \quad E(\alpha\alpha') = E(\alpha)E(\alpha') \tag{8.19}$$

for all $\alpha, \alpha' \in G(H^*)$.

We wish to describe the triple (M, μ, ρ) in such a manner that the decomposition of M into eigenspaces of $\pi(t)$ is more explicit.

Definition 8.5.1 *$\mathcal{BWE}_r$ is the set of all triples (B, W, E), where*

$$B = \{m_1, \ldots, m_r\}$$

is a basis for M,

$$W = (\alpha_1, \ldots, \alpha_r) \in G(H^*) \times \cdots \times G(H^*),$$

and

$$E = (e^i_j) \in \mathrm{M}_r(H).$$

If $(B, W, E) \in \mathcal{BWE}_r$, then $(M, \mu_{B,W})$ is a left H-module which is determined by $a{\cdot}m_i = \alpha_i(a)m_i$ for all $a \in H$ and $1 \leq i \leq r$. Let $\rho_{B,E} : M \longrightarrow M \otimes H$ be the linear map determined by $\rho(m_j) = m_i \otimes e^i_j$ for all $1 \leq j \leq r$. We have that $\mu = \mu_{B,W}$, where B is the basis for M initially described for M and α_i is defined by (8.14), and $\rho = \rho_{B,E}$, where E is determined by (8.15). Thus we have proved all but parts e) and f) of the following:

Proposition 8.5.1 *Suppose that the field k contains a primitive n^{th} root of unity. Let H be the group algebra of the cyclic group of order n over k and M be an r-dimensional vector space over k. Suppose $(B, W, E) \in \mathcal{BWE}_r$, where $W = (\alpha_1, \ldots, \alpha_r)$ and $E = (e^i_j)$. Then:*

a) *$(M, \rho_{B,E})$ is a right H-comodule if and only if $E(\epsilon) = I$ and $E(\alpha\alpha') = E(\alpha)E(\alpha')$ for all $\alpha, \alpha' \in G(H^*)$.*

b) *The compatibility condition (2.17)*

$$a_{(1)}\cdot m^{<1>} \otimes a_{(2)}m^{(2)} = (a_{(2)}\cdot m)^{<1>} \otimes (a_{(2)}\cdot m)^{(2)}a_{(1)}$$

for all $a \in H$ and $m \in M$ is satisfied if and only if $e^i_j \neq 0$ implies $\alpha_i = \alpha_j$.

Let $R : M \otimes M \longrightarrow M \otimes M$ be defined by $R(m \otimes n) = m^{<1>} \otimes m^{(2)}\cdot n$ for $m, n \in M$.

c) *R is a solution to the QYBE if and only if*

$$E(\alpha_k)E(\alpha_j)\alpha_k(e^v_j) = E(\alpha_v)E(\alpha_k)\alpha_k(e^v_j)$$

for all $1 \leq j, k, v \leq r$.

d) *R is an invertible operator if and only if $E(\alpha_1), \ldots, E(\alpha_r)$ are invertible matrices.*

Suppose that the compatibility condition of part b) is satisfied. Then:

e) *R is a solution to the QYBE if and only if*

$$E(\alpha_k)E(\alpha_j)\alpha_k(e^v_j) = E(\alpha_j)E(\alpha_k)\alpha_k(e^v_j)$$

for all $1 \leq j, k, v \leq r$.

f) *Suppose that R is invertible. Then R is a solution to the QYBE if and only if $\{E(\alpha_1), \ldots, E(\alpha_r)\}$ is a commuting family of matrices.*

Proof: We need only establish parts e) and f). Suppose that the compatibility condition of part b) holds. We need only show that the equations of parts c) and e) are one in the same. Fix $1 \leq j, k, v \leq r$. If $\alpha_k(e^v_j) = 0$ then these equations are the same. If $\alpha_k(e^v_j) \neq 0$ then $e^v_j \neq 0$. Therefore $\alpha_v = \alpha_j$ by part b) and again these equations are the same. We have shown part e).

Suppose that R is invertible. Then $E(\alpha_1), \ldots, E(\alpha_r)$ are invertible matrices by part d). Thus for fixed $1 \leq k, j \leq r$ necessarily $\alpha_k(e^v_j) = E(\alpha_k)^v_j \neq 0$ for some $1 \leq v \leq r$. Thus part f) follows from part e). This concludes our proof of the proposition. ∎

Observe that the equation of part e) of the preceding proposition is satisfied when $E(\alpha_k)$ and $E(\alpha_j)$ commute for all $1 \leq j, k \leq r$.

Suppose that the hypothesis of Proposition 8.5.1 and the compatibility condition (2.17) hold. Then $e^i_j \neq 0$ implies $\alpha_i = \alpha_j$ by part b) of the proposition. If necessary reorder the basis $B = \{m_1, \ldots, m_r\}$ for M so that there is a sequence $0 = r_0 < r_1 < r_2 < \ldots < r_s = r$ such that $\alpha_{r_{i-1}+1} = \alpha_{r_{i-1}+2} = \cdots = \alpha_{r_i}$ for all $1 \leq i \leq s$ and $\alpha_{r_1}, \ldots, \alpha_{r_s}$ are distinct. Since $e^i_j = 0$ when $\alpha_i \neq \alpha_j$ it follows that

$$E = \begin{pmatrix} E_{r_1} & & \\ & \ddots & \\ & & E_{r_s} \end{pmatrix}$$

is a diagonal array, where $E_{r_i} \in \mathrm{M}_{r_i - r_{i-1}}(H)$ is the sub-matrix of E given by

$$E_{r_i} = (e^u_v)_{r_{i-1} < u,v \leq r_i}.$$

Suppose further that R satisfies the QYBE and that $E(\alpha_{r_j})$ and $E(\alpha_{r_k})$ do not commute. Then $\alpha_{r_k}(e^v_\ell) = 0$ for all $r_{j-1} < v, \ell \leq r_j$. Therefore $E_{r_j}(\alpha_{r_k}) = 0$. We have shown:

Corollary 8.5.1 *Assume the hypothesis of the previous proposition. Suppose further that* (i) *there is a sequence* $0 = r_0 < r_1 < r_2 < \ldots < r_s = r$ *such that* $\alpha_{r_{i-1}+1} = \alpha_{r_{i-1}+2} = \cdots = \alpha_{r_i}$ *for all* $1 \leq i \leq s$ *and* $\alpha_{r_1}, \ldots, \alpha_{r_s}$ *are distinct and* (ii) *the compatibility condition (2.17) is satisfied. Then the following are equivalent:*

a) *R associated with $(M, \mu_{B,W}, \rho_{B,E})$ is a solution to the QYBE.*

b) *For $1 \leq j, k \leq s$ either $E(\alpha_{r_k})$ and $E(\alpha_{r_j})$ commute or $E_{r_k}(\alpha_{r_j}) = 0 = E_{r_j}(\alpha_{r_k})$.*

∎

Suppose that $n = 2$. Then the characteristic of k is not 2. In this case $G(H^*) = \{\epsilon, \eta\}$ where $\eta(t) = -1$. Let $(B, W, E) \in \mathcal{BWE}_r$ where $B = \{m_1, \ldots, m_r\}$, $W = (\alpha_1, \ldots, \alpha_r)$ and $E = (e^i_j)$. Observe that $(M, \rho_{B,E})$ is a right H-comodule if and only if

$$E(\epsilon) = I \quad \text{and} \quad E(\eta)^2 = I.$$

Case 1: $\alpha_1 = \cdots = \alpha_r$.

The compatibility condition (2.17) is satisfied for $\mu_{B,W}$ and $\rho_{B,E}$ and R associated to $(M, \mu_{B,W}, \rho_{B,E})$ is a solution to the QYBE.

Case 2: $\alpha_1, \ldots, \alpha_r$ are not all the same.

By rearranging B if necessary we may assume that $\eta = \alpha_1 = \cdots = \alpha_m$ and $\epsilon = \alpha_{m+1} = \cdots = \alpha_r$ for some $1 \leq m < r$. Since H has basis $\{1, t\}$ there are $\mathcal{A}, \mathcal{B} \in \mathrm{M}_r(k)$ uniquely determined by

$$E = \mathcal{A} + \mathcal{B}t.$$

Thus

$$\mathcal{A} = \frac{1}{2}(E(\epsilon) + E(\eta)) \quad \text{and} \quad \mathcal{B} = \frac{1}{2}(E(\epsilon) - E(\eta)).$$

Thus $\mathcal{A}$ and $\mathcal{B}$ commute if and only if $E(\epsilon)$ and $E(\eta)$ commute. Assume that $E(\epsilon)$ and $E(\eta)$ commute and the compatibility condition (2.17) holds. Then:

a) $\mu_{B,W}$ and $\rho_{B,E}$ satisfy the compatibility condition (2.17) if and only if

$$E = \begin{pmatrix} E_1 & 0 \\ 0 & E_2 \end{pmatrix},$$

where $E_1 \in \mathrm{M}_m(H)$ and $E_2 \in \mathrm{M}_{r-m}(H)$,

b) $(M, \rho_{B,E})$ is a right H-comodule if and only if $\mathcal{B} = I - \mathcal{A}$ and $\mathcal{A}^2 = \mathcal{A}$,

c) R is a solution to the QYBE,

d) R is invertible if and only if $\mathcal{A}^2 - \mathcal{B}^2$ is invertible.

We end this section with a discussion of the category ${}_H\widetilde{\mathcal{QYB}}$ described in Section 3.1.4 when $H = k[G]$. Recall that objects of ${}_H\widetilde{\mathcal{QYB}}$ are triples (M, μ, ρ) where (M, μ) is a left H-module and $\rho : M \longrightarrow M \otimes H$ is a linear map such that (2.17) holds, and (3.3)

$$\begin{aligned} & m^{<1><1>} \otimes m^{<1>(2)} \cdot n^{<1>} \otimes m^{(2)} n^{(2)} \\ = \; & m^{<1>} \otimes {m^{(2)}}_{(1)} \cdot n^{<1>} \otimes {m^{(2)}}_{(2)} n^{(2)} \end{aligned}$$

and (3.4)

$$\begin{aligned} & m^{<1><1>} \otimes (m^{(2)} \cdot n)^{<1>} \otimes (m^{(2)} \cdot n)^{(2)} m^{<1>(2)} \\ = \; & m^{<1>} \otimes ({m^{(2)}}_{(2)} \cdot n)^{<1>} \otimes ({m^{(2)}}_{(2)} \cdot n)^{(2)} {m^{(2)}}_{(1)} \end{aligned}$$

hold for all $m, n \in M$. Observe that (2.17) implies (3.3) and (3.4) when (M, ρ) is a right H-comodule. Recall by Proposition 3.1.1 that if (M, μ, ρ) is an object of ${}_H\widetilde{\mathcal{QYB}}$ then $R_{(\mu,\rho)}$ is a solution to the QYBE.

Suppose that M is a vector space over k with basis $B = \{m_1, \ldots, m_r\}$ and that N is a also a vector space with basis $C = \{n_1, \ldots, n_s\}$ over k. Let $(B, W, E) \in \mathcal{BWE}_r$ and $(C, X, F) \in \mathcal{BWE}_s$. Suppose $W = \{\alpha_1, \ldots, \alpha_r\}, X = \{\beta_1, \ldots, \beta_s\}, E = (e^i_j)$, and $F = (f^k_\ell)$. We define a triple $(B \otimes C, W \otimes X, E \otimes F) \in \mathcal{BWE}_{rs}$, where the tensor symbol is purely formal, as follows. Identify the ordered segment $\{1, \ldots, rs\}$ with the Cartesian product $S = \{(i,j) \mid 1 \leq i \leq r, 1 \leq j \leq s\}$ ordered lexicographically, reading left to right. Then we set

$$B \otimes C = \{m_{(i,j)} \mid (i,j) \in S\},$$

where $m_{(i,j)} = m_i \otimes m_j$,

$$W \otimes X = \{\alpha_{(i,j)} \mid (i,j) \in S\},$$

where $\alpha_{(i,j)} = \alpha_i\beta_j$, and define $E \otimes F \in \mathrm{M}_{rs}(H)$ by

$$(E \otimes F)^{(i,j)}_{(k,\ell)} = f^j_\ell e^i_k$$

for all $(i,j),(k,\ell) \in S$. Let $\mu = \mu_{B\otimes C, W\otimes X}$ and $\rho = \rho_{B\otimes C, E\otimes F}$. Observe that the resulting module structure $(M \otimes N, \mu)$ on the tensor product is given by

$$a{\cdot}(m \otimes n) = a_{(1)}{\cdot}m \otimes a_{(2)}{\cdot}n$$

and the linear resulting linear map $\rho : M \otimes N \longrightarrow (M \otimes N) \otimes H$ is given by

$$\rho(m \otimes n) = (m^{<1>} \otimes n^{<1>}) \otimes n^{(2)}m^{(2)}$$

for $a \in H, m \in M$ and $n \in N$. These are the familiar tensor product definitions given for left QYB H-modules. We will denote the triple $(M \otimes N, \mu_{B\otimes C, W\otimes X}, \rho_{B\otimes C, E\otimes F})$ more informally by $M \otimes N$.

Proposition 8.5.2 *Let $H = k[G]$ be the group algebra of the cyclic group of order n over k and suppose that the field k contains a primitive n^{th} root of unity. Let M be an r-dimensional vector space with basis B over k. Suppose $(B, W, E) \in \mathcal{BWE}_r$, where $W = \{\alpha_1, \ldots, \alpha_r\}$ and $E = (e^i_j)$. Then:*

a) *The compatibility condition (3.3) holds for all $m, n \in M$ if and only if*

$$E(\alpha_u)E(\gamma)\gamma(e^u_k) = E(\alpha_u\gamma)\gamma(e^u_k)$$

for all $1 \leq u, k \leq r$ and $\gamma \in G(H^*)$.

b) *The compatibility condition (3.4) holds for all $m, n \in M$ if and only if*

$$E(\gamma)E(\alpha_k)\gamma(e^u_k) = E(\gamma\alpha_k)\gamma(e^u_k)$$

holds for all $1 \leq k, u \leq r$ and $\gamma \in G(H^*)$.

Now assume that N is an s-dimensional vector space over k with basis C and let $(C, X, F) \in \mathcal{BWE}_s$, where $X = \{\beta_1, \ldots, \beta_s\}$ and $F = (f^k_\ell)$. Then:

c) *The compatibility condition (3.3) holds for $(M \otimes N, \mu_{W\otimes X}, \rho_{E\otimes F})$ if and only if*

$$\begin{aligned}(E(\alpha_u\beta_v)E(\gamma))^i_p(F(\alpha_u\beta_v)F(\gamma))^j_q\gamma(f^v_\ell)\gamma(e^u_k) \\ = E(\alpha_u\beta_v\gamma)^i_pF(\alpha_u\beta_v\gamma)^j_q\gamma(f^v_\ell)\gamma(e^u_k)\end{aligned}$$

for all $\gamma \in G(H^*)$, $1 \le u,i,p,k \le r$, *and* $1 \le v,j,q,\ell \le s$.

d) *The compatibility condition (3.4) holds for* $(M \otimes N, \mu_{W\otimes X}, \rho_{E\otimes F})$ *if and only if*

$$\begin{aligned}(E(\gamma)E(\alpha_k\beta_\ell))^i_p(F(\gamma)F(\alpha_k\beta_\ell))^j_q\gamma(f^v_\ell)\gamma(e^u_k) \\ = \quad E(\gamma\alpha_k\beta_\ell)^i_pF(\gamma\alpha_k\beta_\ell)^j_q\gamma(f^v_\ell)\gamma(e^u_k)\end{aligned}$$

holds for all $\gamma \in G(H^*)$, $1 \le k,i,p,u \le r$, *and* $1 \le \ell,j,q,v \le s$.

Proof: To show part a) we let $B = \{m_1,\ldots,m_r\}$ and set $m = m_i$, $n = m_j$. Then

$$\begin{aligned} m^{<1><1>} \otimes m^{<1>(2)}{\cdot}n^{<1>} \otimes m^{(2)}n^{(2)} &= m_\ell \otimes e^\ell_i{\cdot}m_u \otimes e^i_je^u_k \\ &= m_\ell \otimes \alpha_u(e^\ell_i)m_u \otimes e^i_je^v_k \\ &= m_\ell \otimes m_u \otimes \alpha_u(e^\ell_i)e^i_je^u_k. \end{aligned}$$

On the other hand

$$\begin{aligned} m^{<1>} \otimes m^{(2)}{}_{(1)}{\cdot}n^{<1>} \otimes m^{(2)}{}_{(2)}n^{(2)} &= m_\ell \otimes (e^\ell_j)_{(1)}{\cdot}m_u \otimes (e^\ell_j)_{(2)}e^u_k \\ &= m_\ell \otimes \alpha_u((e^\ell_j)_{(1)})m_u \otimes (e^\ell_j)_{(2)}e^u_k \\ &= m_\ell \otimes m_u \otimes \alpha_u((e^\ell_j)_{(1)})(e^\ell_j)_{(2)}e^u_k. \end{aligned}$$

Therefore the compatibility condition of part a) holds if and only if

$$\alpha_u(e^\ell_i)e^i_je^u_k = \alpha_u((e^\ell_j)_{(1)})(e^\ell_j)_{(2)}e^u_k$$

holds for all $1 \le u,\ell,j,k \le r$. Since $G(H^*)$ spans H^* we conclude this last equation holds if and only if

$$\alpha_u(e^\ell_i)\gamma(e^i_j)\gamma(e^u_k) = \alpha_u((e^\ell_j)_{(1)})\gamma((e^\ell_j)_{(2)})\gamma(e^u_k)$$

holds for all $1 \le u,\ell,j,k \le r$ and $\gamma \in G(H^*)$. Part a) now follows. Parts b) – d) follow in a similar manner. ∎

Corollary 8.5.2 $H = k[G]$ *be the group algebra of the cyclic group of order* n *over* k *and suppose that the field* k *contains a primitive* n^{th} *root of unity. Let* M *be an* r*-dimensional vector space with basis* B *over* k *and* $(B,W,E) \in \mathcal{BWE}_r$. *Assume further that* $E(\gamma)$ *is invertible for all* $\gamma \in G(H^*)$. *Then:*

a) *The compatibility condition (3.3) holds if and only if* $E(\alpha)E(\gamma) = E(\alpha\gamma)$ *holds for all and* $\alpha \in W$ *and* $\gamma \in G(H^*)$.

b) *The compatibility condition (3.4) holds if and only if* $E(\gamma)E(\alpha) = E(\gamma\alpha)$ *holds for all* $\gamma \in G(H^*)$ *and* $\alpha \in W$.

▮

The category ${}_H\widetilde{\mathcal{QYB}}$ is not always closed under tensor products. The following corollary will show how to construct examples. See Exercise 8.5.5. The category ${}_H\widetilde{\mathcal{QYB}}$ is, however, always closed under tensor squares.

Corollary 8.5.3 *Let $H = k[G]$ is the group algebra of the cyclic group of order n over k and suppose that the field k contains a primitive n^{th} root of unity. Let M be a r-dimensional vector space over k with basis B and $(B, W, E) \in \mathcal{BWE}_r$.*

a) *If $(M, \mu_{B,W}, \rho_{B,E})$ is an object of ${}_H\widetilde{\mathcal{QYB}}$ then $M \otimes M$ is an object of ${}_H\widetilde{\mathcal{QYB}}$.*

b) *Suppose that N is an s-dimensional vector space over k with basis C, $(C, X, F) \in \mathcal{BWE}_s$, and $(N, \mu_{C,X}, \rho_{C,F})$ is a left QYB H-module. Assume that $E(\gamma)$ is invertible for all $\gamma \in G(H^*)$ and that $(M, \mu_{B,W}, \rho_{B,E})$ is an object of ${}_H\widetilde{\mathcal{QYB}}$. Then the following are equivalent:*

   i) *$M \otimes N$ is an object of ${}_H\widetilde{\mathcal{QYB}}$.*

   ii) *$E(\gamma)E(\beta) = E(\gamma\beta) = E(\beta)E(\gamma)$ for all $\gamma \in G(H^*)$ and $\beta \in X$.*

The proof is left as an exercise for the reader. ▮

Exercise 8.5.1 Suppose that V is an r-dimensional vector space over the field k. Let $\{\alpha_1, \cdots, \alpha_r\}$ be a basis for V^* and suppose that $E_1, \ldots, E_r \in \mathrm{M}_r(k)$. Show that there is a unique $E \in \mathrm{M}_r(V)$ such that $E(\alpha_i) = E_i$ for all $1 \leq i \leq r$.

Exercise 8.5.2 Suppose that $M = M_1 \oplus \cdots \oplus M_s$ is the direct sum of finite-dimensional subspaces and $R_1, \ldots, R_s \in \mathrm{End}(M)$ satisfy

i) $R_j(M_k) \subseteq M_k$ for all $1 \leq j, k \leq s$, and

ii) for $1 \leq j, k \leq s$ either R_j and R_k commute or $R_j(M_k) = (0) = R_k(M_j)$.

Let E_i be the projection of $M = M_1 \oplus \cdots \oplus M_s$ onto M_i. Show that

a) $R = R_1 \otimes E_1 + \cdots + R_s \otimes E_s$ and $\mathcal{R} = E_1 \otimes R_1 + \cdots + E_s \otimes R_s$ are solutions to the QYBE, and

b) $R^\tau = \mathcal{R}$ and $\mathcal{R}^\tau = R$.

Exercise 8.5.3 Let R and $\mathcal{R}$ be the operators of Exercise 8.5.2.

a) Show that there is a bialgebra A over k generated by grouplike elements $g_1, \ldots, g_r$ and a left QYB A-module structure (M, μ, ρ) on M such that

   i) $\rho(m) = m \otimes g_i$ for all $m \in M_i$,

   ii) $g_i{\cdot}m = R_i(m)$ for all $m \in M$, and

   iii) $\mathcal{R} = R_{(\mu,\rho)}$.

b) Show that $\widetilde{A(\mathcal{R})}$ is generated by grouplike elements.

c) Show that $\widetilde{A(R)}$ is a commutative algebra. [Hint: See Exercise 4.3.3.]

d) Show that $\widetilde{A(R)}$ is generated by grouplike elements if and only if $R_i|_{M_j} = \omega_{i,j} 1_{M_j}$ for all $1 \leq i, j \leq r$ where $\omega_{i,j} \in k$.

e) Show that if $R_i R_j = R_j R_i$ for all $1 \leq i, j \leq r$ and k is algebraically closed then $\widetilde{A(R)}$ is pointed.

f) Determine $\widetilde{A(R)}$ for the R associated to a triple (B, W, E) of the previous section when $n = 2$ and $\mathrm{Dim} M = 3$.

For the remainder of these exercises $H = k[G]$ is the group algebra of the finite cyclic group G of order n over k. We assume that k has a primitive n^{th} root of unity.

Exercise 8.5.4 Let $W \subseteq G(H^*)$ and $\mathcal{G} = (W)$ be the subgroup of $G(H^*)$ which W generates. Let $E \in \mathrm{M}_r(H)$ and suppose that $E(\gamma)$ is invertible for all $\gamma \in G(H^*)$.

a) Suppose that $E(\gamma\alpha) = E(\gamma)E(\alpha)$ for all $\gamma \in G(H^*)$ and $\alpha \in W$ or that $E(\alpha\gamma) = E(\alpha)E(\gamma)$ for all $\alpha \in W$ and $\gamma \in G(H^*)$. Show that the map $\pi : \mathcal{G} \longrightarrow \mathrm{GL}(r, k)$ defined by $\pi(\alpha) = E(\alpha)$ is a group homomorphism.

b) Suppose that $\pi : \mathcal{G} \longrightarrow \mathrm{GL}(r, k)$ is a group homomorphism. Let $\{\gamma_1, \ldots, \gamma_s\}$ be a complete set of coset representatives for $\mathcal{G}$ in $G(H^*)$ and suppose $U_1, \ldots, U_s \in \mathrm{GL}(r, k)$, where $\gamma_1 = e$ and $U_1 = I$.

   i) Show that there is an $E \in \mathrm{M}_r(H)$ such that $E(\gamma_i\alpha) = U_i\pi(\alpha)$ for all $1 \leq i \leq s$ and $\alpha \in \mathcal{G}$.

   ii) For E of part i) show that $E(\gamma\alpha) = E(\gamma)E(\alpha)$ for all $\gamma \in G(H^*)$ and $\alpha \in \mathcal{G}$.

c) Suppose that $r > 3$ is not prime and M is an r-dimensional vector space over k with basis B. Show that there exists a triple $(B, W, E) \in \mathcal{BWE}_r$ such that $(M, \mu_{B,W}, \rho_{B,E})$ satisfies (2.17), (3.3) but not (3.4). Find an example which satisfies (2.17) and (3.4) but not (3.3).

Exercise 8.5.5 Suppose that M is an r-dimensional vector space over k with basis B and $U \in \mathrm{GL}(r, k)$ is not its own inverse.

a) Show there exists a triple $(B, W, E) \in \mathcal{BWE}_r$ such that $W = \{\epsilon\}$ and E satisfies $E(\epsilon) = I$, $E(\gamma) = U$ for $\epsilon \neq \gamma \in G(H^*)$.

b) Show that $(M, \mu_{B,W}, \rho_{B,E})$ is an object of ${}_H\widetilde{\mathcal{QYB}}$.

c) Suppose that N is an s-dimensional vector space over k and that $(C, F, X) \in \mathcal{BWE}_s$. If $X \neq \{\epsilon\}$ and $(N, \mu_{C,X}, \rho_{C,F})$ is a left QYB H-module, show that $M \otimes N$ is not an object of ${}_H\widetilde{\mathcal{QYB}}$.

Exercise 8.5.6 Let $H = k[G]$ be the group algebra of the cyclic group of order $n = 2$ and M be a 3-dimensional vector space over k. Analyze the cases $\eta = \alpha_1 = \alpha_2$ and $\epsilon = \alpha_3$ and $\eta = \alpha_1, \epsilon = \alpha_2 = \alpha_3$ when R is a solution to the QYBE, specifically

a) determining when (M, μ, ρ) is a left QYB H-module, and

b) computing R and determining when R is invertible.

Reconsider the calculations of Section 2.11.2 in light of the analysis of this section.

Exercise 8.5.7 Suppose that $(B, W, E) \in \mathcal{BWE}_r$ and $W = (\alpha_1, \ldots, \alpha_r)$. Assume that the basis $B = \{m_1, \ldots, m_r\}$ for M is ordered so that there is a sequence $0 = r_0 < r_1 < r_2 < \ldots < r_s = r$ such that $\alpha_{r_{i-1}+1} = \alpha_{r_{i-1}+2} = \cdots = \alpha_{r_i}$ for all $1 \leq i \leq s$ and $\alpha_{r_1}, \ldots, \alpha_{r_s}$ are distinct. Let M_{r_i} be the span of $\{m_{r_{i-1}+1}, \ldots, m_{r_i}\}$. Set $\rho = \rho_{B,E}$ and $\mu = \mu_{B,E}$.

a) Show that the compatibility condition (2.17) holds for μ and ρ if and only if $\rho(M_{r_i}) \subseteq M_{r_i} \otimes H$ for all $1 \leq i \leq s$.

For $\alpha \in H^*$ define $R_\alpha \in \mathrm{End}(M)$ by $R_\alpha(m) = \alpha \rightharpoonup m$. Assume that (2.17) holds for μ and ρ.

b) Show that $R_\alpha(M_{r_i}) \subseteq M_{r_i}$ for all $\alpha \in H^*$.

c) Show that R_α and R_β commute for all $\alpha, \beta \in H^*$.

d) Show that R associated with (B, W, E) is described by

$$R = \sum_{i=1}^{s} R_{\alpha_{r_i}} \otimes E_{r_i},$$

where E_{r_i} is the projection of $M = M_{r_1} \oplus \cdots \oplus M_{r_s}$ onto M_{r_i}.

e) For $1 \leq j, k \leq s$ show that $R_{\alpha_{r_j}}$ and $R_{\alpha_{r_k}}$ commute or $R_{\alpha_{r_j}}(M_{r_k}) = (0) = R_{\alpha_{r_k}}(M_{r_j})$.

8.6 Some One-Parameter Solutions

A method for finding QYBE solutions R that has been successful in many instances consists of making some special assumptions on the matrix of coefficients $\left(R^{k,l}_{i,j}\right)$ and then studying the corresponding variety. This is essentially what we did in Section 2.11.2. This method also applies to the one-parameter QYBE.

8.6.1 Some Specific Solutions

Consider one-parameter QYBE solutions

$$X \xrightarrow{R} \mathrm{End}(\otimes^2 M)$$

where $M = \mathbb{C}^2$. Let $\{e_1, e_2\}$ be the standard basis for M and consider the coordinates $R^{k,l}_{i,j}(x)$ of R with respect to this basis. Consider first solutions of the form

$$\begin{pmatrix} R^{1,1}_{1,1} & R^{1,2}_{1,1} & R^{2,1}_{1,1} & R^{2,2}_{1,1} \\ R^{1,1}_{1,2} & R^{1,2}_{1,2} & R^{2,1}_{1,2} & R^{2,2}_{1,2} \\ R^{1,1}_{2,1} & R^{1,2}_{2,1} & R^{2,1}_{2,1} & R^{2,2}_{2,1} \\ R^{1,1}_{2,2} & R^{1,2}_{2,2} & R^{2,1}_{2,2} & R^{2,2}_{2,2} \end{pmatrix} = \begin{pmatrix} 1 & 0 & 0 & 0 \\ 0 & g(x) & h(x) & 0 \\ 0 & h(x) & g(x) & 0 \\ 0 & 0 & 0 & 1 \end{pmatrix}.$$

It is straightforward to compute the corresponding one-parameter QYBE for such an R (which is said to correspond to the XXZ-model [Kulish and Sklyanin, 1982, p. 1613]). It reduces to three functional equations:

$$\begin{aligned} g(z)h(x)h(z+x) + (-g(z+x) + g(x))h(z) &= 0 \\ (g(x)g(z) - 1)h(z+x) + h(x)h(z) &= 0 \\ g(x)h(z)h(z+x) + (-g(z+x) + g(z))h(x) &= 0. \end{aligned}$$

It is well known [Kulish and Sklyanin, 1982, p. 1614], [Tarasov et al., 1983, p. 1069] that a solution to these equations is given by

$$g(x) = \frac{\sin(x)}{\sin(x+q)}, \quad h(x) = \frac{\sin(q)}{\sin(x+q)},$$

where q is an arbitrary constant. This class of solutions is related to the algebraic Bethe Ansatz [Takhtajan, 1984].

Exercise 8.6.1 What is the set Z in Definition 2.4.1 for the above solution?

Now consider those R which correspond to the matrices

$$\begin{pmatrix} f(x) & 0 & 0 & 0 \\ 0 & g(x) & h(x) & 0 \\ 0 & h(x) & g(x) & 0 \\ 0 & 0 & 0 & f(x) \end{pmatrix}.$$

Another straightforward computation shows that such R will satisfy the one-parameter QYBE if the three equations

$$\begin{aligned} (f(x)g(y) - f(y)g(x))h(z) - g(z)h(x)h(y) &= 0 \\ f(y)h(x)h(z) + (g(x)g(z) - f(x)f(z))h(y) &= 0 \\ g(x)h(y)h(z) + (f(y)g(z) - f(z)g(y))h(x) &= 0 \end{aligned}$$

hold. A bit of algebraic manipulation shows that if we set $h(x) = f(x) - g(x)$, so that

$$R(x) = \begin{pmatrix} f(x) & 0 & 0 & 0 \\ 0 & g(x) & f(x) - g(x) & 0 \\ 0 & f(x) - g(x) & g(x) & 0 \\ 0 & 0 & 0 & f(x) \end{pmatrix}, \tag{8.20}$$

these equations reduce to one, viz.

$$(g(x)g(y) - 2f(y)g(x) + f(x)f(y))g(z) - f(x)f(z)g(y) + f(y)f(z)g(x) = 0.$$

Rearranging terms, we can write this equation as

$$(g(x)g(z) - f(x)f(z))g(y) = (2g(x)g(z) - f(x)g(z) - f(z)g(x))f(y). \quad (8.21)$$

Now there are many instances when (8.21) can be solved for y. For example, if $f(x)$ and $g(x)$ are assumed to be polynomials, solutions are easily seen to exist. As observed in [Lambe, 1994], the linear case gives a solution corresponding to the XXX-model.

For another example, ignoring any zeros of g, put (8.21) into the form

$$\lambda(y) = \frac{1 - \lambda(x)\lambda(z)}{2 - \lambda(x) - \lambda(z)} \quad (8.22)$$

where

$$\lambda(x) = \frac{f(x)}{g(x)}.$$

For example, when $f(x) = 1$ and $g(x) = qx$, we obtain

$$\frac{1}{qy} = \frac{q^2xz - 1}{2qxz - z - y},$$

i.e.

$$y = \frac{x + z - 2qxz}{1 - q^2xz}.$$

Note that this solution is *essentially* the same as the classical solution

$$\begin{pmatrix} 1 & 0 & 0 & 0 \\ 0 & \frac{x}{x+q} & \frac{q}{x+q} & 0 \\ 0 & \frac{q}{x+q} & \frac{x}{x+q} & 0 \\ 0 & 0 & 0 & 1 \end{pmatrix}. \quad (8.23)$$

It is a part of the folklore that (8.23) is essentially the only rational two-dimensional one-parameter solution of the form (8.20). To see this, note that by scaling (and ignoring any singularities that might arise), we may say that

$$\begin{pmatrix} f & 0 & 0 & 0 \\ 0 & g & f-g & 0 \\ 0 & f-g & g & 0 \\ 0 & 0 & 0 & f \end{pmatrix}$$

is equivalent to

$$\begin{pmatrix} 1 & 0 & 0 & 0 \\ 0 & \frac{g}{f} & 1-\frac{g}{f} & 0 \\ 0 & 1-\frac{g}{f} & \frac{g}{f} & 0 \\ 0 & 0 & 0 & 1 \end{pmatrix}.$$

Now by assuming the appropriate invertibility and once again ignoring any singularities that might arise, we can make a substitution for x which will bring this matrix into the form of (8.23). Note, however, that in general, the sets Z of Definition 2.4.1 will be different.

More classes of solutions and methods for finding solutions can be found in [Andrews and Baxter, 1986a], [Andrews and Baxter, 1986b], [Andrews and Baxter, 1987], [Bazhanov, 1985], [Cherednik, 1982], [Chudnovsky and Chudnovsky, 1981], [Kulish et al., 1981], [Kulish and Sklyanin, 1982], and the references therein.

Exercise 8.6.2 Let M be an n-dimensional vector space over the field k. Consider the operator

$$R(x) = (f(x) - g(x))\tau_{M,M} + g(x)1_M \otimes 1_M$$

where $f(x)$ and $g(x)$ are two given functions. Prove that R(x) satisfies the one-parameter QYBE if and only if (8.21) holds. This is essentially the Yang family of QYBE solutions [Yang, 1967].

8.6.2 A ρ-Perturbation Example

We end this Chapter with an example of the ρ-perturbation process given in Theorem 7.5.1. Computer algebra was used to generate the example.

We leave the details of the following straightforward proposition to the reader.

Proposition 8.6.1 *Let $M = \mathbb{C}^n$ and let P be a matrix for which $P^t = \exp(t\log(P))$ exists (converges). We have a (parameterized) action μ_t of $E = \mathrm{M}_n(\mathbb{C})$ ($n \times n$-matrices) on M given by*

$$A \cdot v = P^t A P^{-t} v$$

where $A \in E$ and $v \in M$.

Let R be any one-parameter QYBE solution on M. The dual of the parameterized action (M, μ_t) gives a rational comodule structure over the comatrix coalgebra $\mathrm{C}_n(\mathbb{C})$ which extends to a parameterized comodule structure ρ_t^P over $A(R)$ in the sense of Definition 7.5.2 and hence we have a new one-parameter QYBE solution R^P, viz. the corresponding ρ-perturbation of R by ρ^P. ∎

For a simple example, consider the one-parameter QYBE solution

$$R = \begin{pmatrix} 1 & 0 & 0 & 0 \\ 0 & \frac{x}{x+1} & \frac{1}{x+1} & 0 \\ 0 & \frac{1}{x+1} & \frac{x}{x+1} & 0 \\ 0 & 0 & 0 & 1 \end{pmatrix}$$

corresponding to the XXX-magnet model.

Let $P = I + N$ where I is the 2×2-identity matrix and

$$R = \begin{pmatrix} 0 & 1 \\ 0 & 0 \end{pmatrix}.$$

Clearly, we have $\log(P) = P - I = N$ since $N^2 = 0$. Thus,

$$P^x = \begin{pmatrix} 0 & x \\ 0 & 0 \end{pmatrix}$$

and for any matrix

$$A = \begin{pmatrix} a & b \\ c & d \end{pmatrix}$$

we have

$$P^x A P^{-x} = \begin{pmatrix} a + cx & -cx^2 + (d-a)x + b \\ c & -cx + d \end{pmatrix}.$$

It is easy to work out the corresponding parameterized comodule structure over $A(R)$. We have

$$\begin{aligned} \rho_x(e_1) &= \alpha_1^1(x) \otimes e_1 + \alpha_1^2(x) \otimes e_2, \\ \rho_x(e_2) &= \alpha_2^1(x) \otimes e_1 + \alpha_2^2(x) \otimes e_2 \end{aligned}$$

where

$$\begin{aligned} \alpha_1^1(x) &= t_1^1(x) + x t_2^1(x) \\ \alpha_1^2(x) &= -x t_1^1(x) + t_1^2(x) - x^2 t_2^1(x) + x t_2^2(x) \\ \alpha_2^1(x) &= t_2^1(x) \\ \alpha_2^2(x) &= -x t_2^1(x) + t_2^2(x). \end{aligned}$$

The corresponding ρ-perturbation of R is given by

$$R^P(x) = \begin{pmatrix} 1 & \frac{x}{x+1} & -\frac{x}{x+1} & -\frac{x^2}{x+1} \\ 0 & \frac{x}{x+1} & \frac{1}{x+1} & \frac{x}{x+1} \\ 0 & \frac{1}{x+1} & \frac{x}{x+1} & -\frac{x}{x+1} \\ 0 & 0 & 0 & 1 \end{pmatrix}.$$

R^P satisfies the one-parameter QYBE with additive spectral parameter.

9 CATEGORICAL CONSTRUCTIONS AND GENERALIZATIONS OF THE QUANTUM YANG–BAXTER EQUATION

We introduce some categorical constructions related to the QYBE in this Chapter. This is intended to be an introduction to the work found in [Joyal and Street, 1991a], [Joyal and Street, 1991b], [Joyal and Street, 1991c], [Joyal and Street, 1993], [Pareigis, 1996], [Schauenburg, 1992a], [Schauenburg, 1992b] and the references found therein. Our basic reference for category theory is [MacLane, 1988].

9.1 Coends

Let $\mathcal{C}$ be a category. Recall that the opposite category [MacLane, 1988, p. 33] $\mathcal{C}^{op}$ has the same objects as those of $\mathcal{C}$ and morphisms f^{op} in bijective correspondence with morphisms f of $\mathcal{C}$ but such that if $f : A \longrightarrow B$ then $f^{op} : B \longrightarrow A$. The composite $f^{op}g^{op}$ is defined exactly when the composite gf is defined in $\mathcal{C}$ and then $f^{op}g^{op} = (gf)^{op}$. Let $\mathcal{D}$ be another category and let

$$\mathcal{C}^{op} \times \mathcal{C} \xrightarrow{F} \mathcal{D}$$

be a functor.

Definition 9.1.1 *A* wedge *is an object W in $\mathcal{D}$ and a family $\mu_X : F(X,X) \longrightarrow W$ of morphisms such that for all $f : X \longrightarrow Y$ in $\mathcal{C}$ the diagram*

$$\begin{array}{ccc} F(Y,X) & \xrightarrow{F(1_Y,f)} & F(Y,Y) \\ {\scriptstyle F(f,1_X)}\downarrow & & \downarrow{\scriptstyle \mu_Y} \\ F(X,X) & \xrightarrow[\mu_X]{} & W \end{array}$$

commutes.

We use the following notation to denote wedges:

$$\mu : F \xrightarrow{\bullet\bullet} W.$$

Definition 9.1.2 *Let $\mathcal{C}^{op} \times \mathcal{C} \xrightarrow{F} \mathcal{D}$ be a functor. A* coend *of F, denoted by* $\mathrm{coend}(F)$, *is a universal wedge* $\mathrm{coend}(F) = F \xrightarrow{\bullet\bullet} W$, *i.e. it is a wedge $\mu : F \xrightarrow{\bullet\bullet} W$ with the property that to every wedge $\mu' : F \xrightarrow{\bullet\bullet} W'$ there is a unique arrow $h : W \longrightarrow W'$ such that $\mu'_X = h\mu_X$ for all X.*

As usual, when a coend exists, it is unique up to isomorphism.

Now assume that $\mathcal{D}$ is a small category [MacLane, 1988, p. 22] and $\mathcal{C}^{op} \times \mathcal{C} \xrightarrow{F} \mathcal{D}$ is a functor. For suitable categories $\mathcal{C}$, $\mathrm{coend}(F)$ can be defined explicitly in terms of generators and relations, viz.

$$\mathrm{coend}(F) = \coprod_{X \in OB(\mathcal{C})} F(X,X)/\sim \tag{9.1}$$

where $\sim$ is the smallest equivalence relation defined by

$$F(X_1, f)(y) \sim F(f, X_2)(y)$$

for each $f{:}X_2 \longrightarrow X_1$ in $\mathcal{C}$ and each $y \in F(X_1, X_2)$.

9.2 Quasi-Symmetric Monoidal Categories

We have seen that the category ${}_H\mathcal{QYB}$ over a finite dimensional Hopf algebra H is equivalent to modules over the double $D(H)$. We can rephrase this to say that when H is a finite dimensional Hopf algebra, there is a category equivalence

$${}_H\mathcal{QYB} \simeq \mathcal{M}^{D(H)^*}$$

of ${}_H\mathcal{QYB}$ and the category of comodules over the dual of the double. There is a much more general result that requires some new notation to state.

Definition 9.2.1 *Let $OB(C)$ denote the underlying class of objects of a category C. The category C is said to be* monoidal *if there is a functor*

$$\otimes : OB(C) \times OB(C) \longrightarrow OB(C)$$

and a collection of isomorphisms

$$\alpha_{(X,Y,Z)} : (X \otimes Y) \otimes Z \simeq X \otimes (Y \otimes Z)$$

for all $X, Y, Z \in OB(C)$ which satisfy the so-called pentagon identity:

$$\begin{array}{ccc}
(X \otimes (Y \otimes Z)) \otimes W & \longrightarrow & X \otimes ((Y \otimes Z) \otimes W) \\
\downarrow & & \downarrow \\
((X \otimes Y) \otimes Z) \otimes W) & & X \otimes (Y \otimes (Z \otimes W)) \\
\nwarrow & & \nearrow \\
 & (X \otimes Y) \otimes (Z \otimes W) &
\end{array}$$

Definition 9.2.2 *A* unit object *for a monoidal category C is an object k such that there are natural isomorphisms*

$$X \xrightarrow{\epsilon_X} X \otimes k, \quad X \xrightarrow{\sigma_X} k \otimes X$$

which are compatible with the system of isomorphisms $\{\alpha_{(X,Y,Z)}\}$.

Monoidal categories are quite common. For example, if C is the category of sets and functions, we can take $\otimes$ to be the ordinary Cartesian product. Note that any one point set can be taken to be a unit in this case. Another example is given by the category of all modules over a ring R. In this case we take $\otimes = \otimes_R$ and a unit is given by R itself. The usual convention is to simply write $X \otimes Y \otimes Z$ for either $(X \otimes Y) \otimes Z$ or $X \otimes (Y \otimes Z)$ in a monoidal category and let the context determine whether or not an "associativity" isomorphism $\alpha_{(X,Y,Z)}$ needs to be applied. We will follow this convention.

Definition 9.2.3 *An object X^* in a monoidal category C is said to be a* left dual *to an object X if there are morphisms*

$$X^* \otimes X \xrightarrow{e} k, \quad k \xrightarrow{i} X \otimes X^*$$

such that the composites

$$X \longrightarrow k \otimes X \longrightarrow X \otimes X^* \longrightarrow X \otimes k \longrightarrow X$$

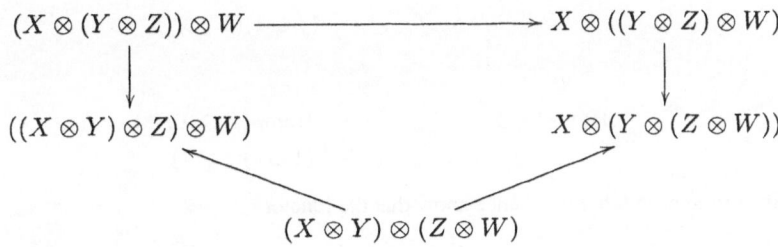

and

$$X^* \longrightarrow X^* \otimes k \longrightarrow X^* \otimes X \otimes X^* \longrightarrow k \otimes X^* \longrightarrow X^*$$

are the identity maps 1_X *and* 1_{X^*} *respectively.*

We leave it to the reader to define the analogous notion of a right dual.

Exercise 9.2.1 Let $\mathcal{C}$ be a category and X, Y, and Z be objects of $\mathcal{C}$.

a) Suppose that (X^*, e, i) is a left dual to X. Prove that the functions

$$\begin{array}{rcl} \mathrm{Hom}_{\mathcal{C}}(Z \otimes X, Y) & \longrightarrow & \mathrm{Hom}_{\mathcal{C}}(Z, X \otimes Y) \\ f & \longmapsto & (f \otimes 1)(1 \otimes i) \end{array}$$

and

$$\begin{array}{rcl} \mathrm{Hom}_{\mathcal{C}}(Z, Y \otimes X^*) & \longrightarrow & \mathrm{Hom}_{\mathcal{C}}(Z \otimes X, Y) \\ g & \longmapsto & (1 \otimes e)(g \otimes 1) \end{array}$$

are inverse to each other and hence show that the functor

$$\begin{array}{rcl} \mathcal{C} & \longrightarrow & \mathcal{C} \\ Y & \longmapsto & Y \otimes X^* \end{array}$$

is right adjoint to the functor

$$\begin{array}{rcl} \mathcal{C} & \longrightarrow & \mathcal{C} \\ Y & \longmapsto & Y \otimes X \end{array}$$

b) Using a) show that if an object X in a monoidal category has a left dual X^*, then the object X^* and the maps e and i are unique up to canonical isomorphism.

Definition 9.2.4 *A monoidal category* $\mathcal{C}$ *is said to be* rigid *if every object* X *has a left and right dual.*

Exercise 9.2.2 Let k be a field.

a) Prove that the category of finite-dimensional vector spaces over k is rigid.

b) Let H be a Hopf algebra over k. Prove that the category of finite-dimensional comodules is monoidal.

c) Using the definitions $M^* = \mathrm{Hom}_k(M, F)$ and

$$\Delta(\alpha)(m) = \alpha(m^{<1>}) \otimes s(m^{(2)}),$$

from [Ulbrich, 1990], show that the category of finitely generated projective right H-comodules is rigid monoidal. Here s is the antipode of H. This uses the isomorphism $M^* \simeq \mathrm{Hom}_H(M, H)$.

Definition 9.2.5 *A monoidal category $\mathcal{C}$ is said to be* quasi-symmetric *if there is a system of isomorphisms*

$$X \otimes Y \xrightarrow{S_{(X,Y)}} Y \otimes X$$

which satisfy the two hexagon identities

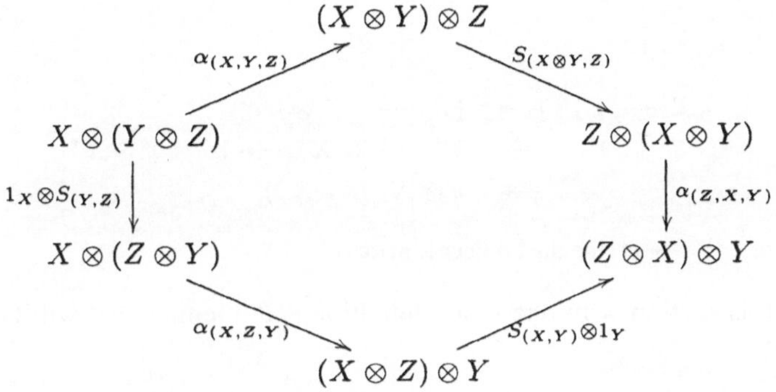

and

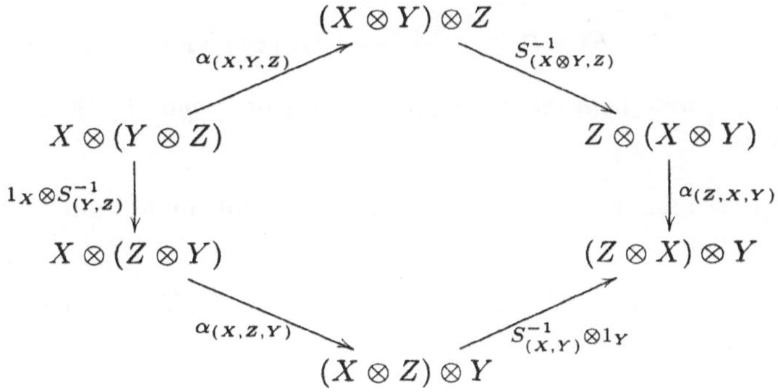

We call S the *symmetry* or *braiding* of $\mathcal{C}$.

If $\mathcal{C}$ is a quasi-symmetric monoidal category and $S_{(Y,X)} = S^{-1}_{(X,Y)}$ then $\mathcal{C}$ is said to be symmetric.

Exercise 9.2.3 Let ${}_k\mathrm{Vec}$ be the category of finite dimensional vector spaces over a field. Prove that ${}_k\mathrm{Vec}$ is a quasi-symmetric monoidal category. The symmetry is given by $X \otimes_k Y \longmapsto Y \otimes_k X$.

Exercise 9.2.4 Prove that if $\mathcal{C}$ is a quasi-symmetric monoidal category and every object has a left dual then it is rigid.

Exercise 9.2.5 Suppose H is a Hopf algebra over a field k. Prove that the full subcategory consisting of the finite-dimensional subobjects of ${}_H\mathcal{QYB}$ is a quasi-symmetric monoidal category.

Exercise 9.2.6 Suppose that $\mathcal{C}$ is a rigid monoidal category. Prove that the correspondence

$$X \longmapsto X^*$$

and

$$\begin{aligned}(X \longrightarrow Y) &\longmapsto (Y^* \longrightarrow Y^* \otimes k \\ &\longrightarrow Y^* \otimes X \otimes X^* \longrightarrow Y^* \otimes Y \otimes X^* \\ &\longrightarrow k \otimes X^* \longrightarrow X^*)\end{aligned}$$

gives a functor $\mathcal{C}^{op} \longrightarrow \mathcal{C}$ (the dual object functor).

We end this section with one more definition and a lemma that will be needed in what follows.

Definition 9.2.6 *Let $\mathcal{C}$ and $\mathcal{D}$ be monoidal categories. A* monoidal functor *is a functor $F : \mathcal{C} \longrightarrow \mathcal{D}$ for which there exists a family of isomorphisms*

$$F(X \otimes Y) \xrightarrow{\eta(X,Y)} F(X) \otimes F(Y)$$

which are compatible with the associativity maps of $\mathcal{C}$ and $\mathcal{D}$. We assume that F preserves units.

It is easy to see that if $F : \mathcal{C} \longrightarrow \mathcal{D}$ is a monoidal functor of rigid categories, then the composite map

$$\begin{aligned}F(X)^* &\longrightarrow F(X)^* \otimes k \longrightarrow F(X)^* \otimes F(k) \longrightarrow F(X)^* \otimes F(X) \otimes F(X^*) \\ &\longrightarrow k \otimes F(X^*) \longrightarrow F(X^*)\end{aligned}$$

allows us to identify $F(X^*)$ with $F(X)^*$.

9.3 Rigid Monoidal Categories and Hopf Algebras

The following theorem from [Deligne and Milne, 1982] and [Saavedra Rivano, 1972] has been generalized in [Ulbrich, 1990]. The purpose of this section is to review this result.

Theorem 9.3.1 *Let k be a field and ${}_k\mathrm{Vec}$ be the category of finite-dimensional vector spaces over k. Let $\mathcal{C}$ be a k-linear, abelian small category and suppose that*

$$G : \mathcal{C} \longrightarrow {}_k\mathrm{Vec}$$

is a faithful, exact, k-linear functor. Then there exists a coalgebra C over k and a category equivalence

$$E : \mathcal{C} \longrightarrow \mathcal{M}^C$$

such that $G = FE$, where $\mathcal{M}^C$ is the category of finite-dimensional right C-comodules and $F : \mathcal{M}^C \longrightarrow {}_k\mathrm{Vec}$ is the forgetful functor.

If in addition, $\mathcal{C}$ is symmetric and rigid, it follows that C is actually a commutative Hopf algebra.

The basic idea in [Ulbrich, 1990] is to drop the symmetry condition and consider the following data. For a vector space V, let $F \otimes V$ be the functor ${}_k\mathrm{Vec} \longrightarrow Sets$ given by $X \longmapsto F(X) \otimes V$ and the set of natural transformations $\mathrm{Nat}(F, F \otimes V)$. One has the functor

$$\begin{array}{rcl} {}_k\mathrm{Vec} & \longrightarrow & Sets \\ V & \longmapsto & \mathrm{Nat}(F, F \otimes V) \end{array}$$

and it is known from [Saavedra Rivano, 1972] that this functor is representable, i.e. there is a k-module H and natural isomorphisms

$$\mathrm{Hom}_k(H, V) \xrightarrow{\ \rho_V\ } \mathrm{Nat}(F, F \otimes V)$$

for all objects V of ${}_k\mathrm{Vec}$. Using this representation, the coproduct, product, unit can be derived. When $\mathcal{C}$ is rigid monoidal, it is also shown in [Ulbrich, 1990] that an antipode can be defined using this data. We have

Theorem 9.3.2 ([Ulbrich, 1990]) *Let $\mathcal{C}$ be a small monoidal category with direct sums and $G : \mathcal{C} \longrightarrow {}_k\mathrm{Vec}$ be a monoidal functor preserving sums. There is a bialgebra H over k and an equivalence $E : \mathcal{C} \longrightarrow \mathcal{M}^H$ such that $G = FE$ where $F : \mathcal{M}^H \longrightarrow {}_k\mathrm{Vec}$ is the forgetful functor. If in addition, $\mathcal{C}$ is rigid, H is a Hopf algebra.*

We will omit the proof and outline the main ideas from [Ulbrich, 1990] in a sequence of exercises. In that case, the representing object mentioned above can be identified as a coend [Schauenburg, 1992b]. Another approach is given in [Majid, 1991b].

Let $G : \mathcal{C} \longrightarrow \mathcal{D}$ be a functor with $\mathcal{C}$ monoidal category and $\mathcal{D}$ a rigid monoidal category. Consider the composite of G and the dual object functor which we denote by G^*. We have a functor

$$\mathcal{C}^{op} \otimes \mathcal{C} \xrightarrow{\ G^* \otimes G\ } \mathcal{D}$$

and hence we can consider

$$H = \mathrm{coend}(G^* \otimes G).$$

Note that the coend formula (9.1) can be reformulated in this special case to read (for suitable $\mathcal{C}$ and $\mathcal{D}$)

$$\mathrm{coend}(G^* \otimes G) = \coprod_{X \in OB(\mathcal{C})} G(X)^* \otimes G(X)/\sim$$

where for $f : X_2 \longrightarrow X_1$ and $\alpha \otimes x\ inG(X_1)^* \otimes G(X_2)$ we have

$$\alpha \otimes G(f)(x) \sim G(f)^*(\alpha) \otimes x.$$

It is shown in [Schauenburg, 1992b] that

Lemma 9.3.1 *Under the hypotheses above, the object* $\mathrm{coend}(G^* \otimes G)$ *represents the functor*

$$\begin{array}{ccc} \mathcal{D} & \longrightarrow & Sets \\ V & \longmapsto & \mathrm{Nat}(G, G \otimes V). \end{array}$$

Exercise 9.3.1 Prove Lemma 9.3.1 by first showing that a natural transformation

$$G \longrightarrow G \otimes V$$

is equivalent to a wedge

$$G^* \otimes G \xrightarrow{\bullet\bullet} V.$$

Exercise 9.3.2 This exercise uses some constructions from [Saavedra Rivano, 1972] and [Ulbrich, 1990]. Assume the hypotheses of Theorem 9.3.2. In parts a) through d) it is not assumed that $\mathcal{C}$ is rigid. The last exercise established natural isomorphisms

$$\mathrm{Hom}_k(H, V) \xrightarrow{\varphi_V} \mathrm{Nat}(G, G \otimes V)$$

where

$$H = \mathrm{coend}(G^* \otimes G).$$

a) Using $\mathrm{Hom}_k(H, H) \xrightarrow{\varphi_H} \mathrm{Nat}(G, G \otimes H)$ we have $(\varphi_H(1_H) \otimes 1)\varphi_H(1_H) : G \longrightarrow G \otimes H \otimes H$. Let

$$\Delta = \varphi_{H \otimes H}^{-1}((\varphi_H(1_H) \otimes 1)\varphi_H(1_H)) : H \longrightarrow H \otimes H.$$

Note also that we have $\mathrm{Hom}_k(H, k) \xrightarrow{\varphi_k} \mathrm{Nat}(G, G \otimes k)$ and that the obvious isomorphism $G(X) \simeq G(X) \otimes k$ gives rise to $e \in \mathrm{Nat}(G, G \otimes k)$. Let

$$\epsilon = \varphi_K^{-1}(e) : H \longrightarrow k.$$

Show that (H, Δ, ϵ) is a coalgebra over k.

b) Using the coproduct just defined, show that the map

$$\rho = \varphi_A(1_A) : G(X) \longrightarrow G(X) \otimes H$$

defines a left H-comodule structure on $G(X)$.

c) Using the comodule structure just defined, show that the map

$$\mathrm{Hom}_k(A \otimes A, V) \xrightarrow{\Phi_V} \mathrm{Nat}(G \otimes G, G \otimes V)$$

given by

$$\Phi_V(\alpha)(x \otimes y) = x^{<1>} \otimes y^{<1>} \otimes \alpha(x^{(2)} \otimes y^{(2)})$$

for $\alpha : H \otimes H \longrightarrow V$, $x \in G(X)$, and $y \in G(Y)$ is an isomorphism.

d) Let

$$\xi : G(X) \otimes G(Y) \simeq G(X \otimes Y) \xrightarrow{\rho} G(X \otimes Y) \otimes H \simeq G(X) \otimes G(Y) \otimes H$$

and

$$m = \Phi_H^{-1}(\xi) : H \otimes H \longrightarrow H.$$

Let

$$\sigma : k \simeq G(k) \xrightarrow{\rho} G(k) \otimes H \simeq H.$$

Prove that $(H, m, \sigma, \Delta, \epsilon)$ is a bialgebra over k.

e) Assume now that $\mathcal{C}$ is rigid. Show that the map given by

$$\begin{aligned} \mathrm{Hom}_k(G(X), G(X) \otimes V) &\simeq \mathrm{Hom}_k(G(X)^*, G(X)^* \otimes V) \\ &\longrightarrow \mathrm{Hom}_k(G(X^*), G(X^*) \otimes V) \end{aligned}$$

is natural and hence corresponds to a map $s : H \longrightarrow H$ under the isomorphism

$$\mathrm{Hom}_k(H, H) \xrightarrow{\varphi_H} \mathrm{Nat}(G, G \otimes H)$$

Prove that s is an antipode for H.

Exercise 9.3.3 Define a category $\mathcal{B}$ whose objects are the non-negative integers $0, 1, \ldots$ and whose morphism sets are

$$\mathrm{Hom}(n, m) = \begin{cases} \emptyset & n \neq m \\ \mathbb{B}_n & n = m \end{cases}$$

where $\mathbb{B}_n$ is the Artin Braid group (2.2.2).

a) Prove that the correspondence

$$(n, m) \longmapsto n + m$$

defines a monoidal structure on $\mathcal{B}$.

b) As a corollary to Theorem 2.2.1, show that any QYBE solution R on M gives rise to a monoidal functor

$$\mathcal{B} \xrightarrow{F_R} {}_k\mathrm{Vec}$$

$$\begin{aligned} F_R(n) &= \otimes^n M \\ \alpha &\longmapsto (\otimes^n M \xrightarrow{\mu_\alpha} \otimes M) \end{aligned}$$

c) Show that the FRT construction is a coend of

$$F^* \otimes F.$$

See [Joyal and Street, 1993], [Joyal and Street, 1991b], [Schauenburg, 1992b], and [Schauenburg, 1992a] for more background.

9.4 Quasi-Symmetric Monoidal Categories and Coquasitriangular Hopf Algebras

We come now to a theorem which was given in a dual formulation in [Majid, 1991b]. This theorem can really be thought of as a continuation of the main theorem in [Ulbrich, 1990]. A good exposition of this material is given in where complete proofs may be found. We have

Theorem 9.4.1 *Let $\mathcal{C}$ be a small quasi-symmetric monoidal category with direct sums and $G : \mathcal{C} \longrightarrow {}_k\mathrm{Vec}$ be a monoidal functor preserving sums. There is a coquasitriangular bialgebra H over k and an equivalence $E : \mathcal{C} \longrightarrow \mathcal{M}^H$ such that $G = FE$ where $F : \mathcal{M}^H \longrightarrow {}_k\mathrm{Vec}$ is the forgetful functor. If in addition, $\mathcal{C}$ is rigid, H is a coquasitriangular Hopf algebra.*

Exercise 9.4.1 Under the hypotheses of Theorem 9.4.1, consider the composite map

$$\begin{aligned} G(X) \otimes G(Y) &\longrightarrow G(X \otimes Y) \longrightarrow G(Y \otimes X) \\ &\longrightarrow G(Y) \otimes G(X) \longrightarrow G(X) \otimes G(Y) \\ &\longrightarrow G(X) \otimes G(Y) \otimes k. \end{aligned}$$

As in the previous exercise, this map corresponds to a map

$$H \otimes H \xrightarrow{r} k.$$

Prove that (H, r) is a coquasitriangular bialgebra.

Exercise 9.4.2 Note that the identity functor ${}_k\mathrm{Vec} \longrightarrow {}_k\mathrm{Vec}$ satisfies the hypotheses of Theorem 9.4.1. What is the associated coquasitriangular bialgebra in this case?

9.5 The Quantum Yang–Baxter Equation in Other Categories

Let $\mathcal{C}$ be any monoidal category. It is straightforward to generalize the braid equation (2.4) in this setting.

Definition 9.5.1 *Let M be an object of $\mathcal{C}$ and R be an endomorphism of $M \otimes M$. R is said to satisfy the braid equation in $\mathcal{C}$ if*

$$(R \otimes 1_M)(1_M \otimes R)(R \otimes 1_M) = (1_M \otimes R)(R \otimes 1_M)(1_M \otimes R)$$

If $\mathcal{C}$ is quasi-symmetric with symmetry S, we can mimic the definition of the QYBE given in Section 2.1 in $\mathcal{C}$ [Lambe and Radford, 1993, p. 258].

Definition 9.5.2 *Let M be an object in $\mathcal{C}$. We say that a morphism $R : M \otimes M \longrightarrow M \otimes M$* satisfies the QYBE in $\mathcal{C}$ with respect to a symmetry S *if and only if*

$$R_{(1,2)} R_{(1,3)} R_{(2,3)} = R_{(2,3)} R_{(1,3)} R_{(1,2)}$$

where

$$R_{(1,2)} = R \otimes 1_M, \quad R_{(2,3)} = 1_M \otimes R$$

and

$$R_{(1,3)} = (1_M \otimes S_{(M,M)})(R \otimes 1_M)(1_M \otimes S_{(M,M)}).$$

There does not seem to be much in the literature on the relationship between these concepts except in the "ordinary" case where $\mathcal{C}$ is the category of vector spaces over a field. We will touch upon one particular case here and leave further investigations to the reader.

9.6 The Category of Graded Modules

Let k be a field and ${}_k\mathcal{GRM}$ be the category of connected graded modules of finite type over k. An M object of ${}_k\mathcal{GRM}$ is a direct sum of finite dimensional vector spaces over k

$$M = \oplus_{i=0}^{\infty} M_i$$

where $M_0 = k$. An element $x \in M_i$ is said to have degree i and we write

$$|x| = i.$$

A map $f : M \longrightarrow N$ in ${}_k\mathcal{GRM}$ is a direct sum of maps

$$f = \oplus_{i=0}^{\infty} f_i$$

where $f_i : M_i \longrightarrow N_{i+k}$ and k is some fixed non-negative integer which we call the degree of f. We also write $|f| = k$ in this case.

There is a monoidal structure on ${}_k\mathcal{GRM}$ given by

$$(M, N) \longmapsto M \otimes N$$

where

$$(M \otimes N)_k = \oplus_{i+j=k} M_i \otimes_k N_j.$$

Given maps $f : M \longrightarrow M'$ and $g : N \longrightarrow N'$ in ${}_k\mathcal{GRM}$, we define

$$\begin{array}{ccc} M \otimes M' & \xrightarrow{f \otimes g} & N' \otimes M' \\ x \otimes y & \longmapsto & (-1)^{|x||g|} f(x) \otimes g(y). \end{array}$$

This is an example of a general sign convention. In dealing with tensor products, whenever one object is transposed with another, a coefficient of minus one to the power of the product of their degrees is acquired. For example, in the definition of $(f \otimes g)(x \otimes y)$, the x is transposed with the g and hence the sign.

In fact, ${}_k\mathcal{GRM}$ is actually a quasi-symmetric category with symmetry

$$S(x \otimes y) = (-1)^{|x||y|} y \otimes x.$$

Exercise 9.6.1 Prove that with the definitions above, $({}_k\mathcal{GRM}, \otimes, S)$ is indeed a quasi-symmetric monoidal category with unit k.

We can easily work out what the QYBE in ${}_k\mathcal{GRM}$ looks like in coordinates in analogy to what we did in Section 2.1.2. We will present this for a degree zero map and leave the general case to the reader. Given a map $R : M \otimes M \longrightarrow M \otimes M$ of degree zero and a homogeneous basis $\{e_i\}$ for M, we write

$$R(e_i \otimes e_j) = R_{i,j}^{k,l} e_k \otimes e_l$$

By applying both sides of the QYBE to basis elements $e_i \otimes e_j \otimes e_k$ and keeping the sign conventions above in context, we obtain the coordinate form of QYBE as

$$(-1)^{|e_{s_2}|(|e_{s_3}|+|e_c|)} R_{j,k}^{s_2,s_3} R_{i,s_3}^{s_1,c} R_{j,k}^{a,b} = (-1)^{|e_{r_2}|(|e_{r_3}|+|e_k|)} R_{i,j}^{r_1,r_2} R_{r_1,k}^{p,r_3} R_{r_2,r_3}^{b,c}.$$

This form of the equation appears in e.g. [Akutsu and Deguchi, 1990] and [Kulish and Sklyanin, 1982].

Exercise 9.6.2 In the current context:

a) Work out the QYBE in coordinates when R has possibly non-zero degree d.

b) Can the graded QYBE solution be transformed into a graded braid equation solution and conversely? What about can be said about the analogous question for a general quasi-symmetric monoidal category.

Appendix A
Prerequisites

A.1 The Ground Ring k and Basic k-Linear Maps

Throughout this text we will be working over a field k. Most definitions can be made over a commutative ring – however the theory we develop would be somewhat limited in this generality.

We will let 1_V denote the identity map of a vector space V over the field k. If V is an algebra over k occasionally 1_V will be used to denote the identity element of V. The meaning of the symbol 1_V should be clear from context. *We will use juxtaposition to denote composition of functions.*

We will use several ways of expressing functions, most often $f : M \longrightarrow N$ which will be written

$$M \xrightarrow{f} N$$

in display mode. Sometimes we will represent f by $(m \longmapsto f(n))$. Linear will mean k-linear and basis will mean linear basis over k unless otherwise specified.

Suppose that M and N are vector spaces over k. We will most often write $\mathrm{Hom}(M, N)$ for the vector space $\mathrm{Hom}_k(M, N)$ of linear maps $f : M \longrightarrow N$. Now assume that $N = k$ and $M^* = \mathrm{Hom}(M, k)$ is the space of linear functionals on M. For a linear functional $\alpha \in M^*$ we write $\alpha(m)$ or $<\alpha, m>$ for the image of $m \in M$ under α.

We will assume that the reader is familiar with basic properties of the tensor product of vector spaces over k. Generally we will omit the subscript k on the tensor product symbol and write $U \otimes V$ instead of the more customary $U \otimes_k V$.

There are certain linear isomorphisms which will play a fundamental role in what follows, namely:

$$
\begin{array}{ll}
k \otimes U \simeq U & \alpha \otimes u \longmapsto \alpha u \\
U \otimes k \simeq U & u \otimes \alpha \longmapsto \alpha u \\
(U \otimes V) \otimes W \simeq U \otimes (V \otimes W) & (u \otimes v) \otimes w \longmapsto u \otimes (v \otimes w) \\
U \otimes V \simeq V \otimes U & u \otimes v \longmapsto v \otimes u,
\end{array}
$$

where U, V, and W are vector spaces over k. These k-linear isomorphisms hold when k is a commutative ring. They will be revisited in more complex categorical settings in Section 9.2. The last isomorphism is very important.

Definition A.1.1 *The linear map* $\tau_{U,V} : U \otimes V \longrightarrow V \otimes U$ *defined by* $\tau_{U,V}(u \otimes v) = v \otimes u$ *for all* $u \in U$ *and* $v \in V$ *is the* twist map.

One last detail. Let I be a set (usually an indexing set). Define a function

$$
\begin{array}{rcl}
I \times I & \xrightarrow{\delta} & Z_2 \\
\delta(i, j) & = & \begin{cases} 1 & i = j \\ 0 & i \neq j, \end{cases}
\end{array}
$$

where $\mathbb{Z}_2 = \{0, 1\}$ is regarded as a semigroup under multiplication inherited from the ring of integers. This function is called a *Kronecker delta*. Note that the Kronecker delta δ is symmetric, that is $\delta(i, j) = \delta(j, i)$. For this reason any number of notations can unambiguously be used to express it: $\delta_{i,j} = \delta_{j,i} = \delta_i^j = \delta_j^i = \delta^{i,j} = \delta^{j,i}$. The notation used will depend upon what seems appropriate in its context. When I is replaced by I^n observe that

$$\delta_{j_1,\ldots,j_n}^{i_1,\ldots,i_n} = \delta_{j_1}^{i_1} \cdots \delta_{j_n}^{i_n}.$$

A.2 Algebras, Coalgebras, and Their Representations

We assume that the reader is familiar with the usual notions of *algebras* and their *modules* over k. There are many good references for these notions which can be found in most mathematics department libraries.

Less familiar perhaps are the notions of *coalgebra* and *comodule* and related structures. We will devote Chapter 1 to developing the theory of these and related objects for treatment of the main topics of this text.

Coalgebras and comodules over a field have a local finiteness character which is not usually found in algebras and modules. Comodules give rise to modules which are locally finite. A module M over and algebra A is said to be *locally finite* if all finitely generated submodules of M are finite-dimensional. It is clear that any module M has a unique maximal locally finite submodule M_{f} which is the sum of all finite-dimensional submodules of M.

A.3 Various Notations Related to the QYBE

There are at least three common notation conventions used in the development of the algebraic theory related to the quantum Yang–Baxter equation (QYBE). The first requires a basis for the various linear spaces involved and expresses relations in terms of *structure constants.* A second is referred to as the (modified) Heyneman-Sweedler (H-S) notation. This notation convention has the advantage of easy formal manipulation of symbolic elements for computation. The third, categorical notation, involves the manipulation of arrows and diagrams and does not involve elements.

We have found that use of a particular notation in a given context is quite often a matter of taste. Since one will find that the extensive literature on the QYBE contains many examples of each of these notations it is a good idea to gain some facility with all three. Of course, an approach using structure constants requires a basis, but it does not require that the basis be infinite-dimensional. All summations (occurring in this text) involving structure constants will be finite.

A.3.1 Structure Constants

Suppose that A is an algebra k and let $B = \{a_i\}$ be a basis for A. Then for every i, j we may write

$$a_i a_j = p_{i,j}^k a_k$$

for some unique *scalars* $p_{i,j}^k \in k$. Here we use the summation convention of *summing over a repeated upper and lower index.* Thus, the right hand side of the equation above is actually a sum over all the values of the index k if A is finite-dimensional. If A is infinite-dimensional, we need to assume that the basis is indexed by some cardinal number and then the sum is over some finite subset of the indexing set (defined for each i, and j). Similar remarks apply to structure constants for other infinite-dimensional vector spaces which might be encountered in the text.

Suppose that M is a left A-module with basis $\{m_j\}$. Then the module action has structure constants given by

$$a_i \cdot m_j = \mu_{i,j}^k m_k.$$

Various axioms take the form of identities which the structure constants must satisfy. For example, the associative law for the left module action A on M

$$a \cdot (b \cdot m) = (ab) \cdot m$$

is expressed as

$$p_{i,j}^s \mu_{s,k}^\ell = \mu_{j,k}^s \mu_{i,s}^\ell. \tag{A.1}$$

The simple verification of this and the other identities presented in this section will be left as an exercise.

The linear dual M^* has a right A-module structure (M^*, μ^T) given by

$$<\alpha{\cdot}a, m> = <\alpha, a{\cdot}m>$$

for all $\alpha \in M^*, a \in A$, and $m \in M$. Now suppose that M is finite-dimensional and let $\{m^i\}$ be the basis for M^* dual to $\{m_i\}$. (This means $<m^i, m_j> = \delta^i_j$ for all i, j in the indexing set.) We use the convention

$$m^i{\cdot}a_j = (\mu^T)^i_{j,k} m^k$$

for expressing this action in terms of structure constants. The reader is left with the exercise of relating the $(\mu^T)^i_{j,k}$'s and the $\mu^k_{i,j}$'s.

When M is a right A-module we use the notation convention

$$m_i{\cdot}a_j = \mu^k_{i,j} m_k$$

for describing structure constants and, when M is finite-dimensional we use the convention

$$a_i{\cdot}m^j = (\mu^T)^j_{i,k} m^k$$

to describe structure constants for the left A-module structure (M^*, μ^T) on M^* defined by $<a{\cdot}\alpha, m> = <\alpha, m{\cdot}a>$ for all $a \in A$, $\alpha \in M^*$, and $m \in M$.

Even though we are not going to define coalgebras and comodules in the appendix, we will still look at linear maps associated with them as a source of exercises in notation conventions. Suppose that C is a coalgebra over k with coproduct $C \xrightarrow{\Delta} C \otimes C$ and counit $\epsilon : C \longrightarrow k$. Suppose that $\{c_i\}$ is a basis for C. There are structure constants $\Delta(c_i) = \Delta^{j,k}_i c_j \otimes c_k$ and $\epsilon(c_i) = \epsilon_i$ determining these maps. Let M be a right comodule M over C with comodule structure map $M \xrightarrow{\rho} M \otimes C$. Suppose $\{m_j\}$ is a basis for M. Then the structure constants for ρ are given by

$$\rho(m_i) = \rho^{j,k}_i m_j \otimes c_k.$$

The comodule axiom $(\rho \otimes 1_C)\rho = (1_M \otimes \Delta)\rho$ is expressed as

$$\rho^{s,j}_i \rho^{k,l}_s = \Delta^{l,j}_s \rho^{k,l}_i.$$

We also leave it as an exercise to see that the counit axiom $(\epsilon \otimes 1_C)\Delta = 1_C = (1_C \otimes \epsilon)\Delta$ is expressed as

$$\begin{aligned} \Delta^{j,k}_i \epsilon_j &= \delta^k_i, \\ \Delta^{j,k}_i \epsilon_k &= \delta^j_i. \end{aligned}$$

Suppose M is a vector space over k with basis $B = \{m_i\}$. Let $f : M \longrightarrow M$ be a linear endomorphism of M. We also write $f(m_i) = f^j_i m_j$. If $R : M \otimes M \longrightarrow M \otimes M$ is linear endomorphism of M then we write

$$R(m_i \otimes m_j) = R^{k,\ell}_{i,j} m_k \otimes m_\ell.$$

Definition A.3.1 *The set of scalars $\{R_{i,j}^{k,\ell}\}$ is the set of B-*coordinates *of R.*

This convention is very natural in many ways. In particular if I is the indexing set for $B = \{m_i\}$ then $I \times I$ indexes the basis $\mathcal{B} = \{m_i \otimes m_j\}$ for $M \otimes M$ according to $m_i \otimes m_j = m_{(i,j)}$ and our convention follows the convention for expressing structure constants for linear maps. Of course this notation extends in the obvious way to maps $f : M \longrightarrow N$ and to higher tensor products.

An important algebra is the endomorphism algebra $A = \mathrm{End}(M)$ of linear endomorphisms of M under function composition. If M has basis $\{m_i\}$ then A has basis $\{m_{i,j}\}$ where

$$m_{i,j} = (m_k \longmapsto \delta_{j,k} m_i) \tag{A.2}$$

Recall that M is a left A-module where $a{\cdot}m = a(m)$ for $a \in \mathrm{End}(M)$ and $m \in M$. To get structure constants for this action we need to write (A.2) in the form

$$m_{i,j}{\cdot}m_k = \delta_{j,k} m_i = \delta_i^r \delta_{j,k} m_r.$$

To summarize:

Proposition A.3.1 *Let M be a vector space over k and let $\{m_i\}$ be a basis for M. Let A be the endomorphism algebra of M (i.e. square matrices over k) with basis $\{m_{i,j}\}$ as above. Let M be a module over A by the action $a = (m \longmapsto a{\cdot}m)$, then the structure constants of this action are given by*

$$m_{(i,j),k}^r = \delta_i^r \delta_{j,k}.$$

∎

As is well known, an A-module structure $\mu : M \otimes A \longrightarrow M$ (or $A \otimes M \longrightarrow M$) is equivalent to a representation $A \longrightarrow \mathrm{End}(M)$. The correspondence is given by

$$\mu \longmapsto (a \longmapsto (m \longmapsto a{\cdot}m)),$$

or $\mu \longmapsto \pi_\mu$, where $a{\cdot}m = \mu(a \otimes m) = \pi_\mu(a)(m)$ for $a \in A$ and $m \in M$. Under this correspondence, the standard action of $A = \mathrm{End}(M)$ on M given above is equivalent to the identity map $1_A : A \longrightarrow \mathrm{End}(M)$.

To determine the structure constants for the endomorphism algebra, note that matrix multiplication satisfies

$$m_{i,j} m_{k,l} = \delta_{j,k} m_{i,l}. \tag{A.3}$$

Thus:

Proposition A.3.2 *With the notation from A.3.1, the endomorphism algebra (matrix algebra) has structure constants*

$$m_{(i,j),(k,l)}^{(r,s)} = \delta_i^r \delta_{j,k} \delta_l^s.$$

∎

Remark A.3.1 We will also work with the matrix associated to a given endomorphism when given a basis for M. Clearly, the matrix associated to $m_{i,j}$ above is the "elementary matrix" $e_{i,j}$ with 1 in the (i,j)-th position and 0 elsewhere.

Throughout the following exercises M and N are vector spaces over the field k.

Exercise A.3.1 Let $\{m_i\}$ be a basis for M over k. Show that $\{m^i\}$ is linearly independent and the map

$$\begin{array}{rcl} M^* & \longrightarrow & M \\ \alpha & \longmapsto & \alpha(m^i)m_i \end{array}$$

is an isomorphism. What is its inverse?

Exercise A.3.2 Using the notation from the exercise above, show that

$$\begin{array}{rcl} M & \longrightarrow & M^{**} \\ x & \longmapsto & (\alpha \longmapsto \alpha(x)) \end{array}$$

is injective. When M is finite-dimensional it is an isomorphism. What is its inverse in this case?

Exercise A.3.3 Show that

$$\begin{array}{rcl} M^* \otimes N & \longrightarrow & \mathrm{Hom}(M,N) \\ \alpha \otimes n & \longmapsto & (m \longmapsto \alpha(m)n) \end{array}$$

is injective. If either M or N is finite dimensional this map is an isomorphism. What is its inverse in this case?

Exercise A.3.4 Using the notation from the exercise above, show that

$$\begin{array}{rcl} M^* \otimes N^* & \longrightarrow & (M \otimes N)^* \\ \alpha \otimes \beta & \longmapsto & (m \otimes n \longmapsto \beta(m)\alpha(n)) \end{array}$$

is injective. If M and N are finite dimensional this map is an isomorphism. What is its inverse in this case?

Remark A.3.2 For finite-dimensional vector spaces the isomorphisms given in the above exercises along with the interchange isomorphism

$$\begin{array}{rcl} M \otimes N & \xrightarrow{\tau_{M,N}} & N \otimes M \\ \tau_{M,N}(m \otimes n) & = & n \otimes m \end{array}$$

give rise to a collection of isomorphisms by composition. For example, we now have $M \otimes N^* \simeq \mathrm{Hom}(M, N)$, $M \otimes N \simeq \mathrm{Hom}(M, N) \simeq N \otimes M$, etc. Whenever we refer to an isomorphism involving finite-dimensional vector spaces without explicitly mentioning the map, it will be one of these compositions.

Exercise A.3.5 Let M and N have bases $\{m_i\}$ and $\{n_j\}$ respectively. Let $f^* : N^* \longrightarrow M^*$ be the dual map (linear transpose) which is defined by $<f^*(\alpha), m> = <\alpha, f(m)>$. What are the structure constants (coordinates) for f^*? Note the way the relationship between f and f^* is expressed in this notation.

A.3.2 Heyneman-Sweedler and H-S Notations

The Heyneman-Sweedler notation is a formal representation of sums arising from coalgebra structure maps and more generally comodule structure maps. It is very useful in tracing complex calculations involving composites of linear maps in an intelligible way. The notation arose first as a means of expressing the coproduct of a coalgebra C over k. Part of the structure of C is a linear map

$$C \xrightarrow{\Delta} C \otimes C,$$

which is called the *coproduct* or *comultiplication map.* The Heyneman-Sweedler notation for $\Delta(c)$ is

$$\Delta(c) = \sum c_{(1)} \otimes c_{(2)}.$$

We will follow the physicists lead and drop the summation symbol and write

$$\Delta(c) = c_{(1)} \otimes c_{(2)}.$$

The Heyneman-Sweedler notation, and its variations, written without the summation symbol we will refer to as H-S notation.

The basic ingredient of a right C-comodule M is a linear map

$$M \xrightarrow{\rho} M \otimes C.$$

There are many variations on the Heyneman-Sweedler notation for coalgebras used to express $\rho(m) \in M \otimes C$. We use

$$\rho(m) = \sum m^{<1>} \otimes m^{(2)}$$

which indicates that the first summand lies on M and the second in C. The H-S notation we use to express $\rho(m)$ is therefore

$$\rho(m) = m^{<1>} \otimes m^{(2)}.$$

We express the coassociative axiom $(\rho \otimes 1_C)\rho = (1_M \otimes \Delta)\rho$ in the H-S notation as

$$m^{<1><1>} \otimes m^{<1>(2)} \otimes m^{(2)} = m^{<1>} \otimes {m^{(2)}}_{(1)} \otimes {m^{(2)}}_{(2)}.$$

On occasion we extend notation to certain operators as well. An important example: if $R : M \otimes M \longrightarrow M \otimes M$ is a linear operator we sometimes write

$$R(m \otimes n) = m_{[1]} \otimes n_{[2]}.$$

A.3.3 Categorical Notation

One notation which does not make use of specific elements in a given object is at times very useful [Yetter, 1990]. The foundations for this approach to notation can be found in [MacLane, 1988]. If A is an algebra with structure map (product) given by $A \otimes A \xrightarrow{p} A$, then the associative law can be expressed by saying the diagram

$$\begin{array}{ccc} A \otimes A \otimes A & \xrightarrow{1_A \otimes p} & A \otimes A \\ {\scriptstyle p \otimes 1_A}\downarrow & & \downarrow{\scriptstyle p} \\ A & \xrightarrow[p]{} & A \otimes A \end{array}$$

commutes, that is

$$p(p \otimes 1_A) = p(1_A \otimes p).$$

In the following exercises, assume that A is an algebra with unit over k, M is a left module over A, C is a coalgebra over k, and N is a right comodule over C. Also assume that the structure maps are as follows.

$$A \otimes A \xrightarrow{p} A, \quad C \xrightarrow{\Delta} C \otimes C$$

$$A \otimes M \xrightarrow{\mu} M, \quad M \xrightarrow{\rho} M \otimes C$$

Exercise A.3.6 Show that the associative law $a{\cdot}(b{\cdot}m) = (ab){\cdot}m$ for a left module action is equivalent to

$$\mu(\mu \otimes 1_M) = p(1_A \otimes \mu)$$

and write out the appropriate diagram. Show that when a basis has been chosen, so that structure constants can be defined, these identities are equivalent to

$$\mu^{s_1}_{j,k}\mu^{r}_{i,s_1} = p^{s_1}_{i,j}\mu^{r}_{s_1,k}.$$

Note that when $M = A$ and the action of A on itself is given by multiplication, this is just the associative law for multiplication.

Exercise A.3.7 Show that the coassociative law for comodule coactions

$$(\rho \otimes 1_C)\rho = (1_N \otimes \Delta)\rho$$

may be expressed as

$$\rho_i^{s_1,j}\rho_{s_1}^{k,l} = \rho_i^{k,s_1}\Delta_{s_1}^{i,j}$$

in terms of structure constants.

A.4 Some Results from Linear Algebra

The discussion of coalgebras and their dual algebras draws from basic linear algebra of the space of functionals on a vector space over a field k. Some of these results can be found in [Heyneman and Radford, 1974]. The rank of tensors is a very useful measure for us in various contexts, particularly in the study of comodules. We begin this section with a discussion of rank.

A.4.1 Rank of Tensors and Endomorphisms

Let U and V be vector spaces over the field k and suppose that $\nu \in U \otimes V$ is not zero. Write $\nu = \sum_{i=1}^{r} u_i \otimes v_i$ where $u_i \in U$ and $v_j \in V$ for $1 \leq i \leq r$.

Definition A.4.1 *The minimal r which arises in all such representations of ν is called the* rank *of ν and is denoted by* Rankν.

We set Rank$0 = 0$.

Lemma A.4.1 *Suppose that U and V are vector spaces over the field k and $0 \neq \nu \in U \otimes V$. Write $\nu = \sum_{i=1}^{r} u_i \otimes v_i$ where $u_i \in U$ and $v_i \in V$. Then the following are equivalent:*

a) $r = \text{Rank}\nu$.

b) $\{u_i\}_{1\leq i\leq r}$ *and* $\{v_i\}_{1\leq i\leq r}$ *are linearly independent.*

Proof: To show that part a) implies part b), we first suppose that some one of the u_i's is linearly dependent on the others. Replacing this particular u_i in the sum representing ν by its realization as a linear combination of the other u_i's we can express ν as a sum of less than r tensors in $U \otimes V$. This contradicts the minimality of r. Thus $\{u_1, \ldots, u_r\}$ is a linearly independent set. The argument for the v_i's is the same.

To show that part b) implies part a) we write $\nu = \sum_{\ell=1}^{\text{Rank}\nu} u'_\ell \otimes v'_\ell$. Assume the hypothesis of part a). Since part a) implies part b) the sets $\{u'_\ell\}_{1\leq \ell\leq \text{Rank}\nu}$ and $\{v'_\ell\}_{1\leq \ell\leq \text{Rank}\nu}$ are linearly independent. Applying $f \in U^*$ to both sides of the equation

$$\sum_{\ell=1}^{\text{Rank}\nu} u'_\ell \otimes v'_\ell = \sum_{i=1}^{r} u_i \otimes v_i$$

we obtain

$$\sum_{\ell=1}^{\mathrm{Rank}\nu} f(u'_\ell)v'_\ell = \sum_{i=1}^{r} f(u_i)v_i.$$

Now a linearly independent subset S of a vector space V can be separated from any given nonzero $s \notin S$ by a functional $f \in U^*$ which takes the value 1 on s and vanishes on S. We conclude, therefore, that the v'_ℓ's and the v_i's have the same span. Since these two sets are linearly independent, it follows that $r = \mathrm{Rank}\nu$. Thus part b) implies part a), and the proof is complete. ∎

The notion of rank can be extended to the tensor product of any finite number of vector spaces *provided* the convention for fully parenthesizing the tensor product is set. For example, let V be a 2-dimensional vector space over k with basis $\{u, v\}$ and consider $\nu = u \otimes v \otimes u + v \otimes u \otimes u \in V \otimes V \otimes V$. Regarding $V \otimes V \otimes V$ as $(V \otimes V) \otimes V$ we compute the rank of $\nu = (u \otimes v + v \otimes u) \otimes u$ as 1, whereas regarding $V \otimes V \otimes V$ as $V \otimes (V \otimes V)$ we compute the rank of $\nu = u \otimes (v \otimes u) + v \otimes (u \otimes u)$ as 2.

For vector spaces U and V over k consider the linear map

$$U^* \otimes V \xrightarrow{\pi} \mathrm{Hom}(U, V)$$

defined by $\pi(u^* \otimes v)(u) = u^*(u)v$. Suppose that $\nu \in U^* \otimes V$ is not zero and write $\nu = \sum_{i=1}^{r} u_i^* \otimes v_i$. Note that $\mathrm{Im}\pi(\nu)$ is in the span of $\{v_1, \ldots, v_r\}$. Therefore

$$\mathrm{Rank}\nu \geq \mathrm{rank}\pi(\nu).$$

If $r = \mathrm{Rank}\pi(\nu)$ then the left tensorands and the right tensorands of this sum form linearly independent sets by Lemma A.4.1. Therefore $\pi(\nu) \neq 0$. Consequently π is one-one.

Now the linear map $\pi(\nu)$ has finite rank. Let $m = \mathrm{rank}\pi(\nu)$ denote the dimension of $\mathrm{Im}\pi(\nu)$. Choose a basis $\{v_1, \ldots, v_m\}$ for $\mathrm{Im}\pi(\nu)$. Then there are $f_1, \ldots, f_m \in U^*$ such that

$$\pi(\nu)(u) = f_1(u)v_1 + \cdots + f_m(u)v_m$$

for all $u \in U$. Since π is one-one we conclude that

$$\pi(\nu) = \sum_{\ell=1}^{m} f_\ell \otimes v_\ell$$

Therefore $\mathrm{Rank}\nu \leq \mathrm{rank}\pi(\nu)$. We now conclude that $\mathrm{Rank}\nu = \mathrm{rank}\pi(\nu)$ which means that $\{f_1, \ldots, f_m\}$ is linearly independent by Lemma A.4.1. We have proved the following:

Proposition A.4.1 *Let U and V be vector spaces over the field k and $U^* \otimes V \xrightarrow{\pi} \mathrm{Hom}(U, V)$ be the map defined by $\pi(u^* \otimes v)(u) = u^*(u)v$. Then:*

a) *π is one-one.*

b) *Suppose that $\nu \in U^* \otimes V$ is not zero and write $\nu = \sum_{i=1}^{\mathrm{Rank}\nu} u_i^* \otimes v_i$. Then $\mathrm{Rank}\nu = \mathrm{rank}\pi(\nu)$ and $\{v_1, \ldots, v_{\mathrm{Rank}\nu}\}$ is a basis for $\mathrm{Im}\pi$.*

∎

As a consequence to the proposition we have the following separation result for finite independent sets of functionals.

Corollary A.4.1 *Let $\{u_1^*, \ldots, u_r^*\}$ be a linearly independent subset of U^* where U is a vector space over the field k. If $\alpha_1, \ldots, \alpha_r \in k$ there is a $u \in U$ such that $u_i^*(u) = \alpha_i$ for $1 \leq i \leq r$.*

Proof: We apply the previous proposition to $U, V = U^*$ and $\nu = \sum_{i=1}^{r} u_i^* \otimes u_i^*$. By Lemma A.4.1 we have that $r = \mathrm{Rank}\nu$. By the previous proposition $\{u_1^*, \ldots, u_r^*\}$ is a basis for $\mathrm{Im}\pi(\nu)$. Therefore $\pi(\nu)(u) = \alpha_1 u_1^* + \cdots + \alpha_r u_r^*$ for some $u \in U$ which gives the result. ∎

Suppose that U is a finite-dimensional vector space over the field k and let $\pi : U^* \otimes U \longrightarrow \mathrm{End}(U)$ be the linear map described in Proposition A.4.1. Since U is finite-dimensional it follows that π is an isomorphism. The evaluation map $ev : U^* \otimes U \longrightarrow k$ defined by $ev(u^* \otimes u) = u^*(u)$ corresponds to the trace map $\mathrm{Tr} : \mathrm{End}(U) \longrightarrow k$. Since the trace map is linear it suffices to show that $\mathrm{Tr}\pi(u^* \otimes u) = u^*(u)$ for non-zero $u^* \in U^*$ and $u \in U$. This is established by extending $\{u\}$ to a basis for U and computing the trace of the resulting matrix representing $\pi(u^* \otimes u)$. Notice in particular that

$$\mathrm{Tr}\pi(\nu) = \sum_{i=1}^{\mathrm{Rank}\nu} u_i^*(u_i)$$

where $\nu = \sum_{i=1}^{\mathrm{Rank}\nu} u_i^* \otimes u_i$.

Throughout the following exercises U and V are vector spaces over the field k.

Exercise A.4.1 Apropos of the proof of Lemma A.4.1, show that the the fact that $\{v_1, \ldots, v_r\}$ is linearly independent *follows* from the fact that $\{u_1, \cdots, u_r\}$ is linearly independent. [Hint: Use the isomorphism $U \otimes V \simeq V \otimes U$ described above.]

Exercise A.4.2 Show that the map $\pi : U^* \otimes V \longrightarrow \mathrm{Hom}(U, V)$ defined by $\pi(u^* \otimes v)(u) = u^*(u)v$ induces a linear isomorphism between $U^* \otimes V$ and $\mathrm{Hom}_f(U, V)$, where the latter is the subspace of $\mathrm{Hom}(U, V)$ consisting of all linear maps $f : U \longrightarrow V$ which have finite rank.

Exercise A.4.3 Let $V_1, \ldots, V_n$ be finite-dimensional vector spaces over the field k. Show that the linear map $\pi : V_1^* \otimes \cdots \otimes V_n^* \longrightarrow (V_1 \otimes \cdots \otimes V_n)^*$ defined by $\pi(v_1^* \otimes \cdots \otimes v_n^*)(v_1 \otimes \cdots \otimes v_n) = v_1^*(v_1) \cdots v_n^*(v_n)$ is an isomorphism.

Exercise A.4.4 Let $0 \neq \nu \in U \otimes V$. Define subspaces L_ν and R_ν of U and V respectively by

$$L_\nu = (1_U \otimes V^*)(\nu) \quad \text{and} \qquad R_\nu = (U^* \otimes 1_V)(\nu).$$

a) Suppose that $\nu = \sum_{i=1}^{\mathrm{Rank}\nu} u_i \otimes v_i$. Show that $\{u_1, \ldots, u_{\mathrm{Rank}\nu}\}$ is a basis for L_ν and that $\{v_1, \ldots, v_{\mathrm{Rank}\nu}\}$ is a basis for R_ν.

b) Show that $\mathrm{Dim}L_\nu = \mathrm{Rank}\nu = \mathrm{Dim}R_\nu$.

c) Suppose that $\{u_1', \ldots, u_{\mathrm{Rank}\nu}'\}$ is a basis for L_ν. Show that $\nu = \sum_{i=1}^{\mathrm{Rank}\nu} u_i' \otimes x_i$ has unique solutions $x_1, \ldots, x_{\mathrm{Rank}\nu} \in U$ and that $\{x_1, \ldots, x_{\mathrm{Rank}\nu}\}$ is a basis for R_ν.

d) Show that $\nu \in L_\nu \otimes R_\nu$ and if $\mathcal{U}$ and $\mathcal{V}$ are subspaces of U and V respectively such that $\nu \in \mathcal{U} \otimes \mathcal{V}$ then $L_\nu \subseteq \mathcal{U}$ and $R_\nu \subseteq \mathcal{V}$.

e) Suppose that $\{\mathcal{U}_i\}_{i \in I}$ and $\{\mathcal{V}_i\}_{i \in I}$ are families of subspaces of U and V respectively indexed by a set I. Show that $\cap_{i \in I}(\mathcal{U}_i \otimes \mathcal{V}_i) = (\cap_{i \in I}\mathcal{U}_i) \otimes (\cap_{i \in I}\mathcal{V}_i)$.

f) Suppose that $\mathcal{U}$ and $\mathcal{V}$ are nonzero subspaces of U and V respectively of the same positive dimension. Show that $\mathcal{U} = L_\nu$ and $\mathcal{V} = R_\nu$ for some $\nu \in U \otimes V$.

Exercise A.4.5 Suppose that V is finite-dimensional. If $\{v_1, \ldots, v_n\}$ is a basis for V we let $\{v^1, \ldots, v^n\}$ denote the corresponding "dual" basis for V^*. (This means that $v^i(v_j) = \delta^i_j$ for $1 \leq i, j \leq n$.)

a) If $\{u_1, \cdots, u_n\}$ and $\{v_1, \ldots, v_n\}$ are bases for V show that

$$\sum_{i=1}^{n} u^i \otimes u_i = \sum_{i=1}^{n} v^i \otimes v_i.$$

(The sum $\sum_{i=1}^{n} u^i \otimes u_i$ is called the *canonical element* of $V^* \otimes V$.)

b) Suppose that $\{v_1, \ldots, v_n\}$ is a basis for V and $T \in \mathrm{End}(V)$. Show that

$$\mathrm{Det}T = \sum_{\sigma \in S_n} \mathrm{sgn}(\sigma)\pi(v^{\sigma 1} \otimes \cdots \otimes v^{\sigma n})(T(v_1) \otimes \cdots \otimes T(v_n)),$$

where π is the linear map of Exercise A.4.3.

A.4.2 Closed Subspaces of U^*

Let U be a vector space over the field k. For a subset $X \subseteq U$ we define $X^\perp \subseteq U^*$ by

$$X^\perp = \{u^* \in U^* \mid u^*(X) = (0)\}$$

and for a subset $Y \subseteq U^*$ we define $Y^\perp \subseteq U$ by

$$Y^\perp = \{u \in U \mid Y(u) = (0)\}.$$

Observe that $X^\perp$ is a subspace of U^* and that $Y^\perp$ is a subspace of U.

There is an asymmetry in the definition of $X^\perp$ and $Y^\perp$. Consider the one-one map $\iota : U \longrightarrow U^{**}$ defined by $\iota(u)(u^*) = u^*(u)$. Note that $Y^\perp = Y^{\perp'} \cap U$, where we identify U with its image under ι in U^{**} and $\perp'$ is defined for $\mathcal{U} = U^*$ and $\mathcal{U}^*$ as $\perp$ is defined for the pair U and U^*.

Lemma A.4.2 *Suppose that U is a vector space over the field k. Let X and Y both be a subspace of U or both be subspaces of U^*. Then:*

a) *If $Y \subseteq X$ then $Y^\perp \supseteq X^\perp$.*

b) $X \subseteq X^{\perp\perp}$.

c) $X^\perp = X^{\perp\perp\perp}$.

Proof: Parts a) and b) are straightforward applications of the definition. To show part c) we first note that $X \subseteq X^{\perp\perp}$ by part b). Therefore $X^\perp \supseteq X^{\perp\perp\perp}$ by part a). But $X^\perp \subseteq X^{\perp\perp\perp}$ by part b) again. This completes the proof of the lemma. ∎

Definition A.4.2 *A subspace X of U or U^* is* closed *if $X = X^{\perp\perp}$.*

Suppose $X \subseteq U$ and $u \in U \backslash X$. Then there is a functional $f \in U^*$ which vanishes on X but does not vanish on u. Therefore *all* subspaces of U are closed. By the previous lemma the closed subspaces of U^* are the subspaces $X^\perp$ of U^* where X runs over the subspaces of U. Furthermore $X \mapsto X^\perp$ describes a bijective inclusion reversing correspondence between the subspaces of U and the closed subspaces of U^*.

Let X and Y be subspaces of U^* where Y is closed and $X \subseteq Y$. Then $X^{\perp\perp} \subseteq Y^{\perp\perp} = Y$. Therefore $X^{\perp\perp} \subseteq Y$. Thus among the closed subsets which contain X there is a unique minimal one which is $X^{\perp\perp}$.

Definition A.4.3 *The subspace $X^{\perp\perp}$ is the* closure *of X, and is denoted by $\overline{X}$.*

By Lemma A.4.2 the closure operation is inclusion preserving and idempotent – that is if $X \subseteq Y$ then $\overline{X} \subseteq \overline{Y}$ and $\overline{(\overline{X})} = \overline{X}$.

Proposition A.4.2 *Suppose that U is a vector space over the field k. Then all subspaces of U^* are closed if and only if U is finite-dimensional.*

Proof: Suppose that U is finite-dimensional. Then the map $U \xrightarrow{\iota} U^{**}$ defined by $\iota(u)(u^*) = u^*(u)$ is an isomorphism. Consequently the notion of closed for U and U^* is the same that for $\mathcal{U} = U^*$ and $\mathcal{U}^* \simeq U$. For any vector space V over k we have noted for the pair V and V^* that all subspaces of V are closed. Therefore all subspaces of $U^* = \mathcal{U}$ are closed.

Now assume that all subspaces of U^* are closed. Choose a basis for U and let $Y \subseteq U^*$ be the subspace of all functionals which vanish on all but finitely many of the basis vectors. Then $Y^\perp = (0)$, so $Y = Y^{\perp\perp} = U^*$. This means that U must be finite-dimensional, and the proposition is proved. ▮

What it means for two subspaces of U^* to have the same closure translates to a very useful "local" condition. For a subspace $X \subseteq U^*$ and a subspace $V \subseteq U$ let $X|_V$ denote the set of restrictions of all $f \in X$ to V. Observe that $X|_V$ is a subspace of V^*.

Proposition A.4.3 *Suppose that U is a vector space over the field k and let X and Y be subspaces of U^*. Then the following are equivalent:*

a) $\overline{X} = \overline{Y}$.

b) $X|_V = Y|_V$ *for all finite-dimensional subspaces V of U.*

Proof: Suppose that $\overline{X} = \overline{Y}$. Then $X^\perp = Y^\perp$ by part c) of Lemma A.4.2. Now let V be a finite-dimensional subspace of U. Since $X^\perp = Y^\perp$ we conclude that $(X|_V)^\perp = (Y|_V)^\perp$. Since V is finite-dimensional, all subspaces of V^* are closed by Proposition A.4.2. Therefore $X|_V = Y|_V$.

Now assume that $X|_V = Y|_V$ for all finite-dimensional subspaces V of U. By symmetry we need only show that $X \subseteq \overline{Y}$. Let $f \in X$ and $v \in Y^\perp$. Set $V = kv$. Then $g|_V = f|_V$ for some $g \in Y$ by assumption. Therefore $f(v) = g(v) = 0$. By definition $f \in Y^{\perp\perp} = \overline{Y}$. This concludes the proof of the proposition. ▮

Definition A.4.4 *Let X and Y be subspaces of U^*. Then Y is a* dense *subspace of X if $Y \subseteq X \subseteq \overline{Y}$.*

If Y is a subspace of X, observe that Y is a dense subspace of X if and only if $\overline{Y} = \overline{X}$ if and only if $Y^\perp = X^\perp$.

We note that Y is a dense subspace of U^* if and only if $Y^\perp = (0)$. To say that Y is a dense subspace of U^* is to say that for any finite-dimensional subspace V of U and $f \in U^*$ there exists a $g \in Y$ such that $g|_V = f|_V$ by Proposition A.4.3.

The subspace constructed in the proof of Proposition A.4.2 is a dense subspace of U^*. We next examine how the operation "$\perp$" relates to intersections and sums of subspaces of U^*.

Proposition A.4.4 *Suppose that U is a vector space over the field k and that $\{V_i\}_{i\in I}$ is a family of subspaces of U. Then:*

a) $(\sum_i V_i)^\perp = \cap_i V_i^\perp$. *Thus the intersection of any family of closed subspaces of U^* is closed.*

b) $\sum_i V_i^\perp = (\cap_i V_i)^\perp$ *provided that* I *is a finite indexing set. Thus the sum of a finite number of closed subspaces of* U^* *is closed.*

Proof: To show part a) we first observe that $\cap_i V_i^\perp \subseteq (\sum_i V_i)^\perp$ by definition. Now $V_j \subseteq \sum_i V_i$ means that $V_j^\perp \supseteq (\sum_i V_i)^\perp$ by part a) of Lemma A.4.2. Therefore $\cap_i V_i^\perp \subseteq (\sum_i V_i)^\perp$ and part a) is established.

We next show part b). Since $V_j \supseteq \cap_i V_i$ means that $V_j^\perp \subseteq (\cap_i V_i)^\perp$ it follows that $\sum_i V_i^\perp \subseteq (\cap_i V_i)^\perp$ in any case.

Conversely, since we are assuming that I is finite, we may suppose that $I = \{1, \ldots, r\}$ for some $r > 0$. Let $\pi_i : U \longrightarrow U/V_i$ be the projection and consider the map $U \xrightarrow{\pi} U/V_1 \oplus \cdots \oplus U/V_r$ defined by $\pi(u) = \pi_1(u) \oplus \cdots \oplus \pi_r(u)$ for $u \in U$. The transpose of the induced inclusion $U/\mathrm{Ker}\pi \longrightarrow U/V_1 \oplus \cdots \oplus U/V_r$ is an onto map $(U/V_1)^* \oplus \cdots \oplus (U/V_r)^* \longrightarrow (U/\mathrm{Ker}\pi)^*$. Now Let V be any subspace of U and $\rho : U \longrightarrow U/V$ be the projection. Then ρ^* is one-one and $\mathrm{Im}\rho^* = V^\perp$. Therefore we have a map $V_1^\perp \oplus \cdots \oplus V_r^\perp \longrightarrow (\mathrm{Ker}\pi)^\perp$ given by $f_1 \oplus \cdots \oplus f_r \longmapsto f$, where $f(u) = f_1(u) + \cdots + f_r(u)$ for $u \in U$. As $\mathrm{Ker}\pi = \cap_{i=1}^r V_i$ we have that $(\cap_{i=1}^r V_i)^\perp \subseteq V_1^\perp + \cdots + V_r^\perp$. This concludes the proof of the proposition. ▮

Part b) of the preceding proposition may very well fail to be true when the indexing set I is not finite.

Let U and V be vector spaces over the field k, and consider the linear map $U^* \otimes V^* \xrightarrow{i} (U \otimes V)^*$ defined by $i(u^* \otimes v^*)(u \otimes v) = u^*(u)v^*(v)$. Then i is one-one. This is easily seen from Lemma A.4.1 and Corollary A.4.1. Under this identification we notice that $U^* \otimes V^*$ is a dense subspace of $(U \otimes V)^*$.

Lemma A.4.3 *Suppose that* U *and* V *are vector spaces over the field* k *and let* $X \subseteq U^*$ *and* $Y \subseteq V^*$ *be subspaces. Regard* $X \otimes Y$ *as a subspace of* $(U \otimes V)^*$. *Then* $(X \otimes Y)^\perp = X^\perp \otimes V + U \otimes Y^\perp$.

Proof: Let $\nu \in (X \otimes Y)^\perp$. Then we may write $\nu = \sum_{i=1}^r u_i \otimes v_i + \sum_{j=1}^s u'_j \otimes v'_j$ where $u_1, \ldots, u_r \in X^\perp$ and $\{u'_1, \ldots, u'_s\}$ is linearly independent modulo $X^\perp$. For a fixed j_0 choose an $f \in U^*$ which vanishes on $X^\perp$ and all of the u'_j's except for u'_{j_0} on which f takes the value 1. By definition $f \in \overline{X}$. By Proposition A.4.3 there exists a $g \in X$ which agrees with f on the u_i's and the u'_j's. Thus for $h \in Y$ we compute $0 = (g \otimes h)(\nu) = h(v'_{j_0})$ which means that $v'_{j_0} \in Y^\perp$. We have shown that $\nu \in X^\perp \otimes V + U \otimes Y^\perp$. Therefore $(X \otimes Y)^\perp \subseteq X^\perp \otimes V + U \otimes Y^\perp$. Since the other inclusion follows by definition, the lemma is proved. ▮

Proposition A.4.5 *Suppose that* U *and* V *are vector spaces over the field* k *and let* $f : U \longrightarrow V$ *be linear. Let* $\mathcal{U}$ *be a subspace of* U *and* $\mathcal{V}$ *be a subspace of* V. *If* $F = f^*$ *is the transpose of* f *then:*

a) $F(\mathcal{V}^\perp) = (f^{-1}(\mathcal{V}))^\perp$. *Thus* F *takes a closed subset of* V^* *to a closed subset of* U^*.

b) $F^{-1}(\mathcal{U}^{\perp}) = f(\mathcal{U})^{\perp}$. *Thus the preimage of a closed subset of U^* under F is a closed subset of V^*.*

c) *If I is a subspace of V^* and J is a subspace of U^* then $F(I) \subseteq J$ implies that $f(J^{\perp}) \subseteq I^{\perp}$.*

Proof: We first show part a). Consider the commutative diagram

$$\begin{array}{ccc} U & \xrightarrow{\ f\ } & V \\ \downarrow{\scriptstyle \pi} & & \downarrow{\scriptstyle \pi'} \\ U/f^{-1}(\mathcal{V}) & \xrightarrow[\ \overline{f}\ ]{} & V/\mathcal{V} \end{array}$$

where π and π' are the projections. Notice that $\overline{f}$ is one-one. Taking transposes yields the commutative diagram

$$\begin{array}{ccc} U^* & \xleftarrow{\ F\ } & V^* \\ {\scriptstyle \pi^*}\uparrow & & \uparrow{\scriptstyle \pi'^*} \\ (U/f^{-1}(\mathcal{V}))^* & \xleftarrow{\ \overline{F}\ } & (V/\mathcal{V})^* \end{array}$$

where $\overline{F}$ is onto. Thus $\mathrm{Im}\pi^* = \mathrm{Im}F\pi'^*$ and $F(\mathcal{V}^{\perp}) = f^{-1}(\mathcal{V})^{\perp}$ follows.

To show part b) we let $v^* \in V^*$ and observe that $F(v^*) \in \mathcal{U}^{\perp}$ if and only if $v^*(f(u)) = F(v^*)(u) = 0$ for all $u \in \mathcal{U}$ if and only if $v^* \in f(\mathcal{U})^{\perp}$. To show part c) we note for $v^* \in I$ and $u \in J^{\perp}$ that

$$v^*(f(u)) = F(v^*)(u) \in F(I)(u) \in J(u) = (0).$$

Therefore $f(u) \in I^{\perp}$ by definition. ∎

We end this section with two other useful observations on transposes and closed sets.

Corollary A.4.2 *Suppose that U and V are vector spaces over the field k and that $U \xrightarrow{f} V$ is linear. Let $F = f^*$ be the transpose of f. Then for a subspace I of V^**

a) $F(\overline{I}) = \overline{F(I)}$, *and*

b) $F(I)^{\perp} = f^{-1}(I^{\perp})$.

Proof: $F(\overline{I})$ is closed by part a) of the preceding proposition. Therefore $F(I) \subseteq F(\overline{I})$ means that $\overline{F(I)} \subseteq F(\overline{I})$. Now $I \subseteq F^{-1}(\overline{F(I)})$ means that $\overline{I} \subseteq F^{-1}(\overline{F(I)})$ by part b) of the preceding proposition. Therefore $F(\overline{I}) \subseteq \overline{F(I)}$, and part a) follows.

We use part a) of the corollary and part a) of the preceding proposition to calculate that

$$F(I)^{\perp} = \overline{F(I)}^{\perp} = F(\overline{I})^{\perp} = F(I^{\perp\perp})^{\perp} = f^{-1}(I^{\perp}).$$

∎

Throughout the following exercises U and V are vector spaces over the field k.

Exercise A.4.6 Show that all finite-dimensional subspaces of U^* are closed.

Exercise A.4.7 Show that $\iota : U^* \otimes V^* \longrightarrow (U \otimes V)^*$ defined by $\iota(u^* \otimes v^*)(u \otimes v) = u^*(u)v^*(v)$ is an isomorphism if and only if either U or V is finite-dimensional.

Exercise A.4.8 Apropos of part b) of Proposition A.4.4, show that the sum of closed subspaces of U^* may not be closed.

Exercise A.4.9 Suppose that $f : U \longrightarrow V$ is linear and let $F : V^* \longrightarrow U^*$ be the transpose of f. Let X and Y be subspaces of V^*. Show that if X is a dense subspace of Y then $F(X)$ is a dense subspace of $F(Y)$.

Exercise A.4.10 Suppose that $V_1, \ldots, V_n$ are vector spaces over the field k and $I_1, \ldots, I_n$ are dense subspaces of $V_1, \ldots, V_n$ respectively. Let $\nu, \nu' \in V_1 \otimes \cdots \otimes V_n$. Show that $\nu = \nu'$ if and only if

$$(f_1 \otimes \cdots \otimes f_n)(\nu) = (f_1 \otimes \cdots \otimes f_n)(\nu')$$

for all $f_i \in I_i$, where $1 \leq i \leq n$. [Hint: Use the identification $V_1 \otimes \cdots \otimes V_n = V_1 \otimes (V_2 \otimes \cdots \otimes V_n)$ and write $\nu - \nu' = \sum_{i=1}^{r} u_i \otimes w_i$ where $u_i \in V_1$ and $w_i \in V_2 \otimes \cdots \otimes V_n$.]

A.4.3 Cofinite Subspaces and Continuous Linear Maps

Suppose that $\mathcal{U}$ is a subspace of U.

Definition A.4.5 *The* codimension *of* $\mathcal{U}$ *in* U, *denoted by* $\mathrm{coDim}\mathcal{U}$, *is* $\mathrm{Dim}(U/\mathcal{U})$.

Definition A.4.6 *The subspace* $\mathcal{U}$ *of* U *is a* cofinite subspace *of* U *if* $\mathrm{coDim}\mathcal{U}$ *is finite.*

Therefore $\mathcal{U}$ is a cofinite subspace of U if and only if there is a finite dimensional subspace K of U such that $\mathcal{U} + K = U$. Consequently if $\mathcal{U}$ is a cofinite subspace of U then any subspace $\mathcal{V}$ of U containing $\mathcal{U}$ is a cofinite subspace of U as well.

Suppose that $f : C \longrightarrow D$ is linear. If $\mathcal{U}$ and $\mathcal{V}$ are subspaces of U and $\mathcal{U}$ is a cofinite subspace of $\mathcal{V}$, observe that $f(\mathcal{U})$ is a cofinite subspace of $f(\mathcal{V})$. If $\mathcal{U}$ and $\mathcal{V}$ are subspaces of V on the other hand and $\mathcal{U}$ is a cofinite subspace of $\mathcal{V}$, then $f^{-1}(\mathcal{U})$ is a cofinite subspace of $f^{-1}(\mathcal{V})$.

Let $\mathcal{U}$ be a subspace of U and $\iota : \mathcal{U} \longrightarrow U$ be the inclusion map. The restriction map $U^* \longrightarrow \mathcal{U}^*$ is defined by $\mathrm{Res}|^U_{\mathcal{U}}(u^*) = u^*|_{\mathcal{U}}$. Observe that $\mathrm{Res}|^U_{\mathcal{U}} = \iota^*$ and is onto.

Since $\mathrm{KerRes}|_{\mathcal{U}}^{U} = \mathcal{U}^{\perp}$ there is an induced isomorphism $U/\mathcal{U}^{\perp} \simeq \mathcal{U}^*$. In particular $\mathcal{U}^{\perp}$ is a cofinite subspace of U^* if and only of $\mathcal{U}$ is a finite-dimensional subspace of U.

Lemma A.4.4 *Suppose that U is a vector space over the field k and I and J are subspaces of U^* such that $I \subseteq J$. If I is a cofinite closed subspace of U^* then J is a cofinite closed subspace of U^* as well.*

Proof: We have noted that J is a cofinite subspace of U^* whenever I is a cofinite subspace of U^*. Suppose that I is a closed cofinite subspace of U^*. Then $I = \mathcal{U}^{\perp}$ for some finite-dimensional subspace $\mathcal{U}$ of U^*. Let $\iota : \mathcal{U} \longrightarrow U$ be the inclusion map and $F = \mathrm{Res}|_{\mathcal{U}}^{U}$. Then $F = \iota^*$. Since $J \supseteq I = \mathrm{Ker}F$ we have that $J = F^{-1}(F(J))$. But $F(J)$ is a closed subspace of $\mathcal{U}^*$ since $\mathcal{U}$ is finite-dimensional by Proposition A.4.2. By part b) of Proposition A.4.5 J is closed. ∎

Proposition A.4.6 *Suppose that U is a vector space over the field k and let $U \xrightarrow{\iota} U^{**}$ be the one-one linear map defined by $\iota(u)(u^*) = u^*(u)$. Let $f \in U^{**}$. Then the following are equivalent:*

a) $f \in \mathrm{Im}\iota$.

b) $f(I) = (0)$ *for some closed cofinite subspace of U^*.*

Proof: If $f = \iota(u)$ for some $u \in U$ then $f(I) = (0)$ where $I = (ku)^{\perp}$. Since I is a closed subspace of U^*, part a) implies part b). Conversely, suppose that $f(I) = (0)$ for some closed cofinite subspace I of U^*. Then $\mathrm{Ker}f$ is a closed cofinite subspace $\mathcal{U}^{\perp}$ of U^* by Lemma A.4.4. Since $\mathcal{U} = ku$ for some $u \in U$ it follows that f and $\iota(u)$ have the same kernel. Therefore $f = \alpha\iota(u) = \iota(\alpha u)$ for some $\alpha \in k$. Part b) implies part a) and the proposition is proved. ∎

There is a natural characterization of maps of dual spaces which are transposes in terms of closed sets.

Proposition A.4.7 *Suppose that U and V are vector spaces over the field k and that $F : V^* \longrightarrow U^*$ is a linear map. Then the following are equivalent:*

a) $F = f^*$ *for some linear map $f : U \longrightarrow V$.*

b) *If I is a closed subspace of V^* then $F^{-1}(I)$ is a closed subspace of U^*.*

c) *If I is a cofinite closed subspace of V^* then $F^{-1}(I)$ is a closed subspace of U^*.*

Proof: Part a) implies part b) by part b) of Proposition A.4.5. Part b) clearly implies part c). To complete the proof we will show that part c) implies part a).

Assume that the hypothesis of part c) holds. Let $\mathcal{U}$ be a finite-dimensional subspace of U and let $I = \mathcal{U}^{\perp}$. Then I is a closed cofinite subspace of U^*. Therefore

$F^{-1}(I) = \mathcal{V}^{\perp}$ for some subspace $\mathcal{V}$ of V by assumption. Now $\mathcal{V}$ must be finite-dimensional since $\mathcal{V}^{\perp}$ is a cofinite subspace of V^*.

We next observe that F induces a map of quotient spaces

$$V^*/\mathcal{V}^{\perp} \xrightarrow{F_{\mathcal{U}}} U^*/\mathcal{U}^{\perp}.$$

Consider the composite

$$\mathcal{V}^* \simeq V^*/\mathcal{V}^{\perp} \xrightarrow{F_{\mathcal{U}}} U^*/\mathcal{U}^{\perp} \simeq \mathcal{U}^*$$

where the isomorphisms on the ends are induced by the restriction maps $\mathrm{Res}|^V_{\mathcal{V}}$ and $\mathrm{Res}|^U_{\mathcal{U}}$ respectively. Since $\mathcal{U}$ and $\mathcal{V}$ are finite-dimensional, this composite is $f^*_{\mathcal{U}}$ for some linear map $\mathcal{U} \xrightarrow{f_{\mathcal{U}}} \mathcal{V}$. Notice that $\mathrm{Res}|^U_{\mathcal{U}} F = f^*_{\mathcal{U}} \mathrm{Res}|^V_{\mathcal{V}}$. Thus

$$F(v^*)(u) = v^*(f_{\mathcal{U}}(u))$$

for all $v^* \in V^*$ and $u \in \mathcal{U}$. From this last equation we see that if $\mathcal{U} \subseteq \mathcal{U}'$ and $\mathcal{U}'$ is also a finite-dimensional subspace of U then $f_{\mathcal{U}'}|_{\mathcal{U}} = f_{\mathcal{U}}$. Therefore there is a (unique) linear map $f : U \longrightarrow V$ such that $f|_{\mathcal{U}} = f_{\mathcal{U}}$ for all finite-dimensional subspaces $\mathcal{U}$ of U. This means

$$F(v^*)(u) = v^*(f(u))$$

for all $v^* \in V^*$ and $u \in U$. By definition $F = f^*$. Part c) implies part a), and the proof of the proposition is complete. ∎

Definition A.4.7 *A linear map $F : V^* \longrightarrow U^*$ which satisfies any of the three equivalent conditions of the preceding proposition is said to be* continuous.

Observe that the one-one linear map

$$\mathrm{Hom}(U, V) \longrightarrow \mathrm{Hom}(V^*, U^*) \quad (f \longmapsto f^*)$$

induces a linear isomorphism between the space of linear maps from U to V and the space of continuous linear maps from V^* to U^*.

Throughout the following exercises U and V are vector spaces over the field k.

Exercise A.4.11 Let I and J be subspaces of U^* and suppose that $I \subseteq J$. Define $[J : I] = \mathrm{Dim}(J/I)$.

a) Suppose that K is a subspace of U^* and $J \subseteq K$. Show that

$$[K : I] = [K : J] + [J : I].$$

b) Suppose that $[J : I]$ is finite. Show that $[J : I] \geq [\overline{J} : \overline{I}]$.

Exercise A.4.12 Suppose that $F, G : V^* \longrightarrow U^*$ are continuous. Show that if F and G agree on a dense subspace of V^*, then $F = G$.

Exercise A.4.13 Prove that the following are equivalent:

a) All cofinite subspaces of U^* are closed.

b) U is finite-dimensional.

References

[Abe, 1980] Abe, E. (1980). *Hopf algebras.*, volume 74 of *Cambridge Tracts in Mathematics.* Cambridge University Press, Cambridge, UK.

[Akutsu and Deguchi, 1990] Akutsu, Y. and Deguchi (1990). Graded solutions of the Yang-Baxter relation and link polynomials. *J. Phys. A*, 23(11):1861–1875.

[Akutsu et al., 1989] Akutsu, Y., Deguchi, T., and Wadati, M. (1989). Exactly solvable models and knot theory. *Phys. Rep.*, 180(4-5):247–332.

[Andrews, 1989] Andrews, G. E. (1989). Physics, Ramanujan, and computer algebra. In *Computer algebra (New York, 1984)*, volume 113 of *Lecture Notes in Pure and Appl. Math.*, pages 97–109. Dekker, New York.

[Andrews and Baxter, 1986a] Andrews, G. E. and Baxter, R. J. (1986a). Lattice gas generalization of the hard hexagon model. I. Star-triangle relation and local densities. *J. Statist. Phys.*, 44(1-2):249–271.

[Andrews and Baxter, 1986b] Andrews, G. E. and Baxter, R. J. (1986b). Lattice gas generalization of the hard hexagon model. II. The local densities as elliptic functions. *J. Statist. Phys.*, 44(5-6):713–728.

[Andrews and Baxter, 1987] Andrews, G. E. and Baxter, R. J. (1987). Lattice gas generalization of the hard hexagon model. III. q-trinomial coefficients. *J. Statist. Phys.*, 47(3-4):297–330.

[Artin, 1947] Artin, E. (1947). Theory of braids. *Ann. Math.*, 48:101–126.

[Baxter, 1982] Baxter, R. (1982). *Exactly Solved Models in Stastistical Mechanics.* Academic Press, London, UK.

[Bazhanov, 1985] Bazhanov, V. V. (1985). Trigonometric solutions of triangle equations and classical Lie algebras. *Phys. Lett. B*, 159(4,5,6):321–324.

[Bergman, 1978] Bergman, G. M. (1978). The diamond lemma for ring theory. *Adv. Math.*, 29:178–218.

[Birman, 1975] Birman, J. (1975). *Braids, links and mapping class groups*, volume 82 of *Ann. of Math. Studies.* Princeton Univ. Press.

[Chari and Pressley, 1995] Chari, V. and Pressley, A. (1995). *A guide to quantum groups. Corrected reprint of the 1994 original.* Cambridge University Press, Cambridge, UK.

[Cherednik, 1982] Cherednik, I. V. (1982). On the properties of factorized S matrices in elliptic functions. *Soviet J. Nuclear Phys.*, 36(2):320–324 (1983).

[Chudnovsky and Chudnovsky, 1981] Chudnovsky, D. V. and Chudnovsky, G. V. (1981). The construction of factorized S-matrices – The relationship between the Baxter model and "Zamolodchikov algebras". *Phys. Lett. B*, 98(1,2):83–87.

[Cotta-Ramusino et al., 1993] Cotta-Ramusino, P., Lambe, L., and Rinaldi, M. (1993). *Construction of quantum groups and the Yang-Baxter equation with spectral parameter*, pages 171–186. Plenum, New York.

[D'Ariano et al., 1985] D'Ariano, G. M., Montorsi, A., and Rasetti, M. G., editors (1985). *Integrable systems in statistical mechanics*, volume 1 of *Series on Advances in Statistical Mechanics*. World Scientific Publishing Co., Singapore.

[Deligne and Milne, 1982] Deligne, P. and Milne, J. (1982). *Tannakian Categories*, volume 900 of *Lecture Notes in Math.* Springer-Verlag, Berlin, Heidelberg, New York.

[Drinfel'd, 1987] Drinfel'd, V. G. (1987). Quantum groups. In *Proceedings of the International Congress of Mathematicians, Vol. 1, 2 (Berkeley, Calif., 1986)*, pages 798–820, Providence, RI. Amer. Math. Soc.

[Drinfel'd, 1990] Drinfel'd, V. G. (1990). Almost cocommutative Hopf algebras. *Leningrad Math. J.*, 1(1):321–342.

[Eckmann and Hilton, 1962] Eckmann, B. and Hilton, P. (1962). Group like structures in general categories I. Multiplications and comultiplications. *Math. Annalen*, 145:227–255.

[Faddeev and Takhtadzhan, 1987] Faddeev, L. and Takhtadzhan, L. (1987). *Hamiltonian Methods in the Theory of Solitons*. Springer-Verlag, Berlin-Heidelberg-New York.

[Faddeev, 1990] Faddeev, L. D. (1990). Lectures on quantum inverse scattering method. In *Nankai Lectures Math. Phys.*, pages 23–70. World Sci. Publishing, Teaneck, NJ.

[Faddeev, 1995] Faddeev, L. D. (1995). Algebraic aspects of Bethe-ansatz. *Internat. J. Modern Phys. A*, 10(13):1845–1878.

[Faddeev et al., 1988] Faddeev, L. D., Reshetikhin, N. Y., and Takhtadzhan, L. A. (1988). Quantization of Lie groups and Lie algebras. In *Algebraic analysis, Papers Dedicated to Prof. Mikio Sato on the Occasion of his Sixtieth Birthday*, volume I, pages 129–139. Academic Press, Boston, MA.

[Faddeev et al., 1989] Faddeev, L. D., Reshetikhin, N. Y., and Takhtadzhan, L. A. (1989). Quantum groups. In *Braid group, knot theory and statistical mechanics*, volume 9 of *Adv. Ser. Math. Phys.*, pages 97–110. World Scientific Publishing Co. Inc., Teaneck, NJ.

[Faddeev et al., 1990] Faddeev, L. D., Reshetikhin, N. Y., and Takhtadzhan, L. A. (1990). Quantization of Lie groups and Lie algebras. *Leningrad Math. J.*, 1(1):193–225. Translation from Algebra Anal. 1, No.1, 178-206 (1989).

[Gelaki, 1996] Gelaki, S. (1996). On pointed ribbon Hopf algebras. *J. Algebra*, 181(3): 760–786.

[Gugenheim, 1962] Gugenheim, V. K. A. M. (1962). On extensions of algebras, coalgebras and Hopf algebras. I. *Amer. J. Math.*, 84:349–382.

[Heyneman, 1966] Heyneman, R. G. (1966). Unpublished result.

[Heyneman and Radford, 1974] Heyneman, R. G. and Radford, D. E. (1974). Reflexivity and coalgebras of finite type. *J. Algebra*, 28:215–246.

[Heyneman and Sweedler, 1969] Heyneman, R. G. and Sweedler, M. E. (1969). Affine Hopf algebras I. *J. Algebra*, 13:192–241.

[Heyneman and Sweedler, 1970] Heyneman, R. G. and Sweedler, M. E. (1970). Affine Hopf algebras II. *J. Algebra*, 16:271–297.

[Hietarinta, 1993a] Hietarinta, J. (1993a). The complete solution to the constant quantum Yang-Baxter equation in two dimensions. In Clarkson, P. A., editor, *Applications of analytic and geometric methods to nonlinear differential equations*, volume 413 of *NATO ASI Ser., Ser. C, Math. Phys. Sci.*, pages 149–154. Kluwer Academic Publishers, Boston, Dordrecht, London.

[Hietarinta, 1993b] Hietarinta, J. (1993b). Solving the two-dimensional constant quantum Yang-Baxter equation. *J. Math. Phys.*, 34(5):1725–1756.

[Hietarinta, 1993c] Hietarinta, J. (1993c). The upper triangular solutions to the three-state constant quantum Yang-Baxter equation. *J. Phys. A*, 26(23):7077–7095.

[Hlavatiý, 1992] Hlavatiý, L. (1992). On solutions of the Yang–Baxter equation without additivity. *J. Physics a: Math. Gen.*, 25:1395–1397.

[Jacobson, 1962] Jacobson, N. (1962). *Lie algebras*, volume 10 of *Interscience Tracts in Pure and Appl. Math.* Interscience Publishers, a Division of John Wiley and Sons, New York-London.

[Jenks and Sutor, 1992] Jenks, R. D. and Sutor, R. S. (1992). *Axiom. The scientific computation system.* Springer-Verlag, Berlin, Heidelberg, New York.

[Jimbo, 1985] Jimbo, M. (1985). A q-difference analogue of $U(g)$ and the yang–baxter equation. *Lett. Math. Phys.*, 10:63–69.

[Jones, 1990] Jones, V. F. R. (1990). Baxterization. *Internat. J. Modern Phys. B*, 4(5):701–713.

[Joyal and Street, 1991a] Joyal, A. and Street, R. (1991a). The geometry of tensor calculus I. *Adv. Math.*, 88(1):55–112.

[Joyal and Street, 1991b] Joyal, A. and Street, R. (1991b). An introduction to Tannaka duality and quantum groups. In *Category theory (Como, 1990)*, volume 1488 of *Lecture Notes in Math.*, pages 413–492. Springer, Berlin.

[Joyal and Street, 1991c] Joyal, A. and Street, R. (1991c). Tortile Yang-Baxter operators in tensor categories. *J. Pure Appl. Algebra*, 71(1):43–51.

[Joyal and Street, 1993] Joyal, A. and Street, R. (1993). Braided tensor categories. *Adv. Math.*, 102(1):20–78.

[Kauffman, 1991] Kauffman, L. H. (1991). *Knots and Physics*, volume 1 of *Series on Knots and Everything*. World Scientific, Singapore.

[Kauffman and Lins, 1994] Kauffman, L. H. and Lins, S. (1994). *Temperley-Lieb Recoupling Theory and Invariants of 3-manifolds*, volume 134 of *Annals of Mathematics Studies*. Princeton University Press, Princeton, NJ.

[Kauffman and Radford, 1995] Kauffman, L. H. and Radford, D. E. (1995). Invariants of 3-manifolds derived from finite dimensional Hopf algebras. *J. Knot Theory Ramifications*, 4(1):131–162.

[Koorwinder, 1990] Koorwinder, T. (1990). Orthogonal polynomials in conection with quantum groups. *NATO ASI Series*, 294:257–292.

[Kulish and Sklyanin, 1982] Kulish, P. and Sklyanin, E. (1982). Solutions of the Yang-Baxter equation. *J. Sov. Math.*, 19:1596–1620. Translation from Zap. Nauchn. Semin. Leningr. Otd. Mat. Inst. Steklova 95, 129-160 (1980).

[Kulish et al., 1981] Kulish, P. P., Reshetikhin, N. Y., and Sklyanin, E. K. (1981). Yang–Baxter equation and representation theory: I. *Lett. Math. Phys.*, 5:393–403.

[Lambe, 1994] Lambe, L. (1994). On the quantum Yang-Baxter equation with spectral parameter I. *Proc. Lond. Math. Soc.*, 68(1):31–50.

[Lambe, 1996] Lambe, L. (1996). The 1996 Adams Lectures at Manchester University: New computational methods in algebra and topology. Part I: Solving the Quantum Yang-Baxter Equation.

[Lambe and Radford, 1993] Lambe, L. A. and Radford, D. E. (1993). Algebraic aspects of the quantum Yang-Baxter equation. *J. Algebra*, 154(1):228–288.

[Larson and Radford, 1988] Larson, R. G. and Radford, D. E. (1988). Semisimple cosemisimple Hopf algebras. *Am. J. Math.*, 110(1):187–195.

[Larson and Towber, 1991] Larson, R. G. and Towber, J. (1991). Two dual classes of bialgebras related to the concepts of "quantum groups" and "quantum lie algebras". *Comm. in Alg.*, 19:3295–3345.

[Liguori and Mintchev, 1992] Liguori, M. and Mintchev, M. (1992). Spectral parameters of the quantum Yang–Baxter equation. *Physics Letters B*, 275:371–374.

[MacLane, 1988] MacLane, S. (1988). *Categories for the working mathematician. 4th corrected printing*, volume 5 of *Graduate Texts in Mathematics*. Springer-Verlag, Berlin, Heidelberg, New York.

[Majid, 1990a] Majid, S. (1990a). Physics for algebraists: Non-commutative and non-cocommutative Hopf algebras by a bicrossproduct construction. *J. Algebra*, 130(1):17–64.

[Majid, 1990b] Majid, S. (1990b). Quasitriangular Hopf algebras and Yang-Baxter equations. *Internat. J. Modern Phys. A*, 5(1):1–91.

[Majid, 1991a] Majid, S. (1991a). Doubles of quasitriangular Hopf algebras. *Comm. Algebra*, 19(11):3061–3073.

[Majid, 1991b] Majid, S. (1991b). Reconstruction theorems and rational conformal field theories. *Internat. J. Modern Phys. A*, 6(24):4359–4374.

[Majid, 1994] Majid, S. (1994). Algebras and Hopf algebras in braided categories. In Bergen, J. e. a., editor, *Advances in Hopf algebras, Conference, August 10-14, 1992, Chicago, IL, USA*, volume 158 of *Lect. Notes Pure Appl. Math.*, pages 55–105. Marcel Dekker, New York, NY.

[Michaelis, 1980] Michaelis, W. (1980). Lie coalgebras. *Adv. in Math.*, 38:1–54.

[Montgomery, 1993] Montgomery, S. (1993). *Hopf Algebras and Their Actions on Rings*, volume 82 of *Regional Conference Series in Mathematics*. AMS, Providence, RI.

[Montgomery and Smith, 1990] Montgomery, S. and Smith, S. P. (1990). Skew derivations and $\mathcal{U}_q(sl(2))$. *Israel J. Math.*, 72:158–156.

[Nichols and Zoeller, 1989] Nichols, W. D. and Zoeller, M. (1989). A Hopf algebra freeness theorem. *Am. J. Math.*, 111(2):381–385.

[Pareigis, 1981] Pareigis, B. (1981). A noncommutative noncocommutative hopf algebra in "nature". *J. Algebra*, 70(2):356–374.

[Pareigis, 1996] Pareigis, B. (1996). Reconstruction of hidden symmetries. *J. Algebra*, 183(1):90–154.

[Radford, 1971] Radford, D. E. (1971). A free rank 4 Hopf algebra with antipode of order 4. *Proc. Amer. Math. Soc.*, 30:55–58.

[Radford, 1973] Radford, D. E. (1973). Coreflexive coalgebras. *J. Algebra*, 26:512–535.

[Radford, 1977] Radford, D. E. (1977). Operators on Hopf algebras. *Amer. J. Math.*, 99:139–158.

[Radford, 1985] Radford, D. E. (1985). The structure of Hopf algebras with a projection. *J. Algebra*, 92:322–347.

[Radford, 1992] Radford, D. E. (1992). On the antipode of a quasitriangular Hopf algebra. *J. Algebra*, 151(1):1–11.

[Radford, 1993a] Radford, D. E. (1993a). Irreducible representations of $\mathcal{U}_q(g)$ arising from $\mathrm{Mod}^{\bullet}_{C^{\frac{1}{2}}}$. In *Quantum deformations of algebras and their representations (Ramat-Gan, 1991/1992; Rehovot, 1991/1992)*, volume 7 of *Israel Math. Conf. Proc.*, pages 143–170. Bar-Ilan Univ., Ramat Gan.

[Radford, 1993b] Radford, D. E. (1993b). Minimal quasitriangular Hopf algebras. *J. Algebra*, 157(2):285–315.

[Radford, 1993c] Radford, D. E. (1993c). Solutions to the quantum Yang-Baxter equation and the Drinfel'd double. *J. Algebra*, 161(1):20–32.

[Radford, 1994a] Radford, D. E. (1994a). On Kauffman's knot invariants arising from finite-dimensional Hopf algebras. In *Advances in Hopf algebras (Chicago, IL, 1992)*, volume 158 of *Lecture Notes in Pure and Appl. Math.*, pages 205–266. Dekker, New York.

[Radford, 1994b] Radford, D. E. (1994b). Solutions to the quantum Yang-Baxter equation arising from pointed bi-algebras. *Trans. Am. Math. Soc.*, 343(1):455–477.

[Radford, 1998] Radford, D. E. (1998). Generalized double crossproducts associated with the quantized enveloping algebras. *Comm. in Alg.*

[Radford and Towber, 1993] Radford, D. E. and Towber, J. (1993). Yetter-Drinfel'd categories associated to an arbitrary bialgebra. *J. Pure Appl. Algebra*, 87(3):259–279.

[Reshetikhin, 1991] Reshetikhin, N. (1991). Invariants of links and 3-manifolds related to quantum groups. In Satake, I., editor, *Proc. Int. Congr. Math., Kyoto, August 21–29, 1990*, volume II, pages 1373–1375, Tokyo. Springer-Verlag.

[Reshetikhin and Turaev, 1991] Reshetikhin, N. Y. and Turaev, V. G. (1991). Invariants of 3-manifolds via link polynomials and quantum groups. *Invent. Math.*, 103(3):547–597.

[Rosso, 1988] Rosso, M. (1988). Finite dimensional representations of the quantum analog of the enveloping algebra of a complex simple Lie algebra. *Commun. Math. Phys.*, 117(4):581–593.

[Saavedra Rivano, 1972] Saavedra Rivano, N. (1972). *Catégories Tannakiennes*, volume 265 of *Lecture Notes in Mathematics*. Springer-Verlag, Berlin-Heidelberg-New York.

[Schauenburg, 1992a] Schauenburg, P. (1992a). *On coquasitriangular Hopf algebras and the quantum Yang–Baxter equation*, volume 67 of *Algebra-Berichte*. R. Fischer, Munich, Germany.

[Schauenburg, 1992b] Schauenburg, P. (1992b). *Tannaka duality for arbitrary Hopf algebras*, volume 66 of *Algebra-Berichte*. R. Fischer, Munich, Germany.

[Serre, 1992] Serre, J.-P. (1992). *Lie algebras and Lie groups. 1964 lectures, given at Harvard University, 2nd ed.*, volume 1500 of *Lecture Notes in Mathematics*. Springer-Verlag, Berlin-Heidelberg-New York.

[Shafarevich, 1974] Shafarevich, I. R. (1974). *Basic Algebraic Geometry*, volume 213 of *Die Grundlehren der mathematischen Wissenschaften*. Springer-Verlag, Berlin-Heidelberg-New York.

[Shnider and Sternberg, 1993] Shnider, S. and Sternberg, S. (1993). *Quantum groups. From coalgebras to Drinfel'd algebras. A guided tour*, volume II of *Graduate Texts in Mathematical Physics*. International Press, Cambridge, MA.

[Singer, 1972] Singer, W. M. (1972). Extension theory for connected Hopf algebras. *J. Algebra*, 21:1–16.

[Sklyanin, 1982] Sklyanin, E. K. (1982). Quantum version of the method of inverse scattering problem. *J. Sov. Math.*, 19:1546–1595. Translation from Zap. Nauchn. Semin. Leningr. Otd. Mat. Inst. Steklova 95, 55-128 (1980).

[Sklyanin, 1991] Sklyanin, E. K. (1991). Quantum inverse scattering method: Selected topics. *Preprint Series in Theoretical Physics, Univ. Helsinki*, HU-TFT-91-51:1–36.

[Soibelman and Vaksman, 1988] Soibelman, Y. S. and Vaksman, L. L. (1988). Algebra of functions on the quantum group $SU(2)$. *Func. Anal. Appl.*, 22:170–181.

[Sweedler, 1966] Sweedler, M. E. (1966). Personal Communication with R. Heynaman.

[Sweedler, 1969] Sweedler, M. E. (1969). *Hopf algebras*. W. A. Benjamin, Inc., New York, NY.

[Taft, 1971] Taft, E. J. (1971). The order of the antipode of finite-dimensional Hopf algebra. *Proc. nat. Acad. Sci. USA*, 68:2631–2633.

[Taft, 1972] Taft, E. J. (1972). Reflexivity of algebras and coalgebras. *Amer. J. Math.*, 94:1111–1130.

[Taft, 1982] Taft, E. J. (1982). Noncocommutative sequences of divided powers. In *Conf. Proc. in Lie algebras and related topics*, volume 993 of *Lecture Notes in Math.*, pages 203–209. Springer-Verlag, Berlin, Heidelberg, New York.

[Taft and Wilson, 1974] Taft, E. J. and Wilson, R. (1974). On antipodes in pointed Hopf algebras. *J. Algebra*, 29:27–32.

[Takeuchi, 1971] Takeuchi, M. (1971). Free Hopf algebras generated by coalgebras. *J. Math. Soc. Japan*, 23(4):561–582.

[Takeuchi, 1992] Takeuchi, M. (1992). Hopf algebra techniques applied to the quantum group $U_q(sl(2))$. In *Deformation theory and quantum groups with applications to mathematical physics*, volume 134 of *Contemp. Math.*, pages 309–323. Amer. Math. Soc., Providence, RI.

[Takhtajan, 1984] Takhtajan, L. A. (1984). Integral models in classical and quantum field theory. In Ciesielski, Z. and aw Olech, C., editors, *Proc. Int. Congr. Math.*,

Warsaw, August 16–24, 1983, pages 1331–1346, Amsterdam-New York. North-Holland Publishing Co.

[Takhtajan, 1990a] Takhtajan, L. A. (1990a). Introduction to quantum groups. In *Quantum groups (Clausthal, 1989)*, volume 370 of *Lecture Notes in Phys.*, pages 3–28. Springer, Berlin.

[Takhtajan, 1990b] Takhtajan, L. A. (1990b). Lectures on quantum groups. In *Introduction to quantum group and integrable massive models of quantum field theory (Nankai, 1989)*, Nankai Lectures Math. Phys., pages 69–197. World Sci. Publishing, River Edge, NJ.

[Takhtajan, 1993] Takhtajan, L. A. (1993). Elementary introduction to quantum groups. In *Important developments in soliton theory*, Springer Ser. Nonlinear Dynam., pages 441–467. Springer, Berlin.

[Tarasov et al., 1983] Tarasov, V., Takhtajan, L. A., and Faddeev, L. D. (1983). Local hamiltonians for integrable quantum models on a lattice. *Theoret. and Math. Phys.*, 57(2):1059–1073. Translated from Teoret. Mat. Fiz. 57, no. 2, 163–181 (1983).

[Ulbrich, 1990] Ulbrich, K.-H. (1990). On Hopf algebras and rigid monoidal categories. *Isr. J. Math.*, 72(1/2):252–256.

[Waterhouse, 1979] Waterhouse, W. C. (1979). *Introduction to affine group schemes*, volume 66 of *Graduate Texts in Mathematics*. Springer-Verlag, Berlin, Heidelberg, New York.

[Yang, 1967] Yang, C. N. (1967). Some exact results for the many-body problem in one dimension with repulsive delta-function interaction. *Phys. Rev. Lett.*, 19:1312–1315.

[Yang and Ge, 1989] Yang, C. N. and Ge, M. L., editors (1989). *Braid group, knot theory and statistical mechanics*, volume 9 of *Advanced Series in Mathematical Physics*. World Scientific Publishing Co. Inc., Teaneck, NJ.

[Yang and Ge, 1994] Yang, C. N. and Ge, M. L., editors (1994). *Braid group, knot theory and statistical mechanics. II*, volume 17 of *Advanced Series in Mathematical Physics*. World Scientific Publishing Co. Inc., River Edge, NJ.

[Yetter, 1990] Yetter, D. N. (1990). Quantum groups and representations of monoidal categories. *Math. Proc. Camb. Philos. Soc.*, 108(2):261–290.

[Zamolodchikov and Zamolodchikov, 1975] Zamolodchikov, A. B. and Zamolodchikov, A. B. (1975). Factorized S-matrices in two dimensions as the exact solutions of certain relativistic quantum field theory models. *Ann. Physics*, 120(2):253–291.

About the Authors

Dr. Larry A. Lambe received his PhD from the University of Illinois at Chicago in 1980 and has written numerous papers in the area of algebraic topology, homological algebra, QYBE and symbolic computation. He has lectured around the world on these topics annually since 1988. He is on the Board of Editors of the Journal of Symbolic Computation. In September of 1996, he was awarded the degree of PhD Honorus Causa by Stockholm University and in April, 1997, he was made an Honorary Professor in the Mathematics Department of the University of Wales at Bangor, Wales Gwynedd.

Dr. David E. Radford received his PhD from the University of North Carolina at Chapel Hill in 1970. He has written numerous papers in the areas of Hopf algebras and quantum groups. Dr. Radford is Professor of Mathematics at the University of Illinois at Chicago where he has been on the faculty since 1976.

Index

Other *Mathematics and Its Applications* titles of interest:

P.H. Sellers: *Combinatorial Complexes. A Mathematical Theory of Algorithms.* 1979, 200 pp. ISBN 90-277-1000-7

P.M. Cohn: *Universal Algebra.* 1981, 432 pp. ISBN 90-277-1213- 1 (hb), ISBN 90-277-1254-9 (pb,

J. Mockor: *Groups of Divisibility.* 1983, 192 pp. ISBN 90-277-1539-4

A. Wwarynczyk: *Group Representations and Special Functions.* 1986, 704 pp. ISBN 90-277-2294-3 (pb), ISBN 90-277-1269-7 (hb)

I. Bucur: *Selected Topics in Algebra and its Interrelations with Logic, Number Theory and Algebraic Geometry.* 1984, 416 pp. ISBN 90-277-1671-4

H. Walther: *Ten Applications of Graph Theory.* 1985, 264 pp. ISBN 90-277-1599-8

L. Beran: *Orthomodular Lattices. Algebraic Approach.* 1985, 416 pp. ISBN 90-277-1715-X

A. Pazman: *Foundations of Optimum Experimental Design.* 1986, 248 pp. ISBN 90-277-1865-2

K. Wagner and G. Wechsung: *Computational Complexity.* 1986, 552 pp. ISBN 90-277-2146-7

A.N. Philippou, G.E. Bergum and A.F. Horodam (eds.): *Fibonacci Numbers and Their Applications.* 1986, 328 pp. ISBN 90-277-2234-X

C. Nastasescu and F. van Oystaeyen: *Dimensions of Ring Theory.* 1987, 372 pp. ISBN 90-277-2461-X

Shang-Ching Chou: *Mechanical Geometry Theorem Proving.* 1987, 376 pp. ISBN 90-277-2650-7

D. Przeworska-Rolewicz: *Algebraic Analysis.* 1988, 640 pp. ISBN 90-277-2443-1

C.T.J. Dodson: *Categories, Bundles and Spacetime Topology.* 1988, 264 pp. ISBN 90-277-2771-6

V.D. Goppa: *Geometry and Codes.* 1988, 168 pp. ISBN 90-277-2776-7

A.A. Markov and N.M. Nagorny: *The Theory of Algorithms.* 1988, 396 pp. ISBN 90-277-2773-2

E. Kratzel: *Lattice Points.* 1989, 322 pp. ISBN 90-277-2733-3

A.M.W. Glass and W.Ch. Holland (eds.): *Lattice-Ordered Groups. Advances and Techniques.* 1989, 400 pp. ISBN 0-7923-0116-1

N.E. Hurt: *Phase Retrieval and Zero Crossings: Mathematical Methods in Image Reconstruction.* 1989, 320 pp. ISBN 0-7923-0210-9

Du Dingzhu and Hu Guoding (eds.): *Combinatorics, Computing and Complexity.* 1989, 248 pp. ISBN 0-7923-0308-3

Other *Mathematics and Its Applications* titles of interest:

A.Ya. Helemskii: *The Homology of Banach and Topological Algebras.* 1989, 356 pp. ISBN 0-7923-0217-6

J. Martinez (ed.): *Ordered Algebraic Structures.* 1989, 304 pp. ISBN 0-7923-0489-6

V.I. Varshavsky: *Self-Timed Control of Concurrent Processes. The Design of Aperiodic Logical Circuits in Computers and Discrete Systems.* 1989, 428 pp. ISBN 0-7923-0525-6

E. Goles and S. Martinez: *Neural and Automata Networks. Dynamical Behavior and Applications.* 1990, 264 pp. ISBN 0-7923-0632-5

A. Crumeyrolle: *Orthogonal and Symplectic Clifford Algebras. Spinor Structures.* 1990, 364 pp. ISBN 0-7923-0541-8

S. Albeverio, Ph. Blanchard and D. Testard (eds.): *Stochastics, Algebra and Analysis in Classical and Quantum Dynamics.* 1990, 264 pp. ISBN 0-7923-0637-6

G. Karpilovsky: *Symmetric and G-Algebras. With Applications to Group Representations.* 1990, 384 pp. ISBN 0-7923-0761-5

J. Bosak: *Decomposition of Graphs.* 1990, 268 pp. ISBN 0-7923-0747-X

J. Adamek and V. Trnkova: *Automata and Algebras in Categories.* 1990, 488 pp. ISBN 0-7923-0010-6

A.B. Venkov: *Spectral Theory of Automorphic Functions and Its Applications.* 1991, 280 pp. ISBN 0-7923-0487-X

M.A. Tsfasman and S.G. Vladuts: *Algebraic Geometric Codes.* 1991, 668 pp. ISBN 0-7923-0727-5

H.J. Voss: *Cycles and Bridges in Graphs.* 1991, 288 pp. ISBN 0-7923-0899-9

V.K. Kharchenko: *Automorphisms and Derivations of Associative Rings.* 1991, 386 pp. ISBN 0-7923-1382-8

A.Yu. Olshanskii: *Geometry of Defining Relations in Groups.* 1991, 513 pp. ISBN 0-7923-1394-1

F. Brackx and D. Constales: *Computer Algebra with LISP and REDUCE. An Introduction to Computer-Aided Pure Mathematics.* 1992, 286 pp. ISBN 0-7923-1441-7

N.M. Korobov: *Exponential Sums and their Applications.* 1992, 210 pp. ISBN 0-7923-1647-9

D.G. Skordev: *Computability in Combinatory Spaces. An Algebraic Generalization of Abstract First Order Computability.* 1992, 320 pp. ISBN 0-7923-1576-6

E. Goles and S. Martinez: *Statistical Physics, Automata Networks and Dynamical Systems.* 1992, 208 pp. ISBN 0-7923-1595-2

Other *Mathematics and Its Applications* titles of interest:

M.A. Frumkin: *Systolic Computations.* 1992, 320 pp. ISBN 0-7923-1708-4

J. Alajbegovic and J. Mockor: *Approximation Theorems in Commutative Algebra.* 1992, 330 pp. ISBN 0-7923-1948-6

I.A. Faradzev, A.A. Ivanov, M.M. Klin and A.J. Woldar: *Investigations in Algebraic Theory of Combinatorial Objects.* 1993, 516 pp. ISBN 0-7923-1927-3

I.E. Shparlinski: *Computational and Algorithmic Problems in Finite Fields.* 1992, 266 pp. ISBN 0-7923-2057-3

P. Feinsilver and R. Schott: *Algebraic Structures and Operator Calculus.* Vol. I. Representations and Probability Theory. 1993, 224 pp. ISBN 0-7923-2116-2

A.G. Pinus: *Boolean Constructions in Universal Algebras.* 1993, 350 pp. ISBN 0-7923-2117-0

V.V. Alexandrov and N.D. Gorsky: *Image Representation and Processing. A Recursive Approach.* 1993, 200 pp. ISBN 0-7923-2136-7

L.A. Bokut' and G.P. Kukin: *Algorithmic and Combinatorial Algebra.* 1994, 384 pp. ISBN 0-7923-2313-0

Y. Bahturin: *Basic Structures of Modern Algebra.* 1993, 419 pp. ISBN 0-7923-2459-5

R. Krichevsky: *Universal Compression and Retrieval.* 1994, 219 pp. ISBN 0-7923-2672-5

A. Elduque and H.C. Myung: *Mutations of Alternative Algebras.* 1994, 226 pp. ISBN 0-7923-2735-7

E. Goles and S. Martínez (eds.): *Cellular Automata, Dynamical Systems and Neural Networks.* 1994, 189 pp. ISBN 0-7923-2772-1

A.G. Kusraev and S.S. Kutateladze: *Nonstandard Methods of Analysis.* 1994, 444 pp. ISBN 0-7923-2892-2

P. Feinsilver and R. Schott: *Algebraic Structures and Operator Calculus.* Vol. II. Special Functions and Computer Science. 1994, 148 pp. ISBN 0-7923-2921-X

V.M. Kopytov and N. Ya. Medvedev: *The Theory of Lattice-Ordered Groups.* 1994, 400 pp. ISBN 0-7923-3169-9

H. Inassaridze: *Algebraic K-Theory.* 1995, 438 pp. ISBN 0-7923-3185-0

C. Mortensen: *Inconsistent Mathematics.* 1995, 155 pp. ISBN 0-7923-3186-9

R. Abłamowicz and P. Lounesto (eds.): *Clifford Algebras and Spinor Structures.* A Special Volume Dedicated to the Memory of Albert Crumeyrolle (1919–1992). 1995, 421 pp. ISBN 0-7923-3366-7

W. Bosma and A. van der Poorten (eds.), *Computational Algebra and Number Theory.* 1995, 336 pp. ISBN 0-7923-3501-5

Other *Mathematics and Its Applications* titles of interest:

A.L. Rosenberg: *Noncommutative Algebraic Geometry and Representations of Quantized Algebras*. 1995, 316 pp. ISBN 0-7923-3575-9

L. Yanpei: *Embeddability in Graphs*. 1995, 400 pp. ISBN 0-7923-3648-8

B.S. Stechkin and V.I. Baranov: *Extremal Combinatorial Problems and Their Applications*. 1995, 205 pp. ISBN 0-7923-3631-3

Y. Fong, H.E. Bell, W.-F. Ke, G. Mason and G. Pilz (eds.): *Near-Rings and Near-Fields*. 1995, 278 pp. ISBN 0-7923-3635-6

A. Facchini and C. Menini (eds.): *Abelian Groups and Modules*. (Proceedings of the Padova Conference, Padova, Italy, June 23–July 1, 1994). 1995, 537 pp. ISBN 0-7923-3756-5

D. Dikranjan and W. Tholen: *Categorical Structure of Closure Operators*. With Applications to Topology, Algebra and Discrete Mathematics. 1995, 376 pp. ISBN 0-7923-3772-7

A.D. Korshunov (ed.): *Discrete Analysis and Operations Research*. 1996, 351 pp. ISBN 0-7923-3866-9

P. Feinsilver and R. Schott: *Algebraic Structures and Operator Calculus*. Vol. III: Representations of Lie Groups. 1996, 238 pp. ISBN 0-7923-3834-0

M. Gasca and C.A. Micchelli (eds.): *Total Positivity and Its Applications*. 1996, 528 pp. ISBN 0-7923-3924-X

W.D. Wallis (ed.): *Computational and Constructive Design Theory*. 1996, 368 pp. ISBN 0-7923-4015-9

F. Cacace and G. Lamperti: *Advanced Relational Programming*. 1996, 410 pp. ISBN 0-7923-4081-7

N.M. Martin and S. Pollard: *Closure Spaces and Logic*. 1996, 248 pp. ISBN 0-7923-4110-4

A.D. Korshunov (ed.): *Operations Research and Discrete Analysis*. 1997, 340 pp. ISBN 0-7923-4334-4

W.D. Wallis: *One-Factorizations*. 1997, 256 pp. ISBN 0-7923-4323-9

G. Weaver: *Henkin–Keisler Models*. 1997, 266 pp. ISBN 0-7923-4366-2

V.N. Kolokoltsov and V.P. Maslov: *Idempotent Analysis and Its Applications*. 1997, 318 pp. ISBN 0-7923-4509-6

J.P. Ward: *Quaternions and Cayley Numbers*. Algebra and Applications. 1997, 250 pp. ISBN 0-7923-4513-4

E.S. Ljapin and A.E. Evseev: *The Theory of Partial Algebraic Operations*. 1997, 245 pp. ISBN 0-7923-4609-2

Other *Mathematics and Its Applications* titles of interest:

S. Ayupov, A. Rakhimov and S. Usmanov: *Jordan, Real and Lie Structures in Operator Algebras*. 1997, 235 pp. ISBN 0-7923-4684-X

A. Khrennikov: *Non-Archimedean Analysis: Quantum Paradoxes, Dynamical Systems and Biological Models*. 1997, 389 pp. ISBN 0-7923-4800-1

G. Saad and M.J. Thomsen (eds.): *Nearrings, Nearfields and K-Loops*. (Proceedings of the Conference on Nearrings and Nearfields, Hamburg, Germany. July 30–August 6, 1995). 1997, 458 pp. ISBN 0-7923-4799-4

L.A. Lambe and D.E. Radford: *Introduction to the Quantum Yang–Baxter Equation and Quantum Groups: An Algebraic Approach*. 1997, 314 pp. ISBN 0-7923-4721-8